MERCURY CONTAMINATION: A Human Tragedy
Patricia A. D'Itri and Frank M. D'Itri

POLLUTANTS AND HIGH RISK GROUPS
Edward J. Calabrese

METHODOLOGICAL APPROACHES TO DERIVING ENVIRONMENTAL AND OCCUPATIONAL HEALTH STANDARDS
Edward J. Calabrese

NUTRITION AND ENVIRONMENTAL HEALTH—Volume I: The Vitamins
Edward J. Calabrese

NUTRITION AND ENVIRONMENTAL HEALTH—Volume II: Minerals and Macronutrients
Edward J. Calabrese

SULFUR IN THE ENVIRONMENT, Parts I and II
Jerome O. Nriagu, Editor

COPPER IN THE ENVIRONMENT, Parts I and II
Jerome O. Nriagu, Editor

ZINC IN THE ENVIRONMENT, Parts I and II
Jerome O. Nriagu, Editor

CADMIUM IN THE ENVIRONMENT, Parts I and II
Jerome O. Nriagu, Editor

NICKEL IN THE ENVIRONMENT
Jerome O. Nriagu, Editor

ENERGY UTILIZATION AND ENVIRONMENTAL HEALTH
Richard A. Wadden, Editor

FOOD, CLIMATE AND MAN
Margaret R. Biswas and Asit K. Biswas, Editors

CHEMICAL CONCEPTS IN POLLUTANT BEHAVIOR
Ian J. Tinsley

RESOURCE RECOVERY AND RECYCLING
A. F. M. Barton

QUANTITATIVE TOXICOLOGY
V. A. Filov, A. A. Golubev, E. I. Liublina, and N.A. Tolokontsev

ATMOSPHERIC MOTION AND AIR POLLUTION
Richard A. Dobbins

GROUNDWATER POLLUTION MICROBIOLOGY

GROUNDWATER POLLUTION MICROBIOLOGY

Edited by

GABRIEL BITTON
University of Florida

CHARLES P. GERBA
University of Arizona

A WILEY-INTERSCIENCE PUBLICATION

JOHN WILEY & SONS

New York Chichester Brisbane Toronto Singapore

Library of Congress Cataloging in Publication Data:

Groundwater pollution microbiology.

(Environmental science and technology, ISSN 0194-0287)
"A Wiley-Interscience publication."
Includes index.
1. Water, Underground—Microbiology. 2. Water, Underground—Pollution. I. Bitton, Gabriel. II. Gerba, Charles P., 1945-. III. Series.
QR105.5.G76 1984 628.1'6861698 83-14745
ISBN 0-471-09656-3

Printed in the United States of America

10 9 8 7 6 5 4 3 2 1

CONTRIBUTORS

GABRIEL BITTON, Department of Environmental Engineering Sciences, University of Florida, Gainesville, Florida

E. J. BOUWER, Department of Civil Engineering, University of Houston, Houston, Texas

HERMAN BOUWER, U.S. Water Conservation Laboratory, U.S. Department of Agriculture, Phoenix, Arizona

I. CECH, School of Public Health, The University of Texas, Houston, Texas

G. F. CRAUN, Health Effects Research Lab, U.S. EPA, Cincinnati, Ohio

CHARLES P. GERBA, Department of Microbiology, University of Arizona, Tucson, Arizona

C. HAGEDORN, Department of Agronomy, Mississippi State University, Mississippi State, Mississippi

R. HARRIST, The University of Texas School of Public Health, Houston, Texas

B. H. KESWICK, Program in Infectious Diseases and Clinical Microbiology, The University of Texas Medical School at Houston, Houston, Texas

J. C. LANCE, Water Quality and Watershed Research Laboratory, United States Department of Agriculture, Agricultural Research Service, Duran, Oklahoma

FRANKLIN R. LEACH, Department of Biochemistry, Oklahoma State University, Stillwater, Oklahoma

JAY H. LEHR, National Water Well Association, Worthington, Ohio

GAIL E. MALLARD, United States Geological Survey, National Center, Reston, Virginia

P. L. MCCARTY, Department of Civil Engineering, Stanford University, Stanford, California

JAMES F. MCNABB, United States Environmental Protection Agency, Robert S. Kerr Environmental Research Laboratory, Ada, Oklahoma

JOHN J. MONTGOMERY, National Water Well Association, Worthington, Ohio

DAVID M. NIELSEN, National Water Well Association, Worthington, Ohio

Michael James Noonan, Department of Agricultural Microbiology, Lincoln College, Canterbury, New Zealand

B. E. Rittmann, Department of Civil Engineering, University of Illinois at Urbana–Champaign, Urbana, Illinois

John Wilson, United States Environmental Protection Agency, Robert S. Kerr Environmental Research Laboratory, Ada, Oklahoma

SERIES PREFACE

Environmental Science and Technology

The Environmental Science and Technology Series of Monographs, Textbooks, and Advances is devoted to the study of the quality of the environment and to the technology of its conservation. Environmental science therefore relates to the chemical, physical, and biological changes in the environment through contamination or modification, to the physical nature and biological behavior of air, water, soil, food, and waste as they are affected by man's agricultural, industrial, and social activities, and to the application of science and technology to the control and improvement of environmental quality.

The deterioration of environmental quality, which began when man first collected into villages and utilized fire, has existed as a serious problem under the ever-increasing impacts of exponentially increasing population and of industrializing society. Environmental contamination of air, water, soil, and food has become a threat to the continued existence of many plant and animal communities of the ecosystem and may ultimately threaten the very survival of the human race.

It seems clear that if we are to preserve for future generations some semblance of the biological order of the world of the past and hope to improve on the deteriorating standards of urban health, environmental science and technology must quickly come to play a dominant role in designing our social and industrial structure for tomorrow. Scientifically rigorous criteria of environmental quality must be developed. Based in part on these criteria, realistic standards must be established, and our technological progress must be tailored to meet them. It is obvious that civilization will continue to require increasing amounts of fuel, transportation, industrial chemicals, fertilizers, pesticides, and countless other products, and that it will continue to produce waste products of all descriptions. What is urgently needed is a total systems approach to modern civilization through which the pooled talents of scientists and engineers, in cooperation with social scientists and the medical profession, can be focused on the development of order and equilibrium in the presently disparate segments of the human environment. Most of the skills and tools that are needed are already in existence. We surely have a right to hope a technology that has created such manifold environmental problems is also capable of solving them. It is our hope that this Series in Environmental Sciences and Technology will not only serve to make this challenge more explicit to the established professionals, but that it also will help to stimulate the student toward the career opportunities in this vital area.

Robert L. Metcalf
Werner Stumm

PREFACE

Groundwater has historically been considered a reliable and safe source of water, protected from surface contamination by a "living filter," an upper soil mantle that removes pollutants as the water percolates downward through the soil. However, a number of well-documented outbreaks of hepatitis A and viral gastroenteritis traced to contaminated groundwater, as well as numerous reports of chemical contamination, have destroyed the widely held misconception that groundwater is safe from pollution. This has necessitated a reassessment of what is known about the fates of pathogenic microorganisms and chemical pollutants in soil and groundwater.

Less is known about the microbiology of groundwater than about any other part of the biosphere. The significance of microbiology as it relates to groundwater resides in the interactions of pollutants and microorganisms in the subsurface and the effects of such interactions on the quality of groundwater. Determinations of the nature and extent of microbial activity in the subsurface, the movement and fate of pathogens, and the controlling environmental factors are essential if groundwater quality is to be protected.

The purpose of this book is to present a concise review of the available information on groundwater pollution microbiology. It is hoped that this book will be a starting point for those concerned with the quality of our groundwater resources.

GABRIEL BITTON
CHARLES P. GERBA

Gainesville, Florida
Tucson, Arizona
December 1983

CONTENTS

1

GROUNDWATER POLLUTION MICROBIOLOGY: THE EMERGING ISSUE

Gabriel Bitton

Department of Environmental Engineering Science
University of Florida, Gainesville, Florida

Charles P. Gerba

Department of Microbiology
University of Arizona, Tucson, Arizona

1.1. INTRODUCTION

Groundwater is a hidden resource, the study of which was regarded as an "occult" science until recently. Research was focused on groundwater supply, but little interest was expressed with regard to groundwater quality until very recently (1). Yet groundwater supplies 100 million Americans with their drinking water. In rural areas this dependence on groundwater is even more acute: groundwater supplies 90–95 percent of the drinking water of rural populations in the United States (Table 1.1.). Other major users of groundwater are agriculture (irrigation) and industry. Because our reliance on groundwater continues to increase, attention was directed to the potential health problems resulting from our exposure to contaminated groundwater. The sources of microbial and

Table 1.1. Some Facts About Groundwater

Some Parameters	
Age	Few hours to thousands of years
Temperature	Increases with depth ($\sim 3°C/100$ m)
Conductivity	Presence of dissolved ions (e.g., Na^+, Ca^{2+}, Mg^{2+}, Cl^-, SO_4^{2-}, NO_3^- . . .) Concentration: hundreds of mg/liter
Hydrostatic pressure	Increases with depth (1 atm/10 m)
Major Sources of Groundwater Contamination (U.S.A.)	*Domestic Wastewater* 3 billion M^3/year More than 0.6 million wells receive 850 billion gallons/year *Industrial Sources* 100,000 abandoned industrial waste sites 75,000 active industrial landfills 173,000 waste lagoons *Agricultural Practices* Pesticides and fertilizers
Groundwater as a Resource (U.S.A.)	25% of all water used
Groundwater Uses (U.S.A.)	Irrigation 69% Industry 14% Urban and rural drinking 17%
Groundwater for Drinking (U.S.A.)	Supplies the needs of 50% of U.S. population. Supplies 95% of drinking needs in rural areas. U.S. public water systems: 75% rely on groundwater.
Major Public Health Concerns	*Waterborne Diseases:* 54% of diseases occurred in noncommunity water systems relying solely on groundwater. *Inorganic and Organic Chemicals:* Concern about toxic, mutagenic, carcinogenic, and teratogenic compounds. Not well documented.
Groundwater Legislation	Groundwater quality addressed in numerous pieces of legislation. Acts must be coordinated.

chemical contamination of groundwater are numerous, particularly in industrialized countries. These sources include the practice of land disposal of sewage effluents, sludges, and solid wastes; septic tanks effluents; urban runoff; as well as agricultural, mining, and industrial practices (see Chapter 2). The contamintion problem is worrisome because a portion of the U.S. population drinks untreated groundwater. In Florida, the population is almost completely (92 percent) dependent on groundwater to supply its drinking water. Twenty percent of those consumers utilize untreated groundwater (2). As a result, rural households appear to be more exposed to contaminated well water. A recent national survey has shown that two-thirds of all rural households drink water that is unacceptable from chemical and bacteriological standpoints. Moreover, bacterial contamination appeared to be the most common problem in groundwaters from rural areas (3).

Two categories of microbiologists should be actively involved in the field of groundwater contamination: public health microbiologists and microbial ecologists. Public health microbiologists seek to understand the fate of biological pollutants in subsurface environments and the spread of waterborne diseases resulting from groundwater contamination by microbial pathogens. Microbial ecologists are interested in the fate of chemical pollutants in soil and groundwater environments. They are primarily concerned with microbial transformations of organic and inorganic chemicals in groundwater. It is, however, essential that microbiologists grasp the basic concepts of soil science and groundwater hydrology if they wish to understand how microbial pathogens move through soils, how they survive in groundwater, and how active the indigenous microbial populations are in groundwater (see Chapter 2).

1.2. FATE OF CHEMICAL POLLUTANTS IN GROUNDWATER

Groundwater may contain a wide range of organic substances, some of which are of natural (e.g., humic and fulvic acids) or anthropogenic origin. Some of these man-made chemicals are refractory to microbial degradation in soil and may reach groundwater supplies that are used for drinking purposes. Concern has been raised over the toxicity, mutagenicity, teratogenicity, and carcinogenicity of these compounds. Some known carcinogens such as chloroform, carbon tetrachloride, and trichloroethylene have been detected in well water in various regions in the United States. Adequate epidemiological data with regard to exposure to these chemicals is, however, lacking (Chapter 7).

The fate of these trace organics in soil and aquifer materials is governed by physical (solubility, adsorption to surfaces, volatilization), chemical, biological, and environmental factors. Of most concern are organic compounds that are nonvolatile, soluble, and refractory to microbial action (1). One needs to know more about the extent and the mechanisms of microbial degradation of trace organics in subsurface environments. Many of these trace organics adsorb to organic and inorganic surfaces and the extent of their degradation by attached bacteria can be explained via biofilm models (see Chapter 5). Trace organics, at very low concentrations, can be utilized as secondary substrates and their fate can be explained by non-steady-state biofilm models (4,5). Other factors, in addition to very low substrate concentration, may explain the recalcitrance of trace organics in soils and aquifers (6).

Groundwater contamination with inorganic chemicals is also a cause for concern. The most pertinent example is nitrate which can cause infantile methemoglobinemia (see Chapter 7). Nitrogenous compounds are transformed in soils via microbially mediated reactions. Nitrate is formed as a result of these reactions and moves readily through soils to reach groundwater. Chapter 9 focuses on the microbial transformations of nitrogen and on management practices that can minimize groundwater pollution by nitrate following land application of sewage effluents and sludges.

Microorganisms may find in the groundwater environment all the needs (inorganic and organic carbon sources, N, P, S, Ca, Mg, . . . , electron acceptors) necessary for growth and energy generation. Table 1.2 summarizes the environmental factors and growth requirements controlling microbial activity in a subsurface environment. Oxygen is generally absent in deep aquifers although 2–8 mg/liter have been found in water from deep aquifers in Arizona and Nevada (7). In the absence of oxygen, other electron acceptors (e.g., NO_3^-, SO_4^{2-}) may be available in groundwater (8). Bacterial activity in subsurface environments has hardly been addressed. In shallow wells, bacterial concentration was shown to be as high as 10^6 bacteria/ml, using epifluorescence microscopy (9). We have found 10^4 and 10^5 bacteria/ml in wells 20-ft and 450-ft deep, respectively (Bitton, unpublished data). Kinetic analysis, using glutamic acid and glycolate as

Table 1.2. Microbial Growth Requirements in the Subsurface Environment

Parameter	Comments
Bacterial Concentrations	Almost 0–10^6/ml
Environmental Factors	
Temperature:	Increases with depth (3° C/100 m) May preclude growth at high depths
Osmotic pressure:	May have an effect only in saline aquifers
Hydrostatic pressure: pH:	Not likely to exclude microbial growth
Redox potential:	Certain microorganisms may be excluded in very reduced environments
Role of surfaces:	Biodegradation probably occurs on surfaces in the aquifer material
Growth Requirements	
Carbon sources:	*Inorganic C:* Carbonates and bicarbonates *Organic C:* "Humic" substances and wastewater organics some of which are recalcitrant
Mode of utilization of organics:	At very low concentrations organics may be utilized as secondary substrates.
Other elements:	N, P, S, Na, Ca, Mg, . . . may be present in sufficient quantities to allow growth.
Electron acceptors:	O_2: absent in most deep aquifers. However, availability of other electron acceptors such as NO_3^- and SO_4^{2-}
Sampling for Groundwater Indigenous Microorganisms	Methodological difficulties due mostly to contamination during drilling.

substrates, has shown that the V_{max} and the specific activity index was greater for the shallow well (1.5-m depth) bacterial population than for plankton from a stream in Canada (9). Little is known, however, about bacterial activity in deep aquifers and the subsurface microbial populations. The stumbling block is that it is rather difficult to sample subsurface environments for microorganisms (see Chapter 11). Microbial activity in groundwater can be better understood by the use of laboratory microcosms that simulate as best possible the fate of chemical pollutants under subsurface conditions. A review of these laboratory microcosms is covered in Chapter 6.

1.3. FATE OF BIOLOGICAL POLLUTANTS IN GROUNDWATER

The use of nontreated or inadequately treated groundwater has been responsible for many waterborne diseases including gastroenteritis, cholera, hepatitis, typhoid fever, and giardiasis. The causative agents are bacterial and viral pathogens as well as protozoan parasites (see Chapter 7).

Soils generally function as protective or attenuating zones with regard to contamination of groundwater with microbial pathogens. It is thus important to study the factors that control the survival and the transport of these pathogens through soils. This topic is covered in Chapter 4. Bacterial pathogens are retained in soils mainly via straining whereas the much smaller enteric viruses are retained mainly via adsorption (10, 11). Factors that govern the retention of microbial pathogens are cations, pH, soluble organic materials, flow rate, soil type, and the nature of water flow (saturated versus unsaturated flow). Pathogen survival in soils is similarly influenced by climatic and soil factors and by the nature of the microorganism under consideration. Despite this soil barrier, some pathogens may reach groundwater and survive for long periods in the subsurface environment. Septic tanks (Chapter 8) and land disposal of sewage effluents and sludges (Chapter 9) are two important contributors to groundwater pollution by biological agents. Chapter 9 focuses on management practices that could help minimize the pollution burden.

More data should be generated on the persistence of microorganisms of public health significance in groundwater. Chapter 6 gives information on the various microcosms (flasks, dialysis tubing, McFeter-type chambers) that can be used to reach this research goal. The methodological aspects of the determination of microbial biomass and specific biochemicals in groundwater need to be addressed through further research (Chapter 13).

Groundwater pollution can be followed via deliberate injection of suitable microbial tracers. The suitability of a microbial tracer should be based on its retention potential in soils and upon its survival pattern in soil and groundwater. A review of the various microbial tracers (yeasts, bacteria, phages, and viruses) is presented in Chapter 10. In view of the disadvantages of some of the tracers, more needs to be done to find a suitable tracer of groundwater contamination.

Data dealing with various aspects of groundwater pollution microbiology may be analyzed by using a statistical approach. Statistical techniques such as linear regression, synographic computer mapping, or time-series analysis may be used to analyze microbial adsorption and survival in soils or finding the source and route of well contamination with pathogens (Chapter 12).

1.4. RESEARCH NEEDS AND CONCLUSIONS

Within the field of aquatic microbiology, little is known about the microbiological aspects of groundwater pollution in comparison to surface waters. Some of the research needs in groundwater pollution microbiology are summarized in Table 1.3. More needs to be known about the behavior of pathogens in subsurface environments. We need to know more about their persistence and transport in aquifer material. This research should logically lead to the selection of suitable indicator(s) of groundwater contamination with fecal wastes.

Our knowledge of microbial activity in a subsurface environment is very limited. How do indigenous bacterial populations handle trace organics? Is cometabolism important in microbial degradation of trace organics? We also need to know more about the kinetics aspects of bacterial activity in ground-

Table 1.3. Research Areas in Groundwater Pollution Microbiology

1. *Fate of pathogens in subsurface environment*
   Transport of bacterial and viral pathogens through soils.
   Movement in aquifer material.
   Survival pattern in groundwater.

2. *Microbial activity in subsurface environment*
   Characterization of subsurface microbial populations.
   Secondary utilization of substrates at very low concentrations.
   Microbial degradation of trace organics: Mechanism and extent of the phenomenon (Laboratory and field studies).
   Mechanisms of toxification and detoxification of chemicals under groundwater conditions.
   Bacterial activity in groundwater: Kinetic studies and microscopic counting (epifluorescence, autoradiography . . .).

3. *Microorganisms as tracers*
   Search for suitable microbial tracers.
   Use of mutants other than antibiotic-resistant ones.

4. *Methodological problems*
   Improvement of methods for sampling indigenous groundwater microorganisms.
   Microcosms to study the fate of pathogens in groundwater.
   Microcosms to study the fate of chemicals.

5. Use of microorganisms in remedial measures for groundwater cleanup operations.

water. Research in groundwater microbiology was probably hindered by methodological difficulties in sampling indigenous subsurface microbial populations and in studying the fate of biological and chemical pollutants. Research should focus on microcosms that best simulate groundwater conditions.

Probably because it is a hidden resource, groundwater has been abused for many years. The best approach to the protection of this valuable resource is via prevention. Various pieces of legislation have dealt more or less with groundwater protection. This legislation is summarized in Chapter 14. It is worth noting that groundwater has only been recognized as an important part of the U.S. water resources since 1974 (Safe Drinking Water Act of 1974). Legislation pertaining to groundwater pollution needs to be coordinated and a national program needs to be defined and implemented by the regulatory agencies.

Some remedial measures have been proposed for groundwater decontamination. These measures are, however, complex and costly. The best approach to groundwater contamination is clearly stated in the French proverb, "*Il vaut mieux prévenir que guérir.*" (Prevention is better than the cure).

REFERENCES

1. R. A. Freeze, and J. A. Cherry, *Groundwater*, Prentice-Hall, Inc., Englewood Cliffs, N.J., 1979, p. 604.
2. Florida Environ. Service Center, Report No. 4, July/Aug. 1982.
3. F. Troise, and V. Munger Kahn, Eds., *The Groundwater Newsletter*, July 31, 1982. Water Inform. Center, Inc. Syosset, New York (1982).
4. P. L. McCarty, M. Reinhard, and B. E. Rittmann, *Environ. Sci. Technol.* **15,** 40 (1981).
5. B. E. Rittmann, P. L. McCarty, and P. V. Roberts, *Ground Water* **18,** 236 (1980).
6. M. Alexander, *Science* **211,** 132 (1981).
7. I. J. Winograd and F. N. Robertson, *Science* **216,** 1227 (1982).
8. J. F. McNabb and W. J. Dunlap, *Ground Water* **13,** 33 (1975).
9. T. I. Ladd, R. M. Ventullo, P. M. Wallis, and J. W. Costerton, *Appl. Environ. Microbiol.* **44,** 321 (1982).
10. G. Bitton, in G. Bitton and K. C. Marshall, Eds., *Adsorption of Microorganisms to Surfaces*, Wiley-Interscience, New York, 1980, pp. 331–374.
11. C. P. Gerba, C. Wallis, and J. L. Melnick, *J. Irrig. Drain. Div.* **101,** 157 (1975).

2

ELEMENTS OF SOIL SCIENCE AND GROUNDWATER HYDROLOGY

Herman Bouwer

U.S. Water Conservation Laboratory
U.S. Department of Agriculture, Phoenix, Arizona

2.1. INTRODUCTION

The use of groundwater goes back to the early days of mankind. Ingenious groundwater collection systems existed thousands of years ago. However, groundwater has only recently evolved from a "secret and occult substance emanating from the bowels of the earth" to the more clearly understood water resource that it is today. In the United States alone, groundwater supplies about 25 percent of all water used. Sixty-nine percent of the groundwater pumped is for irrigation, 14 percent is for industry, 13 percent is for urban drinking water, and 5 percent is for rural drinking water. About 50 percent of the population in the United States relies on groundwater for drinking and 75 percent of municipal water supply systems use groundwater. Each year, almost one million new wells are drilled. Droughts often bring on a flurry of new well drilling activity. Groundwater resources must be managed wisely in order to minimize contamination and to keep wells from going dry. Ironically, groundwater has been the major source of recent outbreaks of infectious disease in the United States. because its purported pristine purity has been taken for granted and it is not always chlorinated prior to drinking. Groundwater resources must be protected against microbial contamination, wells must be constructed to prevent direct entry of surface water into the groundwater, and the behavior of microorganisms in the underground environment must be better understood. This chapter summarizes aspects of soil science and groundwater hydrology with special reference to the occurrence, movement, and survival of microorganisms.

2.2. VADOSE ZONES AND AQUIFERS

The underground environment typically consists of a vadose zone, one or more aquifers, and some impermeable basement material such as bedrock (Fig. 2.1). The vadose zone is the zone between the surface and the aquifer. It is normally unsaturated. However, it can also contain some saturated or near-saturated portions, such as the topsoil or root zone after a rainfall or irrigation, deeper clay layers that retain more water than the surrounding coarser material, and perching groundwater zones above restricting or perching layers (Fig. 2.1). For this reason, the term *vadose zone* is preferred over the term *unsaturated zone*, which has also been used to define the region above the groundwater. The vadose zone is very important because it forms the link between the surface and the groundwater environments. Unsaturated flow through the vadose zone is important in the downward movement and attenuation of microbial and chemical pollutants from surface sources. Flow in the vadose zone is vital in natural or artificial recharge of groundwater and in the loss of groundwater by evaporation, which can occur directly from the soil surface or via uptake by plants (phreatophytes, for example).

Groundwater-bearing formations or aquifers are essentially saturated. However, entrapped air can also be present, particularly in the region of fluctuating water tables. Thus, groundwater-containing formations should be called groundwater zones rather than saturated zones. The basic distinction between the underground water in the vadose zone and that in the groundwater zone is that the pressure of the water in the vadose zone is less than atmospheric pressure

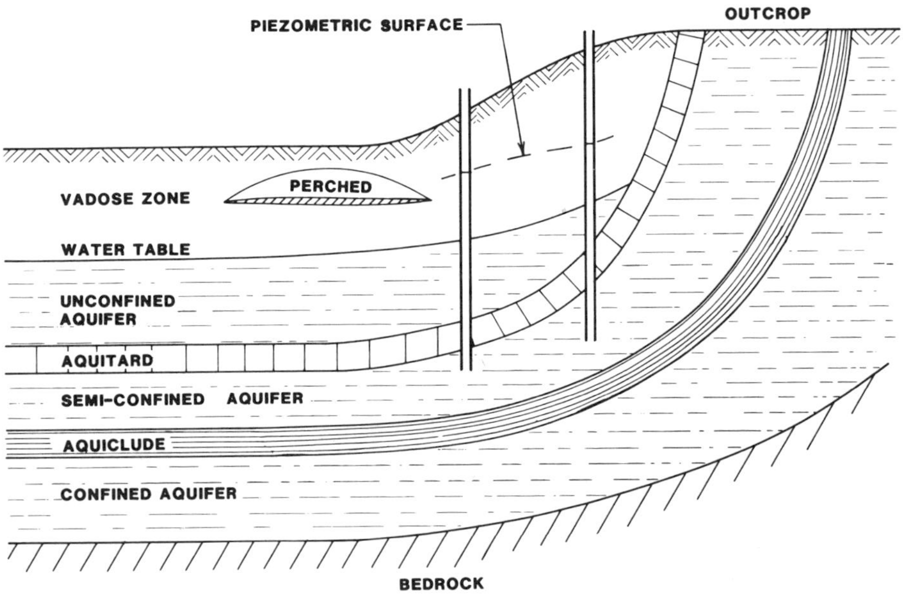

Figure 2.1. Schematic of vadose zone, perched groundwater, unconfined and confined aquifers, water table, and piezometric surface.

(negative gauge pressure), while that in the groundwater zone is greater than atmospheric pressure (positive gauge pressure). Thus, a hole that is dug in the vadose zone will remain dry, whereas a hole dug into the groundwater zone will collect water. Water in the vadose zone can be sampled by applying suction, for example, with a tensiometer-type device that consists of a tube with a porous cup (usually ceramic) at the bottom (see Fig. 2.5). The ceramic is saturated so that it lets water through but no air. Water is then collected in the cup by applying a vacuum to the tube. Other techniques include local pressurization of the air in the vadose zone, which increases the pressure of the water to above atmospheric so that the water will flow into a sampling tube. If the water in the vadose zone is moving downward, samples can also be collected by installing underground plates or troughs to create small bodies of perched groundwater, from which water will flow by gravity into sampling wells or tubes.

Aquifers are groundwater zones of sufficient lateral and vertical extent and of sufficient permeability to produce adequate and sustained flows to wells. Aquifers can be unconfined, confined, or somewhere in between (Fig. 2.1). Unconfined aquifers are like water in a bathtub filled with gravel. The top boundary is the water table (also called the phreatic surface), which can move up or down depending on whether the aquifer is recharged or depleted. For this reason, the water table is also called the free surface. The water table is in contact with atmospheric air through the vadose zone. Thus, the pressure of the groundwater at the water table is equal to atmospheric pressure (zero gauge pressure). Above the water table, the water is at negative gauge pressure and below the water table, the water is at positive gauge pressure. The lower boundary of an unconfined aquifer is a layer of relatively low hydraulic conductivity, called the *impermeable layer* (usually clay or bedrock). Unconfined aquifers yield water (to a pumped well, for example) by a lowering of the water table. Water then drains from the soil pores or rock fractures at the top of the aquifer and is replaced by air through the vadose zone. The volume of water yielded per unit drop of the water table and per unit horizontal area is called the *specific yield* or *storage coefficient*. Values of the specific yield for unconfined aquifers typically are in the range of 0.05–0.25.

Confined aquifers are sandwiched between two essentially impermeable lavers, such as sand between clay layers or sandstone between shale layers (Fig. 2.1). The confining layers are called *aquicludes* if they are essentially impermeable and *aquitards* if they are sufficiently permeable to transmit some water to the confined aquifer or vice versa. Confined aquifers bound by aquitards are called leaky or semiconfined. All water in confined aquifers is at positive gauge pressure and there is no free water table. The pressure condition of the water in a confined aquifer is characterized by the piezometric surface, which is the surface obtained by connecting equilibrium water levels in wells of piezometers reaching into the aquifer (Fig. 2.1). For artesian aquifers, the piezometric surface is above the aquifer itself. If the piezometric surface is above the ground surface, wells tapping the aquifer will be free flowing. Confined aquifers yield water by compression of the aquifer in response to a lowering of the piezometric surface, for example, by pumping water from a well. In contrast to unconfined aquifers, there is no displacement of water by air, and water is yielded by compression of the aquifer. The storage coefficient (yield of groundwater per unit drop of piezometric surface and per unit horizontal area of the aquifer) of confined aquifers is orders of magnitude lower than the specific yield of unconfined aquifers (i.e., on the order of 0.00005–0.01).

Aquifers typically consist of sand and gravel layers and mixes (alluvial or unconsolidated material). Of the consolidated materials, only sandstone is sufficiently permeable to make good aquifers. The other consolidated materials (sedimentary and crystalline rock) must be fractured (shales, granite, basalt, etc.) or have other large pores (solution channels in limestone, for example) to be suitable as aquifers. The fractures, cracks, solution channels, and other large pores in otherwise solid rock are called secondary porosity.

Aquifers may range in thickness from less than 1 m to several hundreds of meters. They may be long and narrow as in small alluvial valleys of they may extend over millions of km^2 and underlie major portions of countries or states. Depth to aquifers may range from 1 m or less to several hundred meters or more.

Unconfined and confined aquifers are the extremes of a whole spectrum of intermediate conditions that can exist in nature. Unconfined and confined aquifers can occur simultaneously as shown in Fig. 2.1. Confining layers may have all sorts of permeability ranges, from essentially zero or much less than that of the aquifer itself to almost the same value as that of the aquifer. Also, confining layers may not be complete and can occur as lenses or as elongated or incomplete layers, thus producing locally confined aquifers that change into semiconfined or unconfined aquifers further away. Alluvial deposits tend to be heterogeneous with old stream beds, flood plains, and oxbows, all horizontally and vertically intertwined. In consolidated materials, secondary porosity (solution channels and fractures) may run in irregular patterns and can be associated with fracture zones and surface lineaments.

2.3. CLASSIFICATION OF SOIL MATERIALS

The soils, sands, or gravels that occur in unconsolidated aquifers, aquicludes, aquitards, and vadose zones are classified according to the size of the individual particles. A commonly used classification scheme is the following system developed by the Soil Survey Staff of the United States Department of Agriculture (28):

Soil Material	Particle Size
Clay	0–2 μm
Silt	2–50 μm
Very fine sand	50–100 μm
Fine sand	0.1–0.25 mm
Medium sand	0.25–0.5 mm
Coarse sand	0.5–1 mm
Very coarse sand	1–2 mm

Particles larger than 2 mm in size are classified as gravel, which is further differentiated as follows [ASTM D 2488; see for example, (27)]:

Soil Material	Particle Size
Fine gravel	0.6–1.9 cm
Coarse gravel	1.9–7.6 cm
Small cobbles	7.6–15.2 cm
Large cobbles	15.2–30.5 cm
Boulders	>30.5 cm

Depending on the relative amounts of sand, silt, and clay in a soil, the texture of the material is classified as sand, silt, loam, clay, or any of the various intermediate classes such as sandy loam, clay loam, and silt loam. The textural classification of a certain soil can be evaluated from the soil textural triangle (Fig. 2.2) prepared by the Soil Survey Staff (28). For example, a soil consisting of 60 percent sand, 30 percent silt, and 10 percent clay is a sandy loam. A soil with 10 percent sand, 60 percent silt, and 30 percent clay is a silty clay loam.

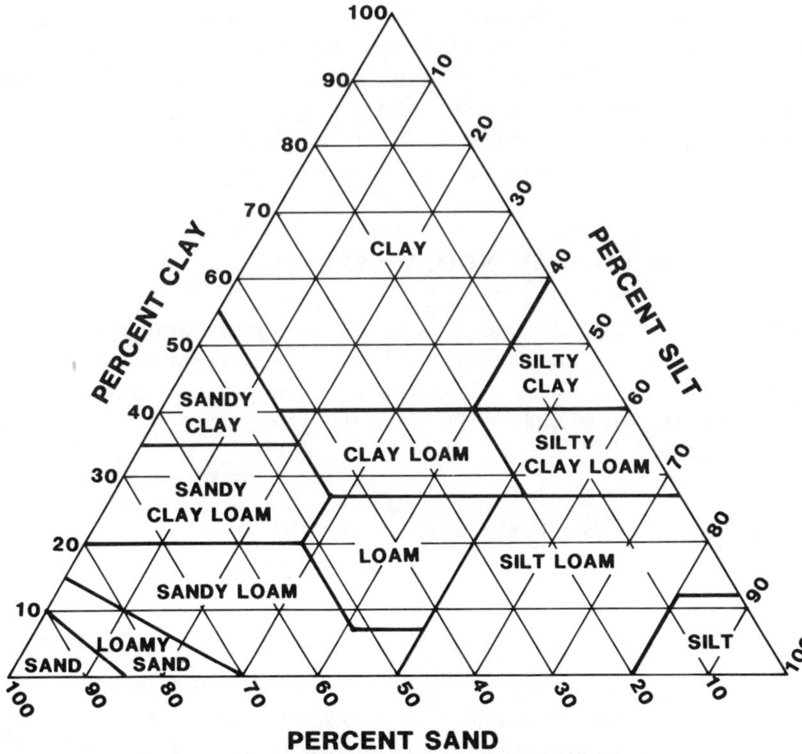

Figure 2.2. Textural triangle for classification of soil.

2.4. POROSITY AND VOLUMETRIC WATER CONTENT

The porosity of a soil or a rock material is the total volume of pores, cracks, fissures, and solution channels per unit volume of material in its natural condition. The porosity is measured by taking an undistrubed sample of the material, drying the sample, weighing it, and dividing the dry weight by the density of the solid material (about 2.65 g/cm^3 for soils and gravels, 2.75 g/cm^3 for limestone and granite, and 3 g/cm^3 for basalt) to get the net volume of the solid phase, V_s. Subtracting this from the total volume V_t of the undisturbed sample gives the porosity as $(V_t - V_s)/V_t$. Typical porosity ranges of underground materials are:

Material	Porosity (percent)
Silts and clays	50–60
Fine sand	40–50
Medium sand	30–40
Coarse sand	20–35
Gravel	20–30

Material	Porosity (percent)
Sand and gravel mixes	10–30
Glacial till	25–45
Dense, solid rock	<1
Fractured and weathered igneous rock	2–10
Permeable recent basalt	2–5
Vesicular lava	10–50
Tuff	30
Sandstone	5–30
Carbonate rock with original and secondary porosity	10–20

The volumetric water content θ of a soil or rock material is the volume of water it contains per unit volume of material in its natural condition. Values of θ are calculated by taking an undisturbed sample and determining its original weight W_w and its weight W_d after drying. If the weights are expressed in grams and the volume V_t of the undisturbed sample in cm^3, θ is calculated as $(W_w - W_d)/V_t$. If the material is saturated with water, there is no air in the pores or cracks and the volumetric water content is equal to the porosity. If the material is unsaturated, the volumetric water content is, of course, less than the porosity.

2.5. CATION-EXCHANGE CAPACITY AND STRUCTURE OF CLAYS

Clay particles are of colloidal size and when wetted they display negatively charged surfaces that attract and adsorb cations. The smaller the cation and the higher its charge, the more strongly it is adsorbed and the more readily it replaces more weakly held cations. Of the more common cations in soil systems, Li^+ is the most weakly held, followed in order of increasing bond strength by Na^+, H^+, K^+, NH^+, Mg^{2+}, Ca^{2+}, and Al^{3+}. Thus, when water with more Ca^{2+} than that in the original soil solution invades a soil or aquifer, Ca^{2+} replaces Na^+, K^+, and the other more weakly held cations in the adsorptive mantle surrounding the clay particles unitl the cations adsorbed to the clay are again in equilibrium with the cations in the solution. The total cation-exchange capacity (milliequivalents of cations adsorbed to the clay particles) of a soil depends on the type and amount of clay in the soil and on the organic matter content. The cation-exchange capacity of clay minerals ranges from around 10 meq/100 g for kaolinite to about 90 meq/100 g for montmorillonite. The cation-exchange capacity of stable organic matter in the soil (humic and fulvic acids and other humus like colloidal substances) may exceed 200 meq/100 g. The cation-exchange capacity of soil materials generally ranges from essentially zero for sands and a few meq/100 g for light-textured soils to about 60 meq/100 g for clay soils. The cation-exchange capacity of a soil is an important parameter in the movement of viruses because clay particles may readily adsorb viruses (see Chapter 4).

The size of the cations adsorbed to the clay particles is important because it determines the distance between the clay particles. For example, Na^+ ions are hydrated, so that they separate the clay particles more than the much smaller Ca^{2+} ions. When clay particles come very close together, they can share adsorbed cations. Also, attractive forces (London–van der Waals forces) can become very strong and hold the particles together to form flocs or aggregates. Thus, Ca-clays have a flocculated structure whereas Na-clays are dispersed and have a deflocculated structure. Ca-clays are friable, easy to work, and have a high hydraulic conductivity. Na-clays are plastic, sticky, hard to work, and have a low hydraulic conductivity. Flocculated clays can change into dispersed clays by passing a solution through them with a higher Na concentration than the original solution in the clay and vice versa. The same applies to other cations, such as K^+ and Mg^{2+}.

The dispersion–flocculation status of a clay is controlled by the sodium adsorption ratio (SAR) of the water in the soil. SAR is defined as $Na^+/[(Ca^{2+} + Mg^{2+})2]^{1/2}$ with the concentrations expressed in meq/liter. The dispersion-flocculation status of a clay is also influenced by the total salt concentration of the soil solution, since a high salt concentration compresses the mantle of adsorbed cations on the clay particles and allows the particles to come closer together and form flocs. The combined effects of SAR and salt concentration of the soil solution on the dispersion–flocculation status of the clay as reflected by the hydraulic conductivity K are illustrated in Fig. 2.3. This graph shows that the

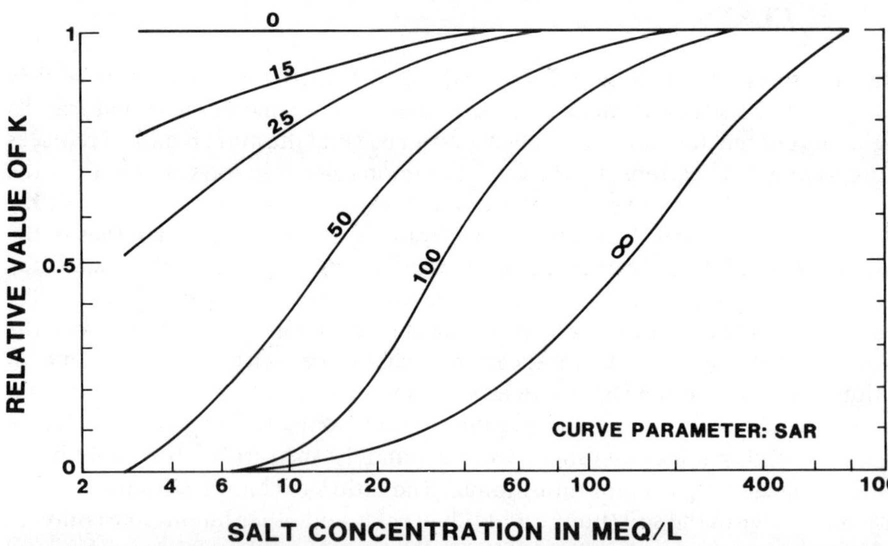

Figure 2.3. Effect of salt concentration and sodium adsorption ratio of soil solution on hydraulic conductivity of Pachappa sandy loam (redrawn from Ref. 19).

cationic composition and salt concentration of the soil solution can have a very pronounced effect on the hydraulic conductivity K of the soil. Thus, a soil or other underground material with flocculated clay behaves like a coarser medium and will allow movement of microorganisms much more readily than the same soil with its clay dispersed.

2.6. SPECIFIC SURFACE

Soils with small particles have a very large internal surface area. This area is expressed as the specific surface, which is the total surface area of the solid particles per g or cm^3 of soil. A medium textured soil can have a specific surface of 100 m^2/g. Organic matter in the soil often has a very large reactive surface and its specific surface may be on the order of 1000 m^2/g. Because soils have such large internal surfaces, they can support the tremendous bacterial populations (up to hundreds of millions per gram of soil in active top soils, less in deeper formations) that grow on these surfaces. These bacterial populations are of great importance in the breakdown of organics in water moving through soils or aquifers, and in other microbiological processes like nitrification, denitrification, and methane formation. The native soil bacteria also may provide competition and therefore a hostile environment for pathogenic organisms that enter the soil environment with polluted water. These antagonistic effects promote the die-off of the pathogens and their removal from the water as it moves through the soil material.

2.7. GROUNDWATER QUALITY AND MICROORGANISMS

Almost all groundwater derives from atmospheric precipitation that infiltrates directly into the soil or seeps into the bottoms of streams or lakes and then continues to move through the vadose zone to the groundwater. Confined aquifers are recharged through aquitards or at their outcrops (Fig. 2.1). The underground residence time or *age* of the resulting *meteoric* groundwater when it is pumped from a well may be 30,000 years or more for deep aquifers in arid areas where the groundwater may date from different climatic periods. Old groundwater can also be pumped in humid areas, for example, from confined aquifers with slowly moving water. Shallow groundwater in humid areas may only be several years old. The groundwater emerging from wet-weather springs may only have been a few hours or days in the underground environment. However, much of the exploitable groundwater typically is on the order of several hundreds of years old.

The chemical composition of groundwater is the result of all the reactions and other processes that have acted upon the water from the moment it condensed in the atmosphere until it is pumped from a well or otherwise leaves the aquifer (e.g., through seeps and springs). Rainwater contains 10–50 mg/liter of dissolved salts. When this water seeps through the soil, it picks up minerals from decayed plants and animals and weathered clay minerals and other soil materials. It also picks up organic compounds, including the very stable humic and fulvic acids that can persist in the water for very long times. By the time the water has moved through the root zone, it can contain several hundred to several thousand mg/liter of dissolved salts. When it moves through the vadose zone, further reactions (ion exchange, precipitation of salt, and solution of salt and weathered minerals) can be expected. Once the water is in the aquifer, additional mineral enrichment occurs as the water is in contact with various minerals for very long times and sometimes at high temperatures. Thus, groundwater typically contains several hundred mg/liter of dissolved salts (mostly sodium, calcium, magnesium, chloride, sulfate, and bicarbonate, plus a wide spectrum of other chemicals in small concentrations or traces; see, for example (12)). High salt concentrations are encountered in old groundwater from the deeper portions of alluvial aquifers in old basins where the water has been in contact with salts and other evaporites. High salt concentrations also occur where the water has been in contact with marine deposits or other rocks with a high salt content, where it has been in aquifers that were associated with old inland seas, and where it has been in coastal aquifers where seawater has intruded inland. Aquifers with marine or otherwise saline water are called *salaquifers*.

Natural groundwater from all but very shallow aquifers is considered free from pathogenic bacteria and viruses. Fortunately for the many consumers of marginally treated groundwater (more than 60 million in the United States alone according to Allen and Geldreich (1)), this is usually true. However, the general absence of pathogens in natural groundwater creates a false sense of security that is periodically brought to the fore when groundwater in aquifers or in distribution systems is contaminated with polluted water and when lack of chlorination or other disinfection causes disease outbreaks (see Chapter 7). Microbial life in aquifers, however, is entirely possible, and nonpathogenic microorganisms native to the aquifer may show up in well discharges.

Surface soils, of course, contain a myriad of microorganisms, but their numbers decrease rapidly below the root zone. Most studies on subsurface microbiological activity have been carried out on surface soils and only meager data are available for deeper regions. Physical and chemical properties of deep subsurface environments were examined by McNabb and Dunlap (18), who concluded that microbiological activity can and does exist in many subsurface regions. Temperatures favorable for microbial life, for example, extend to a depth of about 2000m (assuming a normal temperature increase of 3°C/100m). Water pressures to this depth are not high enough to deter microbial activity and many bacteria can live under the high osmotic pressures of saline water. Carbon sources necessary for microbial life are carbonates and other inorganic carbon that are present in most underground materials. Sedimentary deposits also contain organic carbon, mostly as bitumens and other stable, humus-type substances that are only slowly degraded by bacteria. Other essential elements like N, P, S, and trace elements are also generally present in underground materials. Oxidation–reduction levels often are within the tolerance range of bacteria, except perhaps in deep, isolated formations that may be too reduced for certain microorganisms. Since molecular oxygen is usually absent in deeper regions, anaerobic bacteria will prevail. These bacteria use sulfate, carbon dioxide, nitrate, and simple organic compounds rather that oxygen as electron acceptors in their metabolism.

The study of microbial activity in aquifers and other deep formations is difficult because it is virtually impossible to drill a sterile well. Willis et al. (30) isolated various bacteria in water pumped from a saline aquifer, but concluded that while the bacteria may be native to the aquifer, the possibility of many of the bacterial genera having been introduced by the drilling of the well cannot be precluded. Spring water could give accurate data for biological conditions in the producing aquifers, but these are usually relatively shallow and not indicative of what may happen in deeper formations. Spring water also is easily polluted, so sampling should be done carefully. The best way to study microbial activity in deep formations is to take core samples during drilling. The outer, contaminated portions of the samples are then removed, leaving the inner, untouched portions for bacterial analyses. Despite difficulties of bacterial contamination, the results of such studies often show indigenous bacterial activity, and it is generally accepted that varied and active microfloras exist in deep formations (18). The bacteria include sulfate reducers, denitrifiers, methanogens, and sulfur and hydrocarbon oxidizers (see Chapter 11).

Although bacteria may grow on solid surfaces, they can break loose and show up in the groundwater pumped from the aquifers. Smith et al. (26), for example, reported an average of 1291 denitrifiers/liter in groundwater from 16 wells in the Chalk of London aquifer (the range was 0–17,200 bacteria/liter). Sulfate-reducing bacteria could be detected in water from only two wells and at concentrations of 14 and 200 bacteria/liter. McCabe et al. (17) found that 60 percent of the water samples from 621 wells had a plate count of aerobic bacteria of 10^3–10^5/liter, 7 percent of the samples exceeded a count of 10^6/liter, and 17 percent had a count of zero. While these bacteria normally do not have a direct health significance, they did show that groundwater is not sterile. Coliform bacteria were found in 9 percent of the samples and fecal coliforms in 2 percent. The presence of coliform bacteria was attributed mostly to contamination of the water resulting from poor construction of the wells.

The origin of bacteria in aquifers is unknown. They could have been deposited with the sediments millions of years ago or they may have migrated recently into the formations with meteoric water or during construction of wells. Bacteria generally do not move large distances in fine-textured soil (less than a few meters, for example), but they can migrate much larger distances in coarse-textured or fractured materials (24). Such materials are vulnerable to bacterial contamination by surface water (especially sewage effluent, leachate from garbage dumps, and other polluted water) and animal feces when exposed to the surface without being protected by a relatively fine-textured soil. Pathogenic bacteria, viruses, and other microorganisms not native to the subsurface environment generally do not multiply underground and eventually die. However, they could move considerable distances and survive for a long enough time to be a source of concern around waste disposal areas and other places where pathogens can enter the underground environment (sewage lagoons, sludge drying beds, unlined oxidation ditches, leaky sewer lines, septic tank leach fields, cesspools, garbage dumps, cemeteries, land application systems for wastewater, infiltration of polluted water from streams and lakes, and infiltration of urban runoff).

Two factors significantly affecting the mobility of bacteria and viruses in the underground environment are the size of the water-filled pores including cracks, fissures, and solution channels, and the actual velocity of the water in those pores. Other factors affecting the survival of bacteria in the underground environment are pH, temperature, and oxygen level (see Chapter 4).

2.8. STRAINING AND ADSORPTION OF MICROORGANISMS

The diameter of small bacteria is on the order of 500–1000nm and that of viruses is 30–300nm. While small bacteria may still be adsorbed to soil particles, the larger bacteria are mostly immobilized in soils by physical straining or filtering. Viruses usually are small enough to be adsorbed to clay particles and other reactive surfaces. The isoelectric point of viruses depends on the particular species and ranges from a pH of about 3–7. Below their isoelectric points, the viruses are positively charged and can be adsorbed by the negatively charged surfaces of clay minerals and colloidal organic matter. Above the isoelectric point, viruses are negatively charged and can be immobilized via anion adsorption. However, cation adsorption seems to be the dominant process and virus immobilization in soil generally increases with decreasing pH. Virus immobilization also increases with increasing salt concentration of the soil water and with the concentration of divalent or trivalent cations as opposed to monovalent cations (14). Salt concentrations and di- or trivalent cations tend to compress the thickness of the layer of adsorbed cations around clay particles, thus allowing viruses to come closer to the clay particles and be more strongly adsorbed.

Straining of bacteria can be described by filtration theory as developed for the water treatment industry (see, for example, 11, 13, 22, 25). Filtering starts with the trapping of the larger suspended particles, either at the surface of the soil or at some depth. Individual particles may be blocked in the pores or several particles may interact to form a bridge in the pore that prevents further movement of these particles in the direction of flow. Once movement of the larger suspended particles has been blocked, these particles themselves begin to function as a filter, and trap successively smaller suspended particles until eventually the filter becomes clogged and the hydraulic conductivity and flow rates are reduced to a fraction of their original values. Quantitative aspects of filter clogging were developed by Sakthivadivel and Irmay (25). Ripening of filters by the development of a clogged surface layer is important for the effective filtration of bacteria from the water. Fecal coliform bacteria, for example, penetrated deeper into the soil and reached the groundwater in greater numbers shortly after the start of a new flooding period than after flooding had been continued for some time at the Flushing Meadows Project, which is an experimental groundwater recharge project west of Phoenix, Arizona. The project uses secondary sewage effluent for groundwater recharge with rapid-infiltration basins (5, 6).

Laboratory studies with uniform, spherical materials indicated that bridging occurred when the diameter of suspended particles entering or moving through the medium was more than 0.2 times the diameter of the particles of the medium itself (25). Depending on how the particles in the porous medium were packed, bridging also occurred if the diameter of the suspended material was more than 0.07 times that of the particles in the medium. When the size of the suspended particles was less than 0.07 times the particle size of the medium, the suspended particles moved through the medium without bridging or blocking. From these experiments it can be concluded that bacterial cells can be strained by dispersed clays and fine silts, but can also move readily through coarser materials such as flocculated clays, very fine sands, and coarser soils. Well-sorted sands and gravel materials, fractured rock, and karstic limestone have large pores through which bacteria can travel essentially unrestricted. Normal die-off is the only mechanism that controls the movement and spread of bacteria in such materials (see Chapter 4).

At the Flushing Meadows Project near Phoenix, Arizona, fecal coliform concentrations in secondary sewage effluent were reduced from about 1 million/100 ml to a range of zero to a few hundred/100 ml after percolation through about 1 m of loamy sand, which was the first soil layer below the rapid infiltration basins (4, 5). Beneath the loamy sand were coarse sand and gravel layers to a depth of about 80 m while the groundwater table was at a depth of about 3 m. Water sampled from this unconfined aquifer at depths of 7–10 m contained up to several hundred fecal coliforms/100 ml when sampled directly beneath the basins. The fecal coliform concentration declined as the water moved laterally through the aquifer away from the basins. Fecal coliforms were still readily detected in a monitoring well 30 m away from the basins, but could no longer be detected in a well 90 m from the basins (5). Thus, it appeared that for the particular conditions of this project, the sewage water had to move about 100 m through the aquifer before complete removal of fecal coliform bacteria was obtained. On the other hand, viruses could not be detected in the renovated sewage water sampled at 6–9-m depth immediately below the infiltration basins (10), indicating that the much smaller viruses were readily adsorbed and immobilized in the loamy sand top layer. This study demonstrated that bacteria can be more mobile in the underground environment than viruses.

2.9. REDOX POTENTIAL

An important parameter of the underground environment that determines the type of microbiological processes that take place and the survival of various microorganisms is the oxygen concentration, which is usually measured and expressed in terms of the redox potential Eh. This potential basically expresses the electron availability as it affects the oxidation states of hydrogen, carbon, nitrogen, oxygen, sulphur, manganese, iron, cobalt, and copper in aqueous systems (2). The redox potential is measured as the potential difference in mV between a black platinum electrode (21) and a saturated calomel electrode placed in the system. Gilbert et al. (9), for example, buried black platinum electrodes at depths of 2, 15, 30, 60, and 90 cm to measure Eh values below the bottom of a rapid-infiltration basin intermittently flooded with secondary sewage effluent. Three platinum electrodes were buried at each depth. The reference calomel electrode was fitted into an anchored float that rested on the soil surface during drying. The resulting 15 electrodes were scanned once every hour on a recorder through a pH meter and a timing device. The results typically showed that Eh decreased during flooding and increased during drying (Fig. 2.4). The fastest changes occurred at the 2-cm depth, which also exhibited a marked diurnal cycle during flooding with peak values of Eh at about 1400 hrs. This was probably due to an increase in the dissolved oxygen content of the infiltrating water due to algal activity in the infiltration basin.

Where multiple Eh determinations need to be made in different systems (for example, in various soil columns in the laboratory), the systems can be connected with a calomel bridge to form one continuous reference electrode. The Eh values

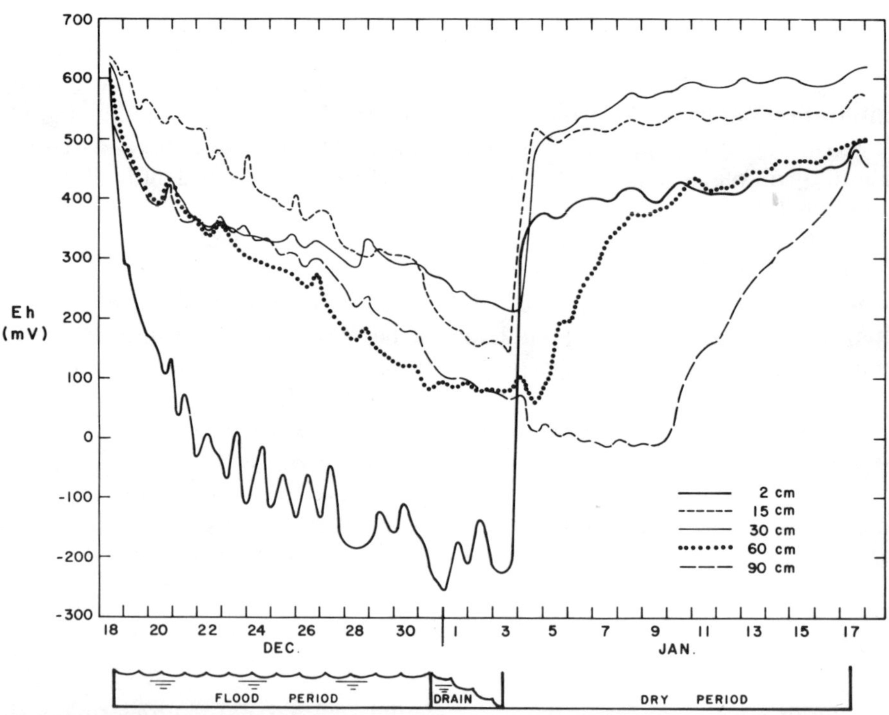

Figure 2.4. Redox potentials at different depths below an infiltration basin intermittently flooded with sewage effluent (9).

for different platinum electrodes in the systems connected by the calomel bridge can then be monitored by using a switch box that connects the electrodes to the calomel electrode in turn and a recorder for monitoring the voltage differences (16).

Redox potentials in the soil are affected by the depth below the surface, the oxygen diffusion rate of the soil as determined by the size and number of air-filled pores which in turn are a function of the soil texture and water content, the pH, and the oxygen sink or demand for oxygen in the soil itself. The latter is affected by the presence of degradable organic carbon and bacterial activity. Oxygen levels in aquatic surface and subsurface environments can also be expressed in terms of the logarithm of the electron concentration pe (see for example, Ref. 29). Values for Eh and pe for various microbiological processes are (at 25°C and pH 7):

	pe	Eh (in mV)
Aerobic respiration	+13.75	+810
Denitrification	+12.65	+750
Sulfate reduction	− 3.75	−220
Methane formation	− 4.13	−240

Bohn (2) reported that nitrate reduction in soil has also been observed at an Eh of +225 mV. Patrick and Mahapatra (20) suggested the following classification of oxygen levels in soils based on their redox potential at pH 7:

Soil Type	Redox Potential
Oxidized soil	>400 mV
Moderately reduced soil	100–400 mV
Reduced soils	−100–100 mV
Highly reduced soils	−300–−100 mV

Since some compounds are more readily degraded and some bacteria die off faster under aerobic rather than anaerobic conditions and vice versa, it is always best if contaminated water in the underground environment moves through both aerobic and anaerobic portions. This is usually achieved in systems where wastewater is intermittently applied to the soil. The upper portion of the vadose zone will then be predominantly aerobic, whereas the deeper portions and the aquifer will be anaerobic.

2.10. WATER CONTENT-PRESSURE HEAD RELATIONS

In unsaturated soils, large pores are empty and water moves only through the small pores which may still be completely filled with water. The water can also move as film flow along the pore walls and around the individual soil particles.

The closer proximity of the water to the solid phase in unsaturated material can be expected to enhance removal of viruses and bacteria from water moving through such media. For this reason, it is always desirable to maintain an unsaturated zone of sufficient extent between the points of entry of water with microorganisms in the underground environment and the underlying groundwater. The more unsaturated the zone is, the better protection against breakthrough of microorganisms can be expected. The same will be true, at least to a certain extent, for consolidated materials where the water moves through cracks, fractures, or other secondary porosity. In these materials, partial saturation also will mean greater contact between the water flowing through the medium and the solid phase.

The water content of unsaturated soils or other underground materials is functionally related to the pressure head of the water (expressed in cm or m of water). The *pressure head* at a certain point simply is the vertical distance between the equilibrium water level in a tube inserted to that point and the point itself (Fig. 2.5). In the groundwater zone, pressure heads are positive and the equilibrium water level in such a tube (called a piezometer) will be above the point (Fig. 2.5 left). In the vadose zone, pressure heads are negative and the water level in the tube will establish itself below the point in question (Fig. 2.5 center). This requires that the end of the tube in the soil be equipped with a porous plug (usually a porous ceramic cup) that will let water through but not air. Negative pressure heads are measured with a tensiometer, which is a vertical tube with a porous cup at the bottom and a vacuum gauge or manometer at the top to measure the negative pressure head in the water at a certain depth (Fig. 2.5 right).

If the water in the vadose zone is completely at rest and in equilibrium with the underlying groundwater table, the pressure head at any point in the vadose zone will be equal to the negative distance of that point above the water table. If the vadose zone consists of a homogeneous soil, the equilibrium water content will be highest near the water table and lowest near the surface (Fig. 2.6). A curve of water content versus height above water table then represents the relation be-

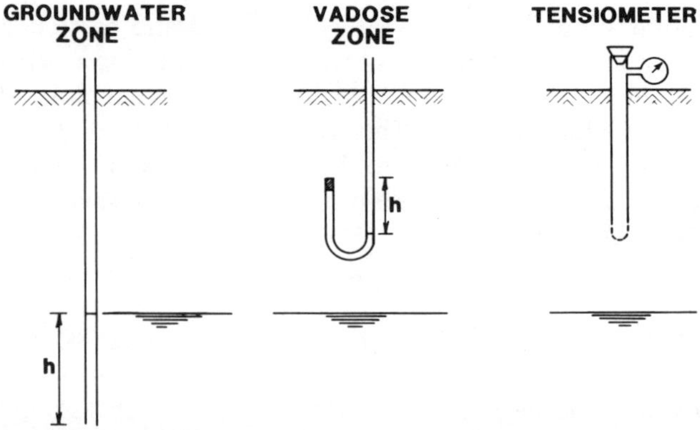

Figure 2.5. Schematic of pressure head measurement in the groundwater zone (left) and in the vadose zone (center and right).

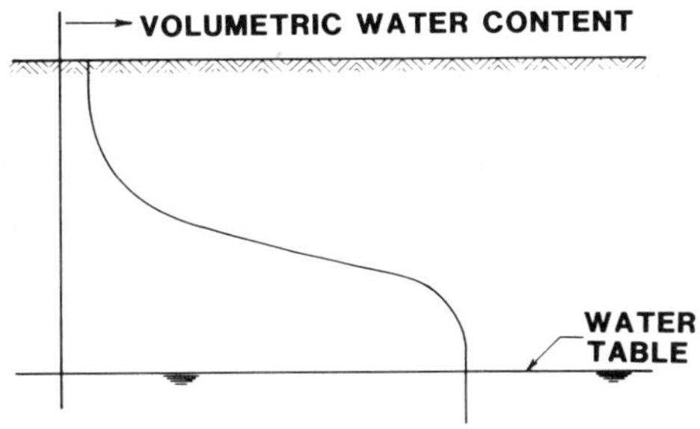

Figure 2.6. Hypothetical water content distribution in a homogeneous vadose zone at static equilibrium.

tween negative pressure head and water content of the soil in the vadose zone. Such curves are called soil water characteristics of the material and are usually plotted with the volumetric water content on the ordinate and pressure head on the abscissa (Fig. 2.7).

True equilibrium conditions never exist in real vadose zones due to vertical movement of water caused by infiltration and evaporation at the surface and by fluctuations of the water table. Also, long periods of time are required for equilibrium conditions to be approached, even if all inputs and outputs of water in the vadose zone have come to a halt. Curves such as in Fig. 2.7 are thus usually determined in the field by simultaneous measurement of negative pressure heads with tensiometers and water contents with, for example, the neutron method. They can also be determined in the laboratory by subjecting the water in an

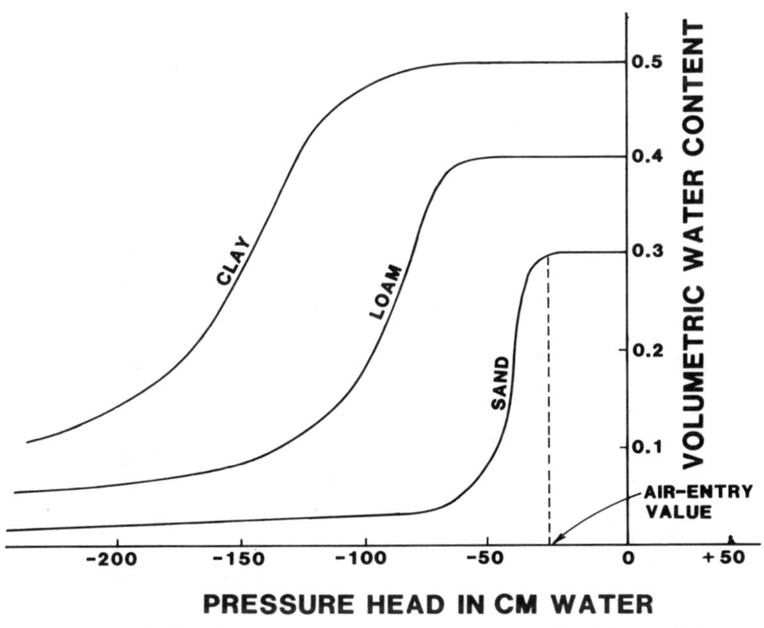

Figure 2.7. Typical relations between water content and pressure head for sand, loam, and clay soils.

undisturbed sample of the soil to various water suctions at the bottom or to various air pressures at the top, and measuring the equilibrium water contents in the sample (23).

Typical (hypothetical) soil water characteristics for a sand, loam, and clay soil are shown in Fig. 2.7. The curves show that finer-textured soils retain more water at zero pressure head (saturated conditions), begin to lose water at lower (more negative) pressure heads, and retain more water at low pressure heads than coarse-textured soils. Thus, a clay layer in an otherwise sandy vadose zone can, at equilibrium water conditions, have a higher water content than the sandy materials above and below it. As a matter of fact, such a clay layer could be essentially saturated, as mentioned at the beginning of the chapter in order to illustrate why vadose zone is a better term than unsaturated zone.

Soil water characteristics are also subject to hysteresis, yielding different curves when water is removed from a wet or saturated soil than when water is added to an unsaturated soil (Fig. 2.8). Also, wetting an unsaturated soil never produces complete saturation even when the pressure heads of the water have become positive. This is because small air pockets become trapped in the wetted zone. The entrapped air disappears very slowly, mainly by solution in the soil water and eventual diffusion back to the atmosphere. Complete saturation can only be obtained on short order if the soil is wetted under vacuum. Because of entrapped air, an unsaturated soil that has been wetted to zero or positive pressure head should not be called a saturated soil. *Rewetted* or *resaturated* soil are better terms.

The curves in Fig. 2.7 show that a decreasing pressure head (starting from zero and going to the left) does not immediately produce a decrease in the soil water content. The soil thus remains essentially saturated until the pressure head reaches the so-called air entry value, whereby the largest pores in the soil are

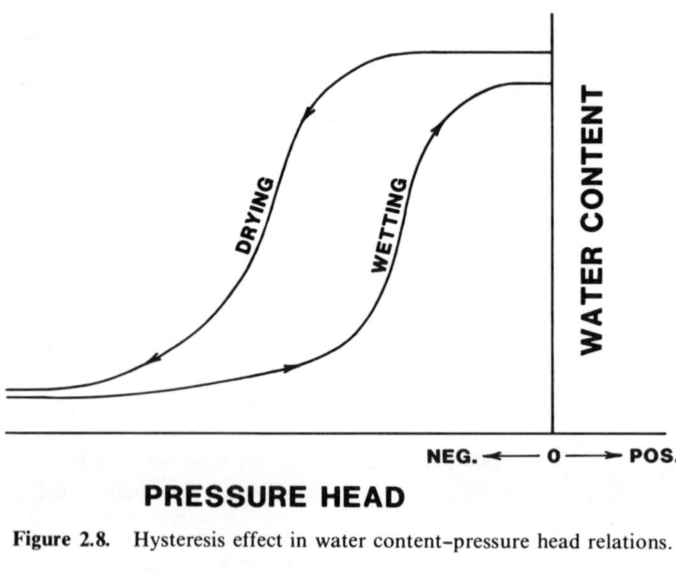

PRESSURE HEAD

Figure 2.8. Hysteresis effect in water content–pressure head relations.

drained and air replaces water. The air entry value ranges from about −20 cm water for coarse sands to −100 cm and less (more negative) for fine-textured soils. If the soil has an air entry value of −30 cm water, this means that the vadose zone for the first 30 cm above the water table is still saturated. This saturated, or essentially saturated, zone is called the *capillary fringe* (see Fig. 2.6). The capillary fringe above a water table is about 20 cm thich for coarse sands (less for very coarse sands or gravels) and 100 cm or more for fine-textured soils.

The top of the capillary fringe rather than the water table itself should be taken as the lower boundary of the unsaturated zone. Thus, where water tables are shallow, the height of the unsaturated zone can be much less than the depth of the water table. Furthermore, laboratory soil columns for studying movement of microorganisms and other processes in unsaturated soil should be long enough to produce the desired length of unsaturated medium. For example, a 60-cm column freely draining at the bottom and filled with sand having an air entry value of −40 cm water only has an unsaturated section of 20 cm. To extend the unsaturated section of such a column, the water at the bottom must be kept at negative pressure head (using, for example, a suction or siphon arrangement).

2.11. WATER MOVEMENT

The flow of water in saturated and unsaturated soils is governed by Darcy's equation. For a system with steady flow and parallel, linear streamlines (Fig. 2.9), Darcy's equation can be written as:

$$v = K\frac{(h_1 + z_1) - (h_2 + z_2)}{L} \qquad [2.1]$$

where v = Darcy velocity of water (length/time); h_1 = pressure head of water at point 1 (length); z_1 = elevation head at point 1 (length); h_2 = pressure head of water at point 2 (length); z_2 = elevation head at point 2 (length); L = distance of flow between points 1 and 2 as measured along streamline (length); K = hydraulic conductivity of soil or aquifer material (length/time).

The pressure head h at a given point in the flow system is the height to which water will rise in a piezometer inserted down to that point. The elevation head of a given point is the vertical distance of that point above an arbitrary horizontal reference plane (dashed line in Fig. 2.9). The sum or pressure head and elevation head at a given point in the flow system is called the total head H. Thus, $h_1 + z_1$ in Equation 2.1 is the total head H_1 at point 1 and $h_2 + z_2$ is the total head H_2 at point 2. The distance L between points 1 and 2 must be measured along the streamline on which the points are located. The ratio $(H_1 - H_2/L)$ is called the *hydraulic gradient* of the flow.

The *hydraulic conductivity* K basically is the factor of proportionality between Darcy velocity and hydraulic gradient. The hydraulic conductivity is a

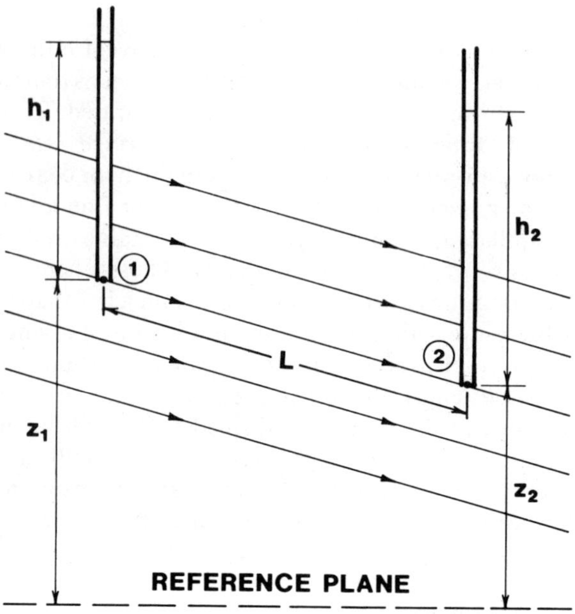

Figure 2.9. Vertical cross section of groundwater flow system with linear, parrallel streamlines.

property of the porous medium. Its value depends on the size and number of pores in the soil or aquifer material. Orders of magnitude for *K* in m/day of saturated granular materials are:

Saturated Granular Material	*K*(m/day)
Clay soils (surface)	0.01–0.2
Deep clay beds	10^{-8}–10^{-2}
Loam soils (surface)	0.1–1
Fine sand	1–5
Medium sand	5–20
Coarse sand	20–100
Gravel	100–1000
Sand and gravel mixes	5–100
Clay, sand, and gravel mixes (till)	0.001–0.1

The hydraulic conductivity of sandstone is considerably less than that of unconsolidated sand with the same grain sizes, due to cementation and higher density of the sandstone. *K* values of other consolidated materials depend entirely on the secondary porosity of the rock (fractures, weathering, and solution channels in carbonate rock). General ranges of *K* in m/day for saturated consolidated materials are:

Saturated Consolidated Material	K (m/day)
Sandstone	0.001–1
Carbonate rock with secondary porosity	0.01–1
Shale	10^{-7}
Dense, solid rock	10^{-5}
Fractured or weathered rock (aquifers)	0.001–10
Fractured or weathered rock (core samples)	almost 0–300
Volcanic rock	almost 0–1000

Darcy's equation applies to both saturated and unsaturated flow. For unsaturated flow, h will be negative and must be measured with tensiometers. Air in unsaturated soils obstructs the flow of water. Thus, the hydraulic conductivity of an unsaturated soil is less than that of the same soil at saturation. The relation between hydraulic conductivity K and the volumetric water content θ, called the *unsaturated hydraulic conductivity characteristic*, can be determined experimentally (3, 7). For a relatively light-textured soil such as a sandy loam, the curve may look like the one in Fig. 2.10. The curve shows

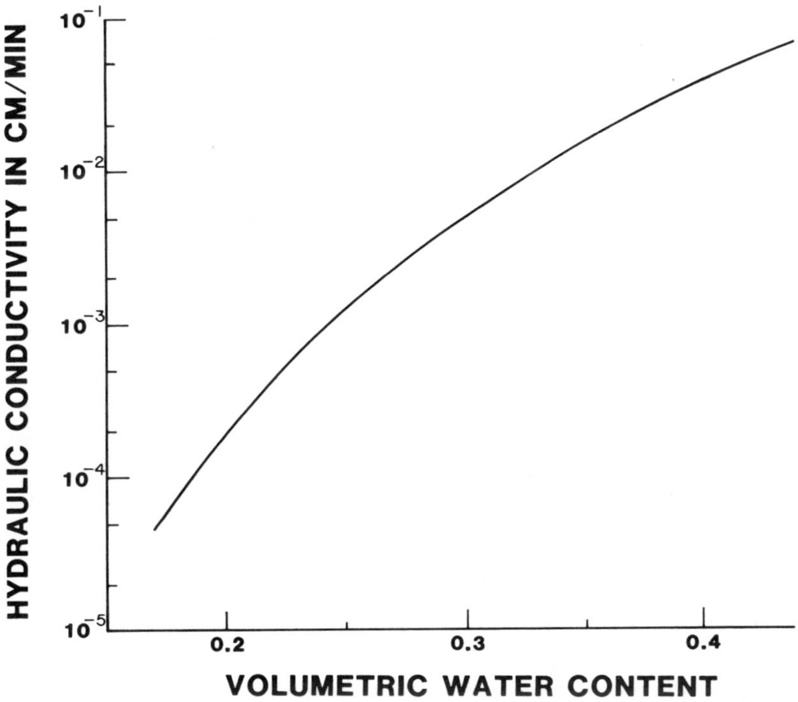

Figure 2.10. Typical relation between hydraulic conductivity and water content for a light soil (sandy loam type, for example).

that K reduces rapidly (several log cycles) with decreasing water content. If the relation between θ and pressure head h is also known, the h–θ and K–θ relations can be combined to yield a curve relating K to h. Since θ–h relations are hysteretic, K–h relations also are different for drying than for wetting. The value of K in Equation 2.1 for unsaturated flow will be the average of the K values at points 1 and 2. To avoid errors, points 1 and 2 for unsaturated flow systems must be taken sufficiently close together so that k_1 belonging to h_1 and K_2 belonging to h_2 are relatively close together and their average can be used without incurring large errors.

The Darcy velocity v is the gross velocity as if the water were moving through the entire porous medium and not only through the water-filled pores. Thus, the volume rate of flow Q through a cross-sectional area A normal to the flow direction is calculated as $Q = vA$. For an aquifer of thickness D, Q in a segment of width W normal to the flow is $Q = vDW = KiDW$, where i is the slope of the water table or piezometric surface in the direction of flow (i is the hydraulic gradient). The product KD is called the transmissivity T of the aquifer and is commonly used in calculations and analyses of flow to wells or other situations of predominantly horizontal flow in aquifers.

2.12. ACTUAL VELOCITY OF WATER

For convective transport of pollutants (including microorganisms) in the underground environment, the actual velocity of the water rather than the Darcy velocity should be considered. The actual or macroscopic velocity of water in a porous medium can be approximated as:

$$v_m = v/\theta \qquad [2.2]$$

where v_m = macroscopic velocity; v = Darcy velocity; θ = volumetric water content (volume fraction.)

This equation applies to saturated as well as unsaturated flow. For saturated flow, θ is equal to the porosity of the material. Equation 2.2 is approximate and assumes that all water in the medium is involved in the flow. Because some of the water is tied up in "dead" pore space, however, Equation 2.2 underestimates the actual velocity of water and the rate of movement of pollutants in the medium. Thus, θ in Equation 2.2 should reflect the *effective* water content of the medium that is involved in the flow.

2.13. DOWNWARD FLOW

Continuous downward flow through the vadose zone from a surface source (waste disposal area, wastewater lagoon, septic tank leach field, and land treatment system) tends to approach a steady-state condition. Water contents

Table 2.1. Values of θ and K from Fig. 2.10 and Actual Downward Velocity Calculated as $v/\theta = K/\theta$

Water content θ (cm^3/cm^3)	Hydraulic conductivity K (cm/min)	(m/day)	Actual velocity v_m (m/day)
0.4	5×10^{-2}	0.72	1.8
0.32	$\times 10^{-2}$	0.144	0.45
0.24	$\times 10^{-3}$	0.0144	0.06
0.18	$\times 10^{-4}$	0.00144	0.008

and pressure heads in the vadose zone (or in each layer thereof if the zone consists of various soils) tend to remain constant with depth. Applying Darcy's equation to such a system of steady, vertical flow shows that the hydraulic gradient is one (gravity flow only) and, hence, the K of the unsaturated soil is equal to the Darcy velocity or average infiltration rate.

Where downward flow rates are too high for proper immobilization of microorganisms, these flow rates can be reduced by decreasing the infiltration rate or seepage rate of the wastewater at the land surface. This will produce longer detention times in the vadose zone and, hence, greater die-off of the pathogenic organisms. Lower downward flow rates also produce lower water contents, which in turn will give better removal of bacteria by more effective straining, and better removal of viruses by closer contact between the viruses and the adsorbing surfaces of the soil particles. Seepage rates from lagoons or disposal areas can be reduced by lining the areas (7). Infiltration rates in land treatment systems can be reduced by lowering application rates and spreading the wastewater over larger areas.

Since a reduction in the downward flow rate will cause a corresponding reduction in the unsaturated K and in the water content θ of the material (see Fig. 2.10), the macroscopic downward velocity in the vadose zone will not decrease in the same proportion as the infiltration rate. Using, for example, the relation between K and θ of Fig. 2.10, various corresponding values of equilibrium water content and downward flow rate (Darcy velocity), which is numerically equal to K because of unit hydraulic gradient, are listed in Table 2.1. Dividing the flow rate by the corresponding θ yields the macroscopic or actual downward velocity of the water, The results show that this actual velocity decreases less rapidly than K, which is equal to the infiltration rate. This behavior should be considered in laboratory or field experiments where removal of microorganisms from downward moving water is studied in relation to the application rate of the water and corresponding residence time in the unsaturated zone.

2.14. LATERAL FLOW

For lateral flow of water through aquifers, the hydraulic gradient in Equation 2.1 is the slope of the water table (unconfined aquifers) or of the piezometric surface (confined aquifers). Thus, Equation 2.1 for this case reduces to $v = Ki$, where i is the slope of the water table or piezometric surface. Dividing the resulting v by the water content or porosity of the aquifer gives an estimate of the actual velocity of the water in the aquifer, which is the more significant one for predicting rates of movement of pollutants and microorganisms. For an unconfined sandy aquifer with $K = 10$ m/day, a water table slope of 1m/km, and a porosity of 20 percent, the actual velocity of the groundwater thus would be $10 \times 0.001/0.2 = 0.05$ m/day or 18.25 m/year.

2.15. ANISOTROPY

If the hydraulic conductivity at a given point in a soil or aquifer is the same in all directions of water flow, the medium is called isotropic. If the hydraulic conductivity depends on the direction of water flow, the medium is called anisotropic. Almost all alluvial soils or aquifers as a whole are more permeable in the horizontal than in the vertical direction because (1) the individual particles or stones are deposited on their flat sides (imbrication) and (2) there is horizontal layering of materials of different texture that makes the hydraulic conductivity of the whole aquifer larger in the horizontal than in the vertical direction. Alluvial deposits that were formed by streams or other moving bodies of water also tend to exhibit anisotropy in the horizontal plane, with a higher hydraulic conductivity in the downstream direction than normal thereto. This is because gravel layers, old stream beds, and other more permeable deposits in these normally very heterogeneous formations tend to be more continuous in the prevailing downstream direction of the old stream channels than normal thereto. Thus, alluvial aquifers as a whole, in addition to being very heterogeneous because they are an intertwined mixture of irregular deposits of gravel, sand, silt, and clay layers, also tend to exhibit three-dimensional anisotropy with the smallest K occurring in the vertical direction, the largest K in the horizontal direction parallel to the predominant flow of water in the ancestral channels, and an intermediate K in the horizontal direction normal to the old prevailing stream flows.

In horizontally isotropic aquifers, lateral flow of groundwater is normal to the contours of the water table or piezometric surface. Thus, the streamlines are orthogonal to the equipotentials. In horizontally anisotropic aquifers, streamlines are no longer orthogonal to equipotential lines and the flow direction depends on the orientation of the principal hydraulic conductivity axes and their angle with the equipotential. This is shown schematically in Fig. 2.11, depicting an alluvial aquifer system with K_x in the direction of the predominant old stream flow larger than K_y normal thereto. For a given point in the aquifer system, an ellipse with principle axes of $K_x^{-1/2}$ and $K_y^{-1/2}$ is constructed (8, 15). If the dashed line through the center of the ellipse represents the equipotential line at that point, the direction of flow in the aquifer at that point is obtained by drawing the hydraulic gradient through the center of the ellipse normal to the equipotential, drawing a tangent to the ellipse where the hydraulic gradient line intersects the ellipse, and drawing the streamline as the line through the center of the ellipse

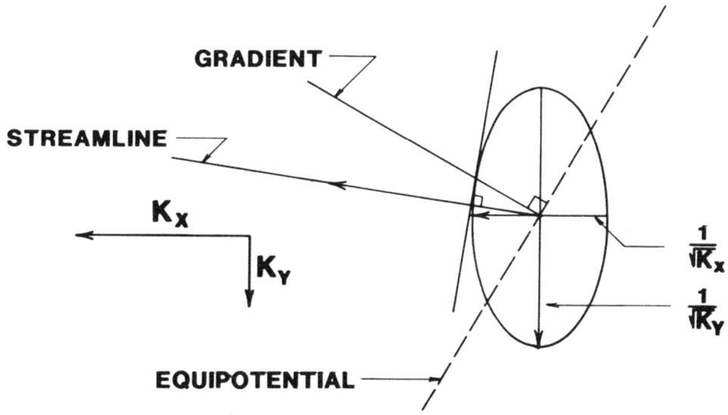

Figure 2.11. Determination of flow direction in horizontally anisotropic aquifer.

normal to the tangent. As shown in the figure, the resulting streamline veers away from the line normal to the equipotential in the direction of the greatest *K* as the groundwater flows along the path of least resistance.

Because of anisotropy and the general heterogeneity of alluvial deposits with old, meandering, and intertwining buried stream channels, lenses and layers of less permeable formations that settled in quieter waters, and the stratified nature of the deposits as a whole, predictions of directional movement of groundwater from water table contours or equipotentials are difficult to make with any accuracy, especially on a small scale. Thus, while simplified models of groundwater flow systems with assumed isotropy and orthogonality between streamlines and equipotentials can be valuable for general hydrologic evaluation of underground flow systems, they must be used with great caution for predicting movement of microorganisms or other pollutants through the aquifer because pollutants may then show up in unexpected places and at unanticipated times. By the same token, surmised streamlines as orthogonals to measured equipotentials cannot be used to conclusively trace an observed incident of bacterial or other pollution of groundwater back to a suspected source. Tracer studies, intensive monitoring of groundwater quality, and analyzing soil and water samples in the vadose zone beneath suspected pollution sources may yield more reliable determinations of sources and convective transport of pollutants in heterogeneous and anisotropic aquifer systems than simplified models.

2.16. DISPERSION AND ATTENUATION

Convective transport of microorganisms in the underground environment is also subject to dispersion. If the organisms enter the soil or aquifer from a line source (polluted river, for example), there will be only longitudinal dispersion, causing the typical *S*-shaped breakthrough curve as some distance from the source (Fig. 2.12). If the organisms enter from a point source, there will also be transverse dispersion, yielding the typical pollution plumes downgradient from the source. The concentration distribution of organisms across the plume is then a probability curve, which becomes wider and flatter with increasing distance from

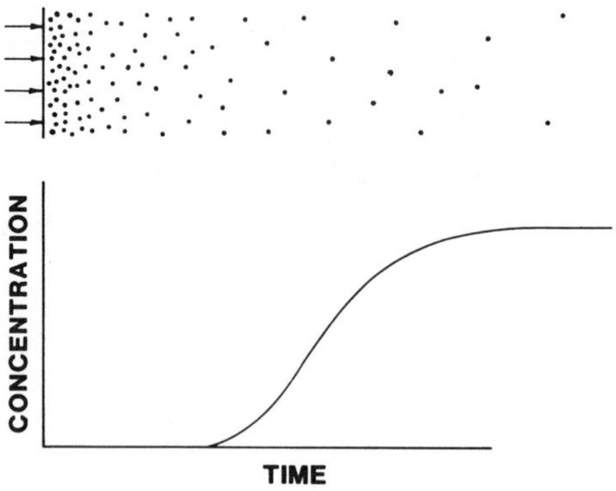

Figure 2.12. Longitudinal dispersion downgradient from line source of contaminants and concentration breakthrough curve at some distance from source.

the source (Fig. 2.13). Straining, adsorption, and die-off will cause a delay in the arrival and a reduction in the concentration of the organisms at points downgradient from the source until eventually no more microorganisms can be detected. The distance from the source where all microorganisms will be removed depends on the type and concentration of microorganisms in the water entering the underground environment, the underground velocities of the water, the type of soil and aquifer materials, and other conditions. It may range from a few meters or less for fine-textured soils and about 100 m for sandy materials to a distance on the order of kilometers for fractured or cavernous rock.

2.17. LAND TREATMENT OF WASTEWATER AND GROUNDWATER MANAGEMENT

The capability of soil–aquifer systems to remove microorganisms and other constituents (e.g., organic carbon, nitrogen, phosphorus, and heavy metals) from wastewater is utilized to provide effective, low-cost, and low-energy treatment of sewage effluent and other wastewater by applying it to the land (see Chapter 9 for more details). The principle of soil–aquifer treatment has, of course, been utilized throughout the ages when human and other wastes were returned to the land. It is still being used today in septic tank and cesspool disposal systems (see Chapter 8). Land treatment is being increasingly used on a larger scale to utilize effluent

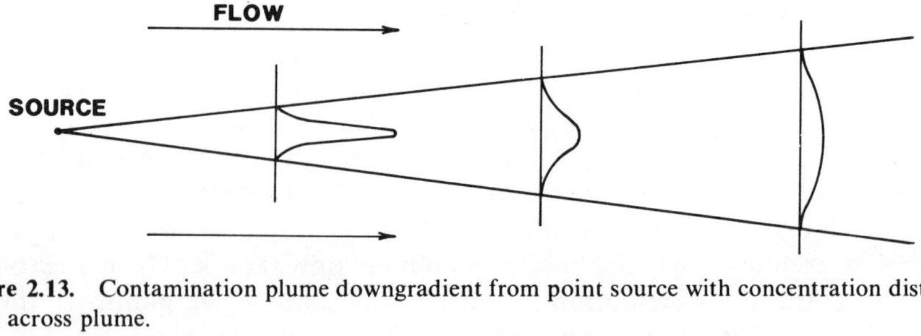

Figure 2.13. Contamination plume downgradient from point source with concentration distributions across plume.

from sewage treatment plants for production of agricultrual crops and groundwater recharge and to minimize pollution of surface water. There are two kinds of land treatment systems for infiltrating sewage into the soil: (1) slow-rate systems whereby the sewage is applied at irrigation-type application rates and most of the effluent water is consumed by crops, and (2) rapid-infiltration systems whereby sewage is applied to infiltration basins for groundwater recharge (Fig. 2.14).

Rapid-infiltration systems require permeable soils (sandy loams to fine sands) to obtain high infiltration rates. Very coarse materials (coarse sands or fine gravels, for example) are not desirable because their specific surface is too small for effective treatment. While the quality of the sewage water is greatly improved as it moves downward through the vadose zone to the groundwater and then laterally through the aquifer, the quality of the resulting *renovated* water often is not as good as that of the native groundwater. Thus, to protect indigenous groundwater against encroachment by renovated sewage water, the renovated water should be removed from the aquifer at some distance from the infiltration system. The desired distance between the infiltration system and the collection facilities depends on the pretreatment and quality of the sewage effluent going into the ground, the hydraulic loading rate and management of the infiltration system, the soil materials in the vadose zone and the aquifer, the depth to the groundwater, and the desired quality of the renovated water. For safety reasons alone, the distance should always be as large as possible. For the sand and gravel system below the Flushing Meadows Project west of Phoenix, Arizona, for example, the distance should be no less than 100 m (6).

Renovated water from rapid-infiltration systems can be removed from aquifers by locating the infiltration system on high ground that drains naturally into a stream or other surface water (Fig. 2.14A). These systems are frequently used to

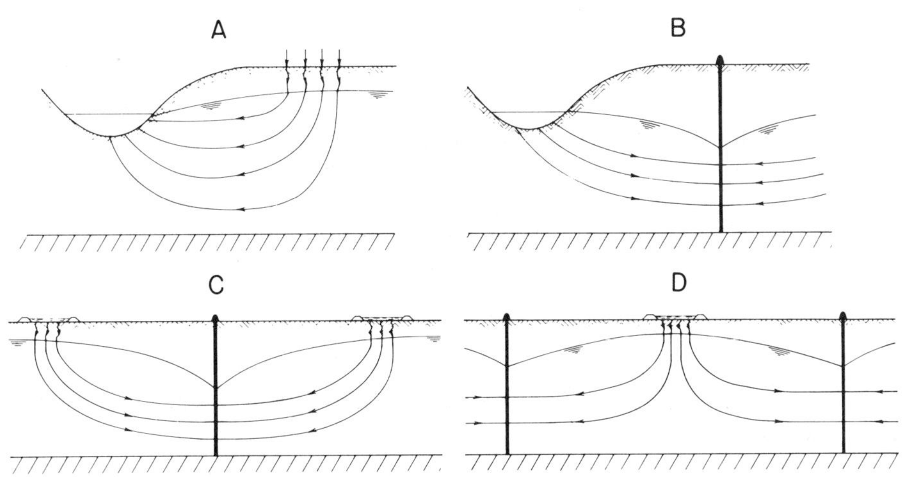

Figure 2.14. Schematic of systems for utilizing portions of aquifers as natural wastewater treatment systems while protecting native groundwater outside the systems.

reduce pollution of the receiving water by giving the wastewater the benefit of soil–aquifer treatment before it enters the stream. The reverse of the system in Fig. 2.14A is the bank-filtration system (Fig. 2.14B), which is used where polluted surface water is a source for municipal water supply. Rather than taking the raw water directly out of the stream or lake, it is pumped from wells some distance from the bank so that the water moves through the aquifer before it goes to the water treatment plant. The aquifer filtration system in this case not only gives significant pretreatment of the water, it also protects the water treatment consumers against accidental or slug discharges of toxic materials in the stream. The concentration wave of such toxicants is damped out in the aquifer by adsorption–desorption reactions. There is also a delay in the arrival of the toxic material at the wells.

Other systems for removal of the renovated water are shown in Fig. 2.14C and D. The two parallel sets of infiltration basins and the line of wells midway between (Fig. 2.14C) are used where the wells should pump renovated water only. This can be desirable where groundwater rights situations restrict pumping of native groundwater or where the renovated water needs to be further treated (for example, disinfection, activated carbon filtration, and possibly reverse osmosis if it is to be used for drinking). Movement of renovated water into the aquifer outside the system of Fig. 2.14C can be avoided by managing infiltration rates and well pumping rates so that the groundwater table below the outside edges of the infiltration strips does not rise higher than the water table in the aquifer outside the system. This can be ascertained by monitoring water levels in observation wells at the periphery of the infiltration system and comparing these levels with those in the aquifer outside the system. In Fig. 2.14, the infiltration basins are concentrated in a cluster that is surrounded by a circle of wells at some distance. In this system, the wells pump a mixture of native groundwater and renovated wastewater. This may be desirable where renovated water needs to be blended with better quality water and where there are no restrictions on withdrawing native groundwater. The basins in Fig. 2.14D can also be arranged in a long strip with lines of wells some distance away on each side. Where the water table is shallow, the renovated water can be collected with underground drains. With the systems in Fig. 2.14, a portion of the aquifer is actually set aside or *zoned* to serve as a natural treatment system for sewage effluent or other wastewater, so that the benefits of soil–aquifer treatment can be obtained without endangering the quality of indigenous groundwater in the rest of the aquifer.

REFERENCES

General sources of information from which material in this chapter was taken and to which reference is made for further detail, are:

H. Bouwer, *Groundwater Hydrology*, McGraw–Hill, New York, New York, 1978.

D. Hillel, *Fundamentals of Soil Physics*, Academic Press, New York, 1980.

Specific references cited in this chapter are:

1. M. J. Allen and E. E. Geldreich, *Ground Water* **13**(1), 45–51 (1975).
2. H. L. Bohn, *Soil Science* **112**, 39–45 (1971).
3. H Bouwer and R. D. Jackson, "Determining Soil Properties," In J. Van Schilfgaarde (Ed.), *Drainage for Agriculture*, ASA Monograph No. 17, Chapter 23, Sect. X. 1974, Am. Soc. Agron., Madison WI, pp, 611–672.
4. H. Bouwer, R. C. Rice, and E. D. Escarcega, *J. Water Pollut. Contr. Fed.* **46**(5), 835–843. (1974).
5. H. Bouwer, J. C. Lance, and M. S. Riggs, *J. Water Pollut. Contr. Fed.* **46**(5), 844–859 (1974),
6. H. Bouwer, R. C. Rice, J. C. Lance, and R. G. Gilbert, *J. Water Poll. Control. Fed.* **52**(10), 2457–2470. (1980).
7. H. Bouwer, *Ground Water* **20**, 531–537 (1982).
8. C. W. Fetter, Jr., *Ground Water Monitoring Review* **1**(3), 28–31 (1981).
9. R. G. Gilbert, J. B. Robinson, and J. B. Miller, *Proc. Intl. Conf. on Land for Waste Management*, Ottawa, Ontario (1974), pp.87–96.
10. R. C. Gilbert, C. P. Gerba, R. C. Rice, and H. Bouwer, *Appl. and Environ. Microbiol*, **32**(3), 333–338. (1976).
11. W. A. Hall, "An Analysis of Some Theories of Sand Filtration," Ph.D. Thesis in Engineering, Univ. of California, Los Angeles. (1952).
12. J. D. Hem, *U.S. Geol. Survey Water Supply Paper 1473*, 2nd ed., (1970), 363 pp.
13. T. Iwasaki, *J. Am. Water Works Assoc.* **29**, 1591–1602. (1937).
14. J. C. Lance and C. P. Gerba, "Virus Removal with Land Filtration," In E. J. Middlebrooks (ed.), *Water Reuse*, Ann Arbor Science Publishers, Inc. Ann Arbor, MI., 641–660 (1982).
15. A. C. Liakopoulos. *Water Resources Res.* **1**, 135–142 (1965).
16. R. S. Linebarger, F. D. Whisler, and J. C. Lance. *Soil Sci. Soc. Am. Proc.* **39**(2), 375–377 (1975).
17. L. J. McCabe, J. M. Symons, R. D. Lee, and G. G. Robeck, *J. Am. Water Works Assoc.* **62,** 670–687 (1970).
18. J. F. McNabb and W. J. Dunlap, *Ground Water* **13**(1), 33–44 (1975).
19. B. L. McNeal, *Proc Soil Sci. Soc. Am.* **32**, 190–193 (1968).
20. W. H. Patrick Jr., and I. C. Mahapatra, *Advan. Agron.* **20**, 323–359. (1968).
21. A. Quispel, *Soil Science* **63,** 265–275 (1949).
22. L. G. Rich, *Unit Operations of Sanitary Engineering*, John Wiley and Sons, Inc., New York, 1961.
23. L. A. Richards, "Physical Condition of Water in Soil," In C. A. Black et al. (Eds.), *Methods of Soil Analysis*, Agronomy Monograph No. 9, Vol. 1, Am. Soc. Agron., Madison, WI, 1965, pp. 128–152.
24. J. C. Romero, *Ground Water*, **8**(2), 37–49. (1970).
25. R. Sakthivadivel, and S. Irmay. Pub. HEL-15-4, Hydraulic Engineering Laboratory, Univ. of California, Berkeley, 1966.
26. D. B. Smith, R. A. Downing, R. A. Monkhouse, R. L. Otlet, and F. J. Pearson, *Water Resources Res.* **12**, 392–404 (1976).

27. Soil Conservation Service, U.S. Dept. of Agriculture., *Drainage of Agricultural Land*, Water Information Center, Inc., Port Washington, New York, 1973, 430 pp.

28. Soil Survey Staff, U.S. Dept. of Agriculture, *Soil Survey Manual*, U.S. Dept of Agriculture Handbook No. 18, 1951, 503 pp.

29. W. Stumm, and J. Morgan, *Aquatic Chemistry*, John Wiley and Sons, Inc., New York, 1976.

30. C. J. Willis, G. H. Elkan, E. Horvath, and K. R. Dail, *Ground Water* **13**(5), 406–409. (1975).

3

SOURCES OF GROUNDWATER POLLUTION

B. H. Keswick

Program in Infectious Diseases and Clinical Microbiology
The University of Texas Medical School at Houston, Texas

3.1. INTRODUCTION

In this chapter we will examine many potential sources of groundwater pollution including microbial as well as chemical contamination (Table 3.1). By examining these sources we will see the problems in protecting groundwater from further pollution.

3.2. IMPACT OF MICROBIAL POLLUTION

There were 264 outbreaks and 62,273 reported cases of waterborne diseases due to contaminated groundwater in the United States from 1946 to 1977 (2). This accounts for about 50 percent of all waterborne illness (Fig. 3.1). Groundwater becomes contaminated when it comes in contact with fecal material or sewage containing pathogenic bacteria, viruses, or parasites of human and animal origin. A *pathogen* can be defined as a microorganism that is capable of producing disease in a host. Bacteria are the most common pathogens in fecal matter and sewage. Their ~ 1–$10\ \mu m$ size makes them intermediate in size between viruses and parasites (Table 3.2). Although normally associated with a host, pathogenic bacteria are able to survive and multiply in the environment, provided that the proper nutrient conditions are present. There are many types of potentially pathogenic bacteria present in feces and sewage including salmonella, shigella, *Vibrio cholerae*, and *Escherichia coli*. *E. coli* was previously considered to be only a harmless nonpathogen. This fallacy should be corrected in the literature on water treatment since enterotoxigenic and enteropathogenic strains exist that can cause illness, particularly traveller's diarrhea.

More than 100 different types of viruses are found in sewage including poliovirus, hepatitis A, echo, coxsackie, rota, adeno, and Norwalk-like viruses. They range in size from approximately 27 nm for poliovirus to about 70 nm for rotavirus. These viruses are capable of causing a variety of illnesses in very low doses (4–6). Viruses have been found to survive for long periods outside their hosts, although they do not replicate in the environment. Techniques for the detection of viruses in large volumes of water have improved greatly over the past decade and should facilitate further study (7).

Unlike trace organics, which will be discussed later, the disease-producing potential of microorganisms is well known and the impact of waterborne illness throughout the world is immense. For example, rotavirus is responsible for 5–25 percent of all the severe diarrhea cases seen in infant children worldwide. In less developed areas, an estimated 5–8 million infant deaths in 1975 alone were attributed to diarrhea (8). Bradley (9) estimates that 5–10 percent of all deaths in tropical developing countries are due to water-related illnesses.

The potential for disease transmission through microbial contamination of groundwater is great cause for concern, especially when it is realized that much of the groundwater used does not receive any form of treatment prior to human consumption.

3.3. SOURCES OF MICROBIAL POLLUTION

3.3.1. Land Disposal of Sewage

Disposal of sewage by land application is a form of groundwater recharge. It is estimated that in the United States (10) that more than 3 billion gallons of sewage effluent from more than 3000 land treatment sites and 15 million septic tanks are applied to land every day (Fig. 3.2). The potential for microbial contamination of groundwater is therefore considerable.

There are several basic types of land treatment facilities: slow-rate, rapid-infiltration, overland flow, wetlands, and subsurface injection. Characteristics for each type of land treatment process are found in Table 3.3 and Fig. 3.3. An increased emphasis on this economically favorable method of sewage disposal has raised concerns about possible groundwater contamination. The possible pathways of contamination are presented in Fig. 3.4.

Even though there have been no reports of disease outbreaks associated with land treatment of sewage wastes, there is a growing number of studies concerning the detection of viruses in groundwater after wastewater application to land or direct groundwater recharge. For example, Wellings et al. (13) demonstrated vertical and lateral movement of viruses in secondary effluent discharged into a cypress dome (Table 3.4). Poliovirus 1, coxsackievirus B4, and echoviruses 7, 11, and 14 were recovered from 3-m (10-ft) deep wells in three of 71 samples. Viruses were also shown to migrate 7–38 m laterally from the application point and to survive at least 28 days. The soil at this site ranged from a top 2-ft layer of black organic soil (4–12 percent clay) to a sandy clay with a permeability of 3×10^{-2} to 3×10^{-6} cm/min. Thus, the viruses moved horizontally as well as vertically and survived many days under ambient conditions, indicating a necessity to evaluate such sites for their aquifer movement and transmission of viruses.

In an earlier study, Wellings et al. (15) recovered viruses from groundwater after spray irrigation of secondary sewage effluent onto an Imolokee sand (little or no silt or clay). Of particular interest in this study was that the viruses survived chlorination, sunlight, spraying, and percolation through 3–6 m (10-20 ft) of sandy soil. Furthermore, a burst of viruses was demonstrated after a period of heavy rains in a sample that had previously been negative. These studies demonstrate that soil type, rainfall, and other factors can affect virus movement into groundwater and that viruses are capable of surviving long periods which, when combined with the ability to move long distances laterally, could lead to wide dispersal through an aquifer.

Vaughn and Landry (16, 17) and Vaughn et al. (18) reported isolations of viruses from four groundwater recharge sites in New York. These sites have soils of coarse sand, fine gravel, and 1–2 percent silt. At the groundwater recharge sites, viruses were recovered at depths up to 11.4 m (37 ft) and at distances up to 45.7 m (150 ft) from the injection point of secondary or tertiary chlorinated effluents. Twenty-four to thirty-three percent of the 100-gallon samples at the four sites were positive for virus with concentrations of 1.3–10.6 PFU/gallon. In order to reach the groundwater, the virus moved through 5.5–9.0 m (18–30 ft) of

Table 3.1. Major Sources of Groundwater Pollution and Types of Pollutants[a]

Source	Type of Pollutant					
	Physical	Inorganic Chemical	Trace Elements	Organic Chemical	Bacteriological	Radiological
Municipal						
Sewer Leakage	Minor	Primary	Secondary	Primary	Primary	Minor
Sewage Effluent	Minor	Primary	Secondary	Primary	Primary	Minor
Sewage Sludge	Minor	Primary	Primary	Primary	Primary	Minor
Urban Runoff	Minor	Secondary	Variable	Primary	Minor	Minor
Solid Wastes	Minor	Primary	Primary	Primary	Secondary	Minor
Lawn Fertilizers	Minor	Primary	Minor	Minor	Minor	Minor
Agricultural						
Evapotranspiration and Leaching	Minor	Primary	Minor	Minor	Minor	Minor
Fertilizers	Minor	Primary	Secondary	Secondary	Minor	Minor
Soil Amendments	Minor	Primary	Minor	Minor	Minor	Minor
Pesticides	Minor	Minor	Minor	Primary	Minor	Minor
Animal Wastes (Feedlots and Dairies)	Minor	Primary	Minor	Secondary	Primary	Minor
Stockpiles	Minor	Primary	Minor	Variable	Variable	Minor

Industrial						
Cooling Water	Primary	Minor	Primary	Minor	Minor	Minor
Process Waters	Variable	Primary	Primary	Variable	Minor	Variable
Storm Runoff	Minor	Secondary	Variable	Primary	Minor	Minor
Boiler Blowdown	Primary	Secondary	Primary	Minor	Minor	Minor
Stockpiles	Minor	Primary	Variable	Variable	Minor	Variable
Water Treatment Plant Effluent	Minor	Primary	Secondary	Minor	Minor	Minor
Hydrocarbons	Secondary	Secondary	Secondary	Primary	Minor	Minor
Tank and Pipeline Leakage	Variable	Variable	Variable	Variable	Minor	Variable
Oilfield Wastes						
Brines	Primary	Primary	Primary	Minor	Minor	Minor
Hydrocarbons	Secondary	Secondary	Secondary	Primary	Minor	Minor
Mining Wastes	Minor	Primary	Primary	Variable	Minor	Variable
Miscellaneous						
Polluted Precipitation and Surface Water	Variable	Variable	Variable	Variable	Variable	Variable
Septic Tanks and Cesspools	Minor	Primary	Minor	Secondary	Primary	Minor
Highway Deicing	Minor	Primary	Minor	Secondary	Minor	Minor
Seawater Intrusion	Primary	Primary	Primary	Minor	Minor	Minor

[a]Courtesy of John Wiley & Sons (1).

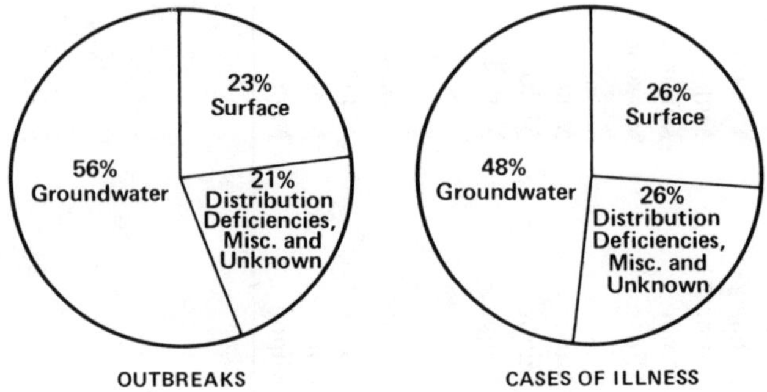

Figure 3.1. Waterborne disease outbreaks in the United States (1971–1977). Courtesy of Elsevier Scientific Publishing (3).

the overlaying soil. Both total coliform and fecal coliform bacteria were also found in these samples. In addition, at the 12 Pines site, virus was demonstrated in groundwater under basins where effluent seeded with virus was applied at a rate of 6–100 cm/hour. The slower infiltration rates were more effective in removing the viruses, suggesting that site management is important for removing viruses.

Schaub and Sorber (19) reported on a study of viruses in groundwater under rapid-infiltration sites at Ft. Devens, Massachusetts. The soil consisted of silty sand and gravel underlaid with bedrock. The groundwater was found to contain viruses at depths of 29 m (95 ft) and lateral distances of 183 m (600 ft) with concentrations of about 8.3 percent of the applied effluent. Fecal streptococci

Table 3.2. Properties of Major Waterborne Bacteria, Viruses, and Parasites

Microorganism	Size (μm)	Infectious Dose (Number of organisms)
Bacteria	1 – 10	
Salmonella typhi		10^5
Shigella dysenteriae		$10^1 - 10^2$
Escherichia coli		$10^6 - 10^8$
Vibrio cholerae		$10^6 - 10^8$
Viruses	0.02 – 0.08	
Enteroviruses (polio, echo, coxsackie)		1 – 10
Rotavirus		Unknown
Norwalk-like virus		Unknown
Hepatitis A		Unknown
Adenovirus		Unknown
Parasites	5 – 20	
Giardia lamblia		10 – 25
Entamoeba hystolitica		Unknown

Table 3.3. Comparison of Design Features for Land Treatment Processes[a]

Feature	Principal Processes			Other Processes	
	Slow Rate	Rapid Infiltration	Overland flow	Wetlands	Subsurface
Application techniques	Sprinkle or surface[b]	Usually surface	Sprinkler or surface	Sprinkler or surface	Subsurface piping
Annual application rate (ft)	2–20	20–560	10–70	4–100	8–87
Field area required (acres)[c]	56–560	2–56	16–110	11–280	13–140
Typical weekly application rate (in.)	0.5–4	4–120	2.5–6[d]	1–25	2–20
Minimum preapplication treatment provided in United States	Primary sedimentation[f]	Primary sedimentation	Screening and grit removal	Primary sedimentation	Primary sedimentation
Disposition of applied wastewater	Evapotranspiration and percolation	Mainly percolation	Surface runoff and evapotranspiration with some percolation	Evapotranspiration, percolation, and runoff	Percolation with some evapotranspiration
Need for vegetation	Required	Optional	Required	Required	Optional

[a]Courtesy of the U.S. EPA (11).
[b]Includes ridge-and-furrow border strip.
[c]Field area in acres not including buffer area, roads, or ditches for 1 Mgal/d (43.8 L/s) flow.
[d]Range for application of screened wastewater.
[e]Range for application of lagoon and secondary effluent.
[f]Depends on the use of the effluent and the type of crop.

Figure 3.2. Location of land treatment sites in the United States (redrawn from Ref. 10).

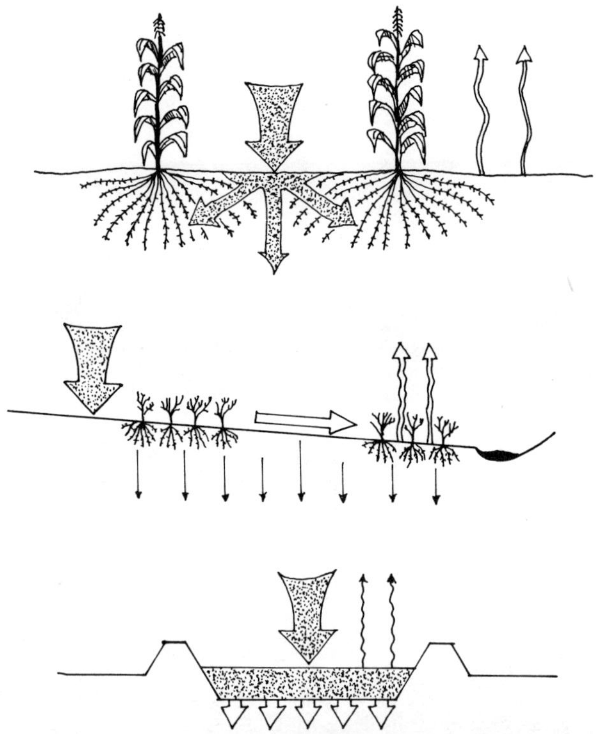

Figure 3.3. Three major types of land treatment sites (redrawn from Ref. 11).

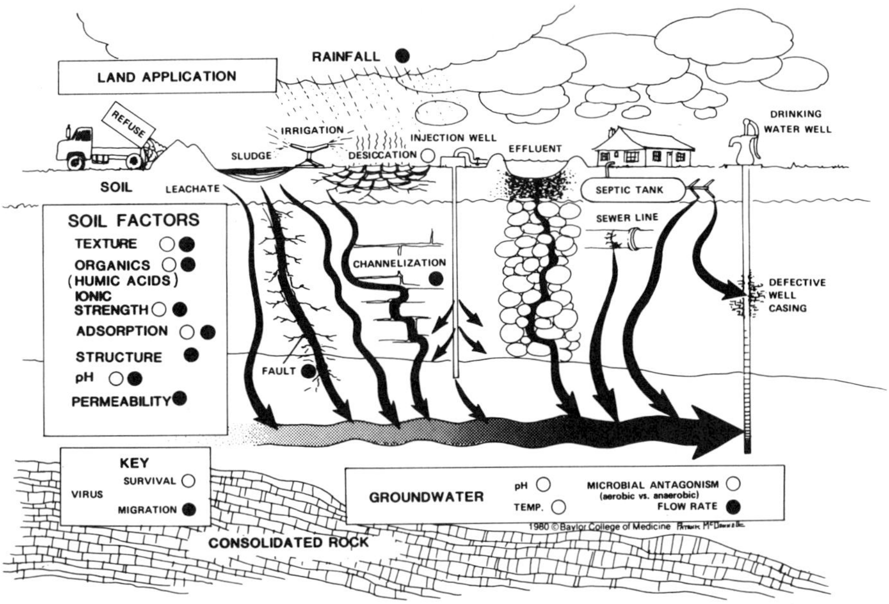

Figure 3.4. Factors affecting the entry, survival, and migration of viruses in groundwater. Courtesy of the American Chemical Society, *Environmental Science and Technology* (12).

were also found in the 28.9-m (95-ft) deep well. Marker f2 phage was applied at this same site; only about 50 percent of the phage was removed, and was detectable for at least 11 days. It was concluded that the site had poor filtration properties, pointing out the need for site-specific evaluation.

At the Vineland, New Jersey rapid-infiltration site (20), primary effluent was applied to Cohansey sand and coarse gravel. Viruses were found at 16.8-m (55-ft) depths and up to 250-m (820-ft) lateral distance in 19 of 40 samples. Polio-, echo-, and coxsackievirus were identified. Total coliforms and fecal coliforms were found consistently at depths up to 6 m beneath the recharge basins. Total coliforms also occasionally occurred at greater depths and downstream. In contrast, no fecal coliforms were found in samples taken below 9.1 m (30 ft) and occurred only once in a shallow well 50 m (165 ft) downgradient. Thus, viruses penetrated deeper into the ground and moved longer distances than the coliforms. The potential for viruses to migrate great distances, as in the previous study, was further demonstrated by Noonan and McNabb (21), who used the phages ϕ X174 and T4 to demonstrate lateral movement of 140 and 911 m, respectively, in New Zealand groundwater in just 96 hrs. From the data, it is evident that the source of pollution may be located some distance from groundwater and may still cause problems. This is especially true in fault zones and in areas where limestone is prevalent.

As mentioned above, bacteria are often found in well water under land

Table 3.4. Isolation of Viruses Beneath Land Treatment Sites[a]

Site Location	Type of Site[b]	Virus Types	Maximum distance Virus Migration (m)	
			Depth	Horizontal
St. Petersburg, Florida	S	Polio 1; coxsackie B4; echo 7	6	–
Gainesville, Florida	S	Coxsackie B4; polio 1, 2	3	7
East Meadow, New York	R	Echo 12; U[c]	11.4	3
Holbrook, New York	R	Echo 6, 21, 24, 25; U	6.1	45.7
Sayville, New York	R	U	2.4	–
12 Pines, New York	R	Polio 2; U	6.4	–
North Masapequa, New York	R	Echo 11, 23; coxsackie A16	9.1	–
Babylon, New York	R	Coxsackie	22.8	408
Ft. Devens, Massachussetts	R	U	28.9	183
Vineland, New Jersey	R	Polio, coxsackie B3; echo	16.8	259
Lake George, New York	R	Phage	45.7	400
Phoenix, Arizona[d]	R	Coxsackie B3	18.3	3
Lubbock, Texas	S	Coxsackie B3	30.5	–
Kerrville, Texas	S	U	1.4	–
Dan Region, Israel	R	Polio 1, 2, 3	31–67	60–270
England	S	Polio 2, 3; coxsackie B4, 5	19.4	–

[a]Courtesy of the American Chemical Society (12) and the American Water Works Association (14).
[b]R = rapid infiltration, S = slow-rate infiltration.
[c]U = unidentified.
[d]Gerba, unpublished data.

treatment sites. Sometimes viruses are found when no bacteria can be detected, suggesting that virus removal by percolation through soil is controlled by different processes than bacterial removal. Parasites are large enough to be removed by filtration through the upper soil profiles. The factors influencing the survival and migration of microorganisms in soils and groundwater are discussed in Chapter 4.

Another way in which groundwater can become contaminated by sewage is through leaks in sewer lines. As pipes age, develop cracks, and allow seepage, it is possible that groundwater could become highly contaminated. This contamination can lead to large outbreaks of illness (22), even though the source of a leak can be difficult to determine.

3.3.2. Land Disposal of Sewage Sludge

Currently sewage treatment plants produce 4.5 billion kg of dry sludge per year in the United States. This total is expected to increase to about 8 billion during this decade (23). Sludge contains viruses, bacteria, and parasites as well as a variety of chemical contaminants. Recently, virus concentrations ranging from 3.8×10^3 to 1.16×10^4 plaque-forming units per liter of sludge have been detected (24). Fecal coliform bacteria are present at levels of approximately 10^5/g sludge (25) whereas the numbers of parasite ova and adults in one study (26) ranged from 0–7712/100 g of solids in anaerobically digested sludge. In general, bacterial pathogens, parasite eggs, and protozoan cysts are effectively destroyed during the thermophilic processes of composting; however, not all enteroviruses are inactivated.

Laboratory studies have demonstrated that viruses survived in sludge up to 35 days in hot and wet conditions in Florida (24) and more than three months in the winter in Cincinnati, Ohio. Similar findings have been made in Denmark (27). Thus, viruses appear capable of survival in drying sludge. In contrast are the results of two field studies (24, 28) where no viruses were detected in leachates under sludge application sites during extensive testing. Bitton and Farrah (24) suggested, based on this information, that enteroviruses are efficiently retained by the sludge soil matrix and so are retarded in their possible movement into groundwater. This restriction, of course, applies in areas where soil conditions and groundwater characteristics are favorable. In other words, it appears that the contamination of groundwater by microorganisms in sludge is unlikely unless the sludge comes into direct contact with groundwater either through natural or man-made processes.

3.3.3. Septic Tanks, Cesspools, and On-Site Facilities

In the United States, three billion cubic meters of wastewater per year are discharged into the soil via individual on-site disposal units including septic tanks (Fig. 3.4) and cesspools. This represents about 20 million housing units and about 30 percent of the U.S. population (29). Since many of these units are improperly installed and poorly maintained, they are subject to failure. The microbiological effects of this failure is demonstrated by a recent survey of rural groundwater where greater than 92 percent of the wells examined contained coliform bacteria (30). Details of this problem are discussed in Chapter 8.

3.3.4. Land Disposal of Solid Wastes

There are more than 20,000 landfill sites in the United States. Many are simple refuse dumps whereas others are sanitary landfills consisting of alternate layers of compacted wastes and soil. (31). In this manner, an estimated 3 kg per capita of wastes are disposed of daily. This accounts for more than 90 percent of the solid wastes generated by municipal and industrial activities. Among these wastes are microbial hazards contained in disposable diapers and animal feces as well as numerous chemical pollutants. Groundwater can be contaminated when rainwater washes these pollutants into an aquifer. These leachates are very important in terms of the chemical pollutants discussed in Section 3.5.3 and have been found to contain viruses and bacteria. Cooper et al. (32) found in lysimeter experiments that the initial numbers of fecal streptococci can be as high as $1 \times 10^9/100$ ml of leachate. Furthermore, the numbers of organisms from "open dumps" were considerably higher than those from "closed dumps." In the same tests, virus recovery was sporadic, with low levels of viruses dectected over a period of 20 weeks. Vaughn, et al. (18) detected viruses at depths of 22.8 m and up to 408 m downstream of a landfill site in New York. Donnelly et al. (33) demonstrated longer survival and greater numbers of fecal streptococci in landfill leachate than most gram-negative rods including coliforms.

This information suggests that poorly managed dump sites in unsuitable locations can contribute to microbial pollution but little is known about the potential impact of this practice. The effects of organic compounds found in these sites on the movement, survival, and genetic stability of microorganisms is also unknown.

3.3.5. Stormwater Recharge and Urban Runoff

Stormwater may contain high levels of bacteria and viruses of human fecal origin in cases where sewage becomes mixed with runoff water. The use of stormwater mixed with sewage for direct recharge should be closely monitored. For example, Vaughn et al. (18) detected viruses at a stormwater recharge site where samples taken at 9-m depths directly in the basin were positive for echoviruses 11 and 23 and coxsackievirus A6. This contamination may have been due to runoff from cesspools in the area. Rainwater is also important to groundwater contamination because of its capacity to increase the mobility of viruses and bacteria in soils. This property has been demonstrated in experiments by Landry et al. (34) and in field studies by Wellings et al. (15), Barrell and Rowland (35), and Vaughn and Landry (16, 17). Therefore, although stormwater runoff does not contain significant levels of pathogens, rainwater may still play an important role in contaminating groundwater by mobilizing organisms in soil at land treatment sites.

3.3.6. Agricultural Practices

Microbiological pollution of groundwater by irrigation of cropland with sewage effluent has been demonstrated (36, 37). At Lubbock, Texas, for example, wastewater is used to irrigate cotton, grass, and other crops. Groundwater is then recycled to a series of surface lakes. Water in lakes at a depth of 30.5 m have been found to contain coxsackie B3 viruses as well as total and fecal coliform bacteria.

Disposal of animal wastes at feedlots is also a potential source of groundwater pollution, especially where shallow groundwater tables are exposed to wastes left in the open. Bacteria and viruses can then leach into groundwater.

3.4. IMPACT OF CHEMICAL POLLUTANTS

Chemical pollutants are most simply divided into two types: inorganic and organic. The effects of inorganic chemicals are better known than those of trace levels of organic chemicals detected in groundwater (38,39). For example, inorganic trace elements that are potentially hazardous and known to be toxic at high enough levels include boron, cadmium, chromium, copper, nickel, mercury, arsenic, molybdenum, vanadium, cobalt, silver, lead, fluorine, aluminum, iron, manganese, titanium, and zinc (40). The study of groundwater contamination by organic chemicals, on the other hand, is in its infancy. Although sampling and analysis is often difficult, especially for low levels, many compounds have been detected, including pesticides, chlorinated-hydrocarbons, and benzene derivatives. As already mentioned, the health significance of many trace organics is unknown and will be discussed in detail in Chapter 7. We now review the sources of these chemical pollutants.

3.5. SOURCES OF CHEMICAL POLLUTANTS

3.5.1. Sewage Effluent Disposal

An unknown number of chemicals are disposed of in sewage (41). Both inorganic and organic compounds can be detected in raw wastewater. This is further complicated by the fact that chlorination during sewage treatment can produce chlorinated hydrocarbons not originally present. Table 3.5 shows a list of chloroorganics found in sewage.

When sewage effluent containing chemical compounds is disposed of by land treatment the potential for groundwater contamination exists. Indeed, there are many cases where a variety of compounds attributable to sewage disposal have been found in groundwater. For example, Baxter et al. (43) detected progressive increases in the concentration of nitrate and chloride in a three-year study in England. In another case, lindane was found at elevated concentrations in groundwater beneath the Roswell, New Mexico (44) site, but within current limits. The potential for chemical pollution of groundwater via sewage disposal should be an important consideration in siting land treatment facilities.

3.5.2. Sludge Disposal

Land application is the most common method of sludge management in the United States. The application of sludges can help maintain soil fertility.

Table 3.5. Chloroorganics Found in Sewage[a]

Compound	
Chloroalkyl acetate	Trichlorobenzene
2-Chlorobenzoic acid	Trichlorocumene
3-Chlorobenzoic acid	Trichlordimethoxybenzene
4-Chlorobenzoic acid	Trichloroethylbenzene
8-Chlorocaffeine	Trichloro-*N*-methylanisole
Chlorocumene	Trichloro-α-methylbenzyl alcohol
Chlorocyclohexane	Trichlorophenol
Chloroethylbenzene	Tichlorophthalate derivative
Chloroform	Trichloromethylstyrene
6-Chloroguanine	Heptachlor
3-Chloro-4-hydroxybenzoic acid	Heptachlor epoxide
4-Chloromandelic acid	1,2,3,4,5,7,7-Heptachloronorbornene
Chloro-α-methylbenzyalcohol	
3-Chloro-2-methyl-but-1-ene	Hexachlorobenzene
4-Chloro-3-methyl-phenol	Hexachloro-1,3-butadiene
2-Chlorophenol	Hexachlorocyclohexane
3-Chlorophenol	Hexachloroethane
4-Chlorophenol	Hexachlorophene
4-Chlorophenylacetic acid	(2- or 1-Napthyl) dichloromethane
4-Chlororesorcinol	Nitrotrichloromethane (chloropicrin)
5-Chlorosalicyclic acid	Octyl chloride
5-Chlorouracil	Pentachlorobiphenyl
5-Chlorouridine	Pentachlorophenol
8-Chloroxanthine	1,1,2,2,2-Pentachloropropane
Dibromochloromethane	Pentachloropentene (isomers)
Dichloroacetate derivative	Pentachlorophenyl methyl ether
Dichloroaniline derivative	1,1,3,3-Tetrachloroacetone
Dichloroaromatic derivative	Tetrachloropropanes (isomers)
o-Dichlorobenzene	Tetrachlorobiphenyl
p-Dichlorobenzene	1,1,1,2-Tetrachloroethane
Dichlorobutane	Tetrachloroethylene
Dichlorobis(ethoxy)benzene	Trihlorobiphenyl
Dichloroethylbenzene	1,1,1-Trichloroacetone
Dichloromethoxytoluene	1,1,1-Trichloroethane
Dichloro-α-methylbenzyl alcohol	1,1,2-Trichloroethane
Dichlorotoluene	Trichloroethane
Hexachloroacetone	(1,1,2) Trichloroethylene
N-Methyltrischloroaniline	Trichlorofluoromethane
Pentachloroacetone	Trichloroheptafluorobutane
Tetrachloroacetone	Trichloropentafluoropropane
Tetrachloroethylstyrene	1,1,2-Trichloro-1,2,2-trifluoroethane
Tetrachloromethoxytoluene	2,4,6-Trichlorophenol
Tetrachlorophenol	Trichloropropanes (isomers)
Tetrachlorophthalate derivative	Vinyl chloride

[a]After Dienzer et al. (42).

However, as with liquid sewage, the potential for groundwater contamination exists. Common mineral elements, such as Al, Fe, Ca, Na, and Si, are present in significant quantities in sludge. Trace elements (Cd, Pb, Hg, As, Cr, Cu, Ni, Zn, B, Se, Mo) as well as nitrogen and phosphorus in both organic and inorganic forms are also present.

A study of cadmium (45) illustrates the problems associated with heavy metals in sludge. It was found that at high loading rates Cd migrated through soil to groundwater, whereas at lower loading rates the Cd remained near the soil surface. Thus, under the right (or wrong) conditions, chemicals in sludge may potentially reach groundwater.

3.5.3. Industrial Wastes

Industrial waste landfills (Fig. 3.5) are the most common source of serious groundwater contamination because of their large number and potential for leaching a variety of hazardous substances that are relatively mobile into the underlying groundwater. (47). The EPA defines hazardous wastes as those that are toxic, corrosive, ignitable, or chemically reactive (46). It was estimated that over 57 million metric tons of hazardous industrial waste was produced in the United States during 1980 alone. The percentages generated by industry classification are presented in Fig. 3.6. Of course, chemical-related industries are the largest producers (46). As little as 10 percent of all this waste was disposed of

Figure 3.5. Locations of commercial hazardous waste management facilities (redrawn from Ref. 46).

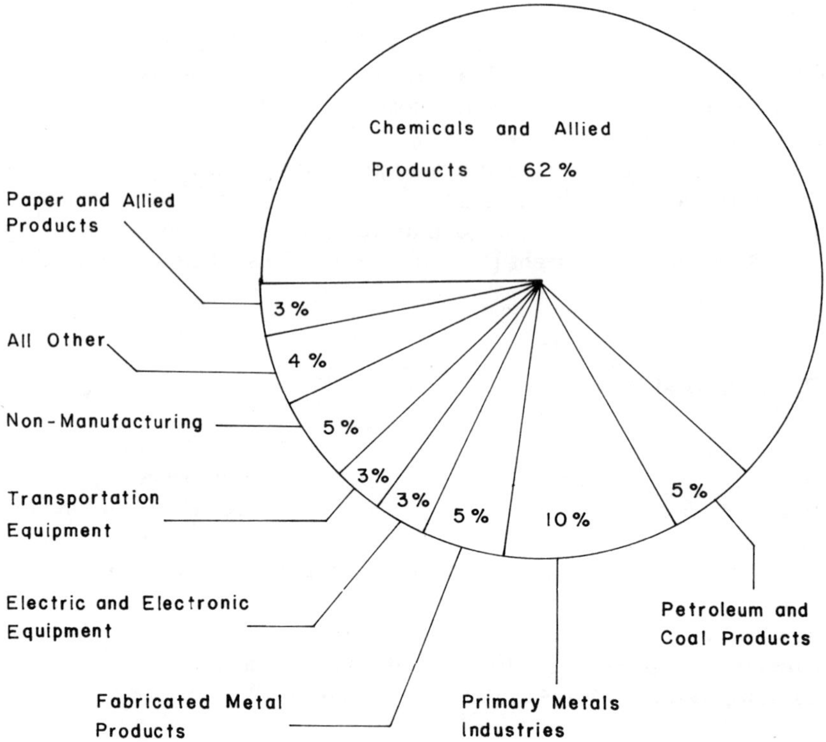

Figure 3.6. Percentage of 1980 hazardous waste generation by standard industrial classification (redrawn from Ref. 46).

properly (Table 3.6). An estimated 32,000–50,000 of the more than 100,000 sites contain hazardous material with 1,000–2,000 posing significant health risks (49). Permanent remedy for these sites could exceed 25 million dollars per site.

The potential for contamination from landfill leachate is greatest in areas with shallow groundwater tables and high rainfalls. The milieu (Table 3.7) of domestic and industrial chemicals received for disposal makes the tasks of

Table 3.6. Commercial Wastes Management for 1980 (Thousand wet metric tons)[a]

Landfill	2699
Land treatment/evaporation	537
Incineration	398
Chemical treatment	2346
Resource recovery	424
Deep-well injection	788
Total	7192

[a]After U.S. EPA (46).

Table 3.7. Chemicals Found in Landfill Sites

Classified

- Acetic acid
- Acetone
- Aldrin
- Alkyl benzene sulfonate
- Allyl ether
- Asbestos
- Benzene hexachloride
- Carbofuran
- Carbon tetrachloride
- Chloroanilic acid
- Chloroform
- Chromic acid
- Cumene
- Hydroxide
- Cyanide
- Cyclohexamine
- Dibromochloropropane
- DDT
- DEF
- Dichlorobenzene
- Dieldrin
- Dichloroethane
- Dichloroethylene
- Dimethyl sulfide
- Dioxane
- Dioxin
- Endosulfan
- Endrin
- Ethanol
- Ethylbenzene
- Ethyl acetate
- Ethyl acetate
- Ethylene dibromide
- Ethylene glycol
- Freon
- Heptane
- Heptachlor
- Hexachlorobenzene
- Hexachlorobutadiene
- Hexachlorocyclopentadiene
- Hexane
- Hydrogen sulfide
- Kepone
- Lindane
- Lithium
- Methacrylic Acid
- Methane gas
- Methyl ethyl ketone
- Methylene chloride
- Mirex
- *o*-Nitoaniline
- Parathion
- Polybrominated biphenyls
- Polychlorinated biphenyls
- Pentachlorophenol
- Phenol
- Sodium
- Styrene
- Tetrachlorobenzene
- Tetrachloroethylene
- Trichloroethane
- Trichloroethylene
- Trimethylsilanol
- Tris (tromethamine)
- Vinyl chloride
- Zinc chloride

Miscellaneous

- Acids
- Alkaline wastes
- Aromatic hydrocarbons
- Chlorinated organics
- Coal tars
- Explosives
- Fly ash
- Herbicides
- Nitrates
- Pesticides
- Petrochemical
- Polynuclear organics
- Solvents
- Toxic metals
- Uranium radiation
- Volatile organics
- Waste oil

[a]After U.S. EPA (48).

sorting out, identifying, and controlling pollutants enormous. It is, however, important that this be done since landfills from ancient times still produce leachates. The wastes buried today may not present themselves for many years.

A recent EPA report (46) listed 168 cases where groundwater was effected by hazardous wastes and 468 cases of well closure due to contamination. Some examples from this report and others illustrate the variety of problems encountered in studying sources of groundwater pollution.

The most infamous case is that of the Love Canal landfill in Niagra Falls, New York (50). Hooker Chemical Company disposed of 21,800 tons of wastes from 1942 to 1952 containing, among others, dioxin, hexachlorobenzene, tetrachloroethylene, chloroform, dichloroethane, benzene, hexachloride, hexachlorobutadiene, dichlorobenzene, trichloroethylene, toluene, and phenol which have migrated from the landfill to surrounding areas. Both ground and surface waters have been effected. The cleanup costs are estimated at 45 million dollars. Three other sites in the area face similar problems and costs. This is a case where the source of contamination is known.

Roux and Althoff (51) described a case history of industrial waste disposal leading to groundwater contamination where the source was not easily identifiable. The primary contaminants were 1,1,1,-trichloroethane and tetrachloroethylene which are cleansing agents with high solubility and mobility in groundwater. In this case, a public water supply and 27 private domestic wells were found to be effected. Two industrial plants adjacent to this area came under suspicion as probable sources of the contaminants. Only careful groundwater mapping revealed that the actual source was only one of the plants. This case points out that it is often difficult to pinpoint a suspected source and to be able to prove liability. Legal aspects of groundwater pollution are discussed in Chapter 14.

Unscrupulous disposal of hazardous wastes has been another serious source of groundwater pollution. In many cases, abandoned dumps, illegal dumps, or incomplete records have hindered assessment of the types of pollutants present and their cleanup. For example, in one case in 1971, a large reputable chemical company contracted a private hauler to dispose of 6000 barrels of liquid chemical wastes at a landfill (46). Instead, the barrels were dumped at an abandoned chicken farm. By early 1974, evidence of contamination in local wells appeared. This resulted in the loss of 148 private wells and a more than $400,000 cost for removing the drums, sampling, analysis, and drilling of new wells. In this case, the source was eventually identified, but in many cases where plants have changed owners or gone out of business, this is impossible. The EPA report (46) cites several examples of illegal and abandoned dumpsites in Michigan, Kentucky, and other states.

In addition to the legal aspects of old, closed, or abandoned dump sites is the disturbing feature that leachate is often produced for years after closure, and contamination may not appear or be detected until well after a site is closed. For example, Dewalle and Chian (52) reported a study of trace organics leaching from a landfill in Delaware, seven years after its closing. Toluene, dichloroethane, butanol, and acetone were present in the highest concentrations in the leachate with toluene, benzene, and trichloroethylene present in significant amounts in one drinking water well. Attempts to limit the spread of the leachate contamination were estimated at over $1.3 million.

Another problem in detecting sources of groundwater pollution which will be discussed in later chapters is that pollution can occur at some distance from the source, especially when aided by rainfall. Murray et al. (53) reports such a case concerning the contamination of a groundwater spring by leachate from a sanitary landfill operation 1.5 km away. This site was situated on karst terrain deemed as "pollution sensitive." Rainfall appeared to be a major contributing factor in promoting migration of the leachate.

In recognizing the potential seriousness of this problem, Congress passed the so-called "Superfund" of $4 billion to aid in the cleanup of hazardous wastes (54). (See Chapter 14.) Clearly, more controls are needed to protect our groundwater from hazardous wastes because the costs of cleanup, if possible at all, are tremendous.

3.5.4. Deep-Well Injection

Industry is faced with the disposal of a multitude of liquid wastes. One alternative method of disposal is the injection of these waters into deep permeable zones (200–4000 m) in order to isolate hazardous substances from the biosphere. Approximately 325 deep wells are in operation around the world (55) and more are under construction. Although generally considered safe, there have been instances of groundwater contamination. The presence of unplugged wells and hydrologic fracturing are perhaps the greatest problems associated with this method.

3.5.5. Agricultural Practices

Two primary groups of potential pollutants are found in agricultural use. These are fertilizers and pesticides. Commercial, inorganic fertilizers contain variable concentrations of N (nitrogen), P (phosphorus) and K (potassium). Nitrogen is the most serious groundwater pollutant. In its NO_3^- form, nitrogen is a common pollutant of groundwater under agricultural lands. In a study in England (56), it was found that high concentrations of nitrate exist in saturated zones of aquifers whose recharge areas are formed by arable farmland. Nitrate contamination is a worldwide problem and in many areas nitrate concentrations exceed the recommended limit for drinking water (see Chapter 7). It is disturbing that in some areas with acceptable NO_3^- levels, NO_3^- is being detected at greater depths and in gradually increasing concentrations.

Shallow groundwater can also become contaminated as a result of leaching of NO_3^- from land application of livestock and poultry wastes. This is often due to poorly constructed wells that make it easy for leachate to reach the groundwater. Cleanup of NO_3^- contamination, although feasible in some cases, is nevertheless expensive and sometimes difficult. Effective prevention appears to be a better approach to the problem (57).

Pesticides contribute a second source of groundwater contamination by agricultural practices (Table 3.8). Leavitt et al. (58) describe a study of atrazine (a widely used pesticide) contamination of groundwater in Nebraska where a tenfold increase was detected in the downward hydraulic gradient direction. Guerrera (59) reported that aldicarb (used for potato pests) could be detected in groundwater even though preliminary tests showed this was unlikely. Nelson et al. (60) reported that large portions of groundwater in the San Joaquin Valley and Santa Anna River Basin are contaminated with 1,2-dibromo-3-chloropropane (DBCP), a pesticide. Drinking water for over 200,000 people is effected and many are consuming water that exceeds California standards. Many pesticides are somewhat insoluble in water, others readily adsorb to soil, and some are even degraded by microorganisms. Todd (1) points out that the impact of pesticides is influenced by the nature of the chemical, rainfall, irrigation rates, and soil characteristics.

Pollution associated with the manufacture and disposal of pesticides and fertilizers has also been reported (46). In Lathrop, California, wastes from fertilizer and pesticide manufacture leached 7–25 feet into the groundwater that serves as the source of drinking water for the area. DBCP, lindane, and DEF as well as radiological contamination have been detected and legal actions have been taken.

In another case, at the Rocky Mountain Arsenal in Colorado (46), a thirty-square mile area of the shallow aquifer was contaminated. This was caused by the disposal of byproducts from the manufacture of pesticides and herbicides from 1943–1957 in unlined holding ponds and has resulted in the temporary abandonment of 64 domestic, stock, and irrigation wells. Aldrin and dieldrin have been detected and an extensive remedial cleanup has been undertaken. Since pesticides and fertilizers are so widely used, often indiscriminately, it is no surprise that their manufacture, use, and disposal may be a major source of groundwater pollution.

3.5.6. Petrochemical Industry Practices

Storage and transhipment of oil, gas, and other petroleum products are potential sources of pollution. Service station storage tanks, pipelines, and storage terminals may all be involved. Thomas and Leis (61) report on groundwater contamination by fuel oil from a company responsible for storage and transhipment. They identified accidental spills, faulty valves, tank overflows, and leaky or broken underground pipes as sources of trouble. Home fuel tanks are also a frequent source of pollution.

Oilfield brines are another source of groundwater pollution produced by the petrochemical industry during drilling operations. Previous disposal practices included storage in lagoons, which resulted in long-term contamination of aquifers in oil-producing areas (62).

Table 3.8. Pesticides in the Environment and their Toxicity (60–62)[a]

	LD_{50} (oral)[b] (g/Kg)	Fatal doses (g/kg)	Chronic Poisoning	Acute Poisoning
DDT	Rat 0.285 Rabbit 0.325	0.4	Not substantiated. Having 648 ppm in their body remained well.	Severe vomiting within 30 min to 1 hr of 5 g. Weakness and numbness of the extremeties. Apprehension and excitement are marked.
Dieldrin	Rat 0.06 Dog 0.068	0.07	Not been established in man. Impair liver function in animals, occasional eliptiform convulsions.	Hyperexcitability, tremors, ataxia, convulsions.
Lindane (benzene hexachloride)	Rat 0.135 Dog 0.12 Rabbit 0.13	0.6	In animals, liver necrosis.	Vomiting and diarrhea, convulsions, circulatory failure.
Malathion	Rat 2.5	0.86	In animals, cholinesterase levels of red blood cells and plasma are reduced markedly.	Headache, tremors, nausea, abdominal cramps, diarrhea, coma, heat block.
Parathion	Rat 0.004	0.0014	Not established in man.	Similar to those of malathion, but more severe and fatal.
2,4-D	Mouse 0.375	0.7	Weakness, fall of blood pressure, muscle damage.	Burning pain, painful and tender muscle fever, paralysis, irreversable fall of blood pressure.
2,4,5,-T	Rat 0.3 Dog 0.1	0.6	Similar to 2,4-D.	Similar to 2,4-D.

[a]Courtesy of the U.S. EPA (48).
[b]LD_{50}: Median lethal dose of a substance that will kill 50% of the experimental subjects.

Present practice consists of injecting these brines, consisting of sodium, calcium, ammonia, boron, chloride, sulfate, trace metals, and dissolved solids, into deep formations geologically isolated from fresh water aquifers. There are over 100,000 such wells in North America. However, as Todd (1) points out, abandoned and unplugged wells and test holes can still provide pathways for brines to rise to overlying aquifers.

3.5.7. Mining Practices

Acid mine drainage from abandoned coal mines is the most important source of groundwater contamination from mining operations. Schubert and Prodan (63) point out that contamination is highest where these materials are disposed of on permeable soils. Groundwater in some areas of Illinois and Appalachia has a very acidic pH (in the range of pH 2.0–4.0) with high concentrations of sulfate, manganese, and trace metals including zinc and cadmium. Olem (64) reviews current studies on the abatement of this problem.

3.5.8. Radioactive Waste Disposal

The use of fissionable materials for power generation and isotopes in industry and medicine necessitates the disposal of wastes generated during ore mining, refining, and actual usage. Burial is the most common means of disposal. Radium is probably of greatest concern in nuclear fuel mining and refining wastes but other isotopes such as U238 and Th230 are also present. Low-level reactor wastes amount to about 40,000–50,000 m^3 per year in the United States. These wastes contain Cs, Sr, and Co among others. Leakage from burial sites where burial containers have deteriorated has been reported (31, 65).

Another problem is that many burial sites have been located in poorly situated sites within groundwater tables. The problem of designing disposal facilities capable of long-term storage of radionucleides is one that needs careful consideration to prevent groundwater contamination.

3.5.9. Miscellaneous Sources

The sources detailed above are not all-inclusive. For example, salt water intrusion is another important potential source of groundwater contamination in areas where groundwater is pumped in excess of its recharge rate (66). Groundwater pollution by heavy metals and pesticides from a dredge soil dump in the Netherlands has also been reported (67). An unusual episode of pollution in France was reported by Aubic (68), where 20,000 bottles of Sauterne were spilled and subsequently polluted a well.

Salt can be an especially troubling pollutant. For example, salt mounds for road deicing have been found to be a problem in areas with shallow groundwater, whereas saltwater intrusion by overpumping of wells is quite serious in others. Laksman (69) points out that approximately two-thirds of the United States is underlain with water containing high amounts of dissolved salts. Pumping wells in excess of recharge rates leads to intrusion of the salt water from below. Careful attention to pumping and recharge is needed to prevent this from occuring.

3.6. SUMMARY

The sources of groundwater pollutants are many and varied. Pollution by disease-causing microorganisms occurs when human and animal wastes containing viruses, bacteria, and parasites come into contact with groundwater and can be responsible for large outbreaks of illness. Chemical pollutants can leach into groundwater from a variety of sources including hazardous waste dumps, sewage land-treatment sites, and injection wells. The myriad of chemicals that may be found in contaminated groundwater make it difficult to study their fate, transport, and health effects. Because of past, often indiscriminate, dumping of hazardous wastes, the number of cases of serious groundwater pollution is likely to increase in the near future. The cleanup of these pollution sources will be extremely expensive.

REFERENCES

1. D. K. Todd, *Groundwater Hydrology*, 2nd ed., John Wiley and Sons, New York, 1980.
2. G. G. Craun, *Groundwater* **17**, 183 (1979).
3. C. P. Gerba and B. H. Keswick. "Survival and Transport of Enteric Viruses and Bacteria in Groundwater," In W. Van Duijvenbooden, P. Glasenbergen, and H. van Lelyveld, Eds., *Quality of Groundwater*, Elsevier, Amsterdam, 1981, pp. 511–515.
4. E. W. Akin, "Minimum Infectious Dose," In *Conference on Microbial Health Considerations of Soil Disposal of Domestic Wastewaters*, U.S. EPA, Norman, Okla., 1981, p. 188.
5. M. Goldfield, "Epidemiological Indicators for Transmission of Viruses in Water," In G. Berg, H. Bodily, E. H. Lennette, J. L. Melnick, and T. G. Metcalf, Eds., *Viruses in Water*, Amer. Public Health Assoc., Washington, D.C., 1976, pp. 70–85.
6. J. L. Melnick "Are Conventional Methods of Epidemiology Appropriate for Risk Assessment of Virus Contamination of Water," In B. P. Sagik and C. A. Sorber, Eds., *Proceedings of the Coference on Risk Assessment and Health Effects of Land Application of Municipal Wastewater and Sludges*, Univ. of Texas at San Antonio, 1978, pp. 61–75.
7. M. D. Sobsey, "Methods for Detecting Enteric Viruses in Water and Wastewater," In G. Berg, H. Bodily, E. H. Lennette, J. L. Melnick, and T. G. Metcalf, Eds., *Viruses in Water*, Amer. Public Health Assoc., Washington, D.C., 1976, pp. 89–127.
8. J. E. Rohde and R. S. Northrup, "Taking Science Where Diarrhea Is," in *Acute Diarrhea in Childhood*, Ciba Foundation Symp. No. 42, *Elsevier/Excerpta Medica*, Amsterdam, 1976, p. 339.

9. D. J. Bradley, "Health Aspects of Water Supplies in Tropical Countries," In R. Feachem, M. McGarry, and D. Mara, Eds., *Water Wastes and Health in Hot Climates*, Wiley-Interscience, New York, 1977, p. 3.
10. W. J. Jewell and B. L. Seabrook. *A History of Land Application as a Treatment Alternative*, EPA 430/9-79-012 U.S. EPA, Washington, D.C., 1979.
11. S. C. Reed, R. W. Crites, R. E. Thomas, and A. B. Hais, *Cost of Land Treatment Systems*, EPA-430/9-75-003, U.S. EPA, Washington, D.C., 1979, p. 5.
12. B. H. Keswick and C. P. Gerba, *Environ. Sci. Technol.* **14**, 1290, (1980).
13. F. M. Welling, A. L. Lewis, C. W. Mountain, and L. V. Pierce, *Appl. Microbiol.* **29**, 1751, (1975).
14. B. H. Keswick and C. P. Gerba, "Fate and Transport of Viruses During Land Treatment of Sewage," In *Water Re-Use II Symposium*, Washington, D.C., 1981. pp. 2261–2281.
15. F. M. Wellings., A. L. Lewis, C. W. Mountain, and L. M. Stark, "Virus Survival Following Wastewater Spray Irrigation of Sandy Soils," In *Virus Survival in Water and Wastewater Systems*, J. F. Malina and B. P. Sagik, Eds., University of Texas at Austin, 1974, pp. 253†260.
16. J. Vaughn and E. F. Landry, "Data Report: An Assessment of the Occurrence of Human Viruses in Long Island Aquatic Systems," Department of Energy and Environment, Brookhaven National Laboratory, Upton, N.Y., 1977.
17. J. Vaughn and E. F. Landry. "The Occurrence of Human Enteroviruses in a Long Island Groundwater Aquifer Recharged with Tertiary Wastewater Effluents," In *State of Knowledge in Land Treatment of Wastewater*, Vol 2, U.S. Government Printing Office, 1978, pp. 233–243.
18. J. Vaughn, E. F. Landry, J. Baranosky, C. A. Beckwith, M. C. Dahl, and D. C. Delihas, *Appl. Environ. Microbiol.* **36**, 47 (1978).
19. S. A. Schaub and C. A. Sorber, *Appl. Environ. Microbiol.*, **33**, 609 (1977).
20. E. L. Koerner and D. A. Haws, *"Long Term Effects of Land Application of* Domestic Wastewater: Vineland, New Jersey, Rapid Infiltration Site," EPA-600/2-79-072, U.S. EPA, Washington, D.C., 1979.
21. M. J. Noonan and J. F. McNabb, "Contamination of Canterbury Groundwater by Viruses," In M. J. Noonan, Ed., *The Quality and Movement of Groundwater in Alluvial Aquifers of New Zealand*, Dept. of Agricultural Microbiology Tech. Publ. No. 2, Lincoln College, Canterbury, New Zealand, 1979.
22. T. H. Hejkal, B. H. Keswick, R. L. LaBell, C. P. Gerba, Y. Sanchez, G. Dreesman, B. Hafkin, and J. L. Melnick, *J. Am. Water Works Assoc.* **74**, 318 (1982).
23. E. W. Akin, H. P. Pahren, W. Jakubowski, and J. B. Lucas, "Health Hazards Associated with Wastewater Effluents and Sludges: Microbiological Considerations," In B. P. Sagik and C. A. Sorber, Eds., *Proceedings of the Conference on Risk Assessment and Health Effects of Land Application of Municipal Wastewater and Sludges*, Univ. Texas at San Antonio, 1978 pp. 9–24.
24. G. Bitton and S. R. Farrah, *ASM News* **46**, 622 (1980).
25. R. L. Edmonds, *Appl. Environ. Microbiol.* **32**, 537 (1976).
26. B. D. Hays, *Water Res.* **11**, 583 (1977).
27. Damgaard-Larsen, K. O. Jensen, E. Lund, and B. Nissen, *Water Res.* **11**, 503 (1977).
28. W. Jakubowski, W. Galke, and G. Love, *Viral and Bacterial Levels Resulting from the Land Application of Digested Sludge*, EPA-600/1-79-015, U.S. EPA, Cincinnati, Ohio, 1979.

29. M. R. Scalf and W. J. Dunlap, *Environmental Effects of Septic Tank Systems*, EPA-600/3-77-096, U.S. EPA, Ada, Oklahoma, 1977. p. 1.

30. S. S. Sandu, W. Warren, and P. Nelson, *Appl. Environ. Microbiol.* **37**, 744. (1979).

31. R. A. Freeze and J. A. Cherry, *Groundwater*, Prentice–Hall, Englewood Cliffs, New Jersey, 1979.

32. R. C. Cooper, J. L. Potter, and C. Leong. *Water Res.* **9**, 733, (1975).

33. J. A. Donnelly, P. V. Scarpino, and D. Brunner, *Abstr. Ann. Mtg. Amer. Soc. Microbiol.* Miami Beach, Fla., 1980. (Abstract Q70)

34. E. F. Landry, J. M. Vaughn, and W. F. Penello, *Appl. Environ. Microbiol.* **40**, 1032 (1980).

35. R. A. E. Barrell and M. G. M. Rowland, *J. Hyg. Camb.* **83**, 143, (1979).

36. S. M. Goyal, B. H. Kerwick, and C. P. Gerba, "Occurrence of Viruses in Groundwater and Soil Beneath Three Application Sites," In *Water Re-Use Symposium II*, Washington, D.C., 1981, pp. 2251–2260.

37. B. E. Moore, *Munic. Wastewater News* **28**, 11, (1980).

38. V. A. Majeti and C. S. Clark, *Potential Health Effects from Persistent Organics in Wastewater and Sludges Used for Land Application*, EPA-600/1-80-025, U.S. EPA, Cincinnati, Ohio, 1980.

39. V. A. Majeti and C. S. Clark, *J. Environ. Eng. Div. ASCE* **107**, 339 (1981).

40. A. C. Chang, A. L. Page, and F. T. Bingham, *J. Water Pollut. Control. Fed.* **53**, 237, (1981).

41. G. F. Lee and R. A. Jones, "Chemical Agents of Health Significance in Municipal Wastewater," In B. P. Sagik and C. A. Sorber, Eds. *Proceedings of the Conference on Risk Assessment and Health Effects on Land Application of Municipal Wastewater and Sludges, Univ. Texas at San Antonio, 1977, pp. 27–58.*

42. M. Deinzer, F. Schaumberg, and J. E. Klein, *Environ. Health Perspect.* **24**, 209, (1978).

43. K. M. Baxter, K. J. Edworthy, H. A. C. Montgomery, and M. J. Beard, "Groundwater Contamination by Sewage and Sewage Effluent Discharge to Aquifers in the U.K.," In W. Van Duijvenbooden, P. Glasenbergen, and H. van Lelyveld, Eds., *Quality of Groundwater*, Elsevier, Amsterdam, 1981, pp. 245–252.

44. E. L. Koerner and D. A. Haws, *Long Term Effects of Land Application of Domestic Wastewater, Roswell, New Mexico, Slow Rate Irrigation Site*, EPA-600/2-79-047, U.S. EPA, Ada, Okalhoma, 1979.

45. R. C. Sidle and L. T. Kardos, *J. Environ. Quality* **6**, 431 (1977).

46. Environmental Protection Agency, *Hazardous Waste Generation and Commercial Hazardous Waste Management Capacity*, SW-894, U.S. EPA, Washington, D.C., 1980.

47. D. E. Lindorff, *Groundwater* **20**, 9 (1981).

48. Environmental Protection Agency, *Damages and Threats Caused by Hazardous Material Sites*, EPA/430/9-8-004, U.S. EPA, Washington, D.C., 1980.

49. D. J. Lennett, *Environment* **22**, 6 (1980).

50. W. J. McDougall and R. A. Fusco, *J. Water Pollut. Control Fed.* **52**, 2914 (1980).

51. P. H. Roux and W. F. Althoff, *Groundwater* **18**, 464 (1980).

52. F. B. DeWalle and S. K. Chian, *J. Am. Water Works Assoc.* **73**, 206 (1981).

53. J. P. Murray, J. V. Rouse, and A. B. Carpenter, *Water Res.* **15**, 745, (1981).

54. J. Trauberman, *Environment* **23**, 25 (1981).

55. L. R. Reeder, "A Review of Three Decades of Deep-Well Injection and the Present State of the Art," In W. Van Duijvanbooden, P. Glasenbergen, and H. van Lelyveld, Eds., *Quality of Groundwater*, Elsevier, Amsterdam, 1981, pp. 373–384.
56. D. B. Oakes and C. P. Young, "The Effect of Farming Practices on Groundwater Quality in the United Kingdom," In W. Van Duijvenbooden, P. Glasenbergen, and H. van Lelyveld, Eds., *Quality of Groundwater*, Elsevier, Amsterdam, 1981, pp. 27–40.
57. I. A. Kahn, *J. Environ. Eng. Div. ASCE* **106**, 331, (1980).
58. G. Wehtje, J. R. C. Leavitt, R. Spalding, L. Miekle, and J. Schepers, "Atrazine Contamination of Groundwater in the Platte Valley of Nebraska: Non-point Sources," In W. Van. Duijvenbooden, P. Glasenbergen, and H. van Lelyveld, Eds., *Quality of Groundwater*, Elsevier, Amsterdam, 1981, pp. 141-146.
59. A. A. Guerrera, *J. Am. Water Works Assoc.* **73**, 190 (1981).
60. S. J. Nelson, M. Isklander, M. Volz, S. Khalilfa, and R. Haberman, "DBCP Contamination of Groundwater in California," In W. Van Duijvenbooden, P. Glasenbergen, and H. van Lilyveld, Eds., *Quality of Groundwater*, Elsevier, Amsterdam, 1981, pp. 169–174.
61. A. Thomas and W. M. Leis, "Ground Water and Surface Water Contaminations by Hydrocarbons: A Case History," In W. Duijvenbooden, P. Glasenbergen, and H. van Lelyveld, Eds., *Quality of Groundwater*, Elsevier, Amsterdam, 1981, pp. 587–592.
62. M. R. Scalf, *Ground Water Pollution in the South Central States*, EPA-R2-73-268, U.S. EPA, Washington, D.C., 1973.
63. J. P. Schubert and P. F. Prodan, "Groundwater Pollution Resulting From Disposal of Pyritic Coal Wastes," In W. Van Duijvenbooden, P. Glasenbergen, and H. van Lelyveld, Eds., *Quality of Groundwater*, Elsevier, Amsterdam, 1981, pp. 319–328.
64. H. Olem, *J. Water Pollut. Control Fed.* **53**, 814 (1981).
65. B. G. Blaylock and C. S. Fore, *J. Water Pollution. Control Fed.* **53**, 861, (1981).
66. R. J. Ireland and F. C. Brassington, "An Investigation of Complex Saline Groundwater Problems in the Permo Triassic Sandstones of North West England," In W. Van Duijvenbooden, P. Glasenbergen, and H. van Lelyveld, Eds., *Quality of Groundwater*, Elsevier, Amsterdam, 1981, pp.697–702.
67. H. N. Kerdijk, "Groundwater Pollution by Heavy Metals and Pesticides from a Dredge Spoil," In W. Van Duijvenbooden, P. Glasenbergen, and H. van Lelyveld, Eds., *Quality of Groundwater*, Elsevier, Amsterdam, 1981, pp. 279–286.
68. J. Aubic. "Accidental Water Pollution Caused by 20,000 Bottles of 'Sauternes'," In W. Van Duijvenbooden, P. Glasenbergen, and H. van Lelyveld, Eds., *Quality of Groundwater*, Elsevier, Amsterdam, 1981, pp. 643–648.
69. B. T. Laksman, *Water and Sewage Works* **94**, 98 (1979).

4

MICROBIAL POLLUTANTS: THEIR SURVIVAL AND TRANSPORT PATTERN TO GROUNDWATER

Charles P. Gerba

Department of Microbiology
University of Arizona,
Tucson, Arizona

Gabriel Bitton

Department of Environmental Engineering Sciences
University of Florida,
Gainesville, Florida

4.1. INTRODUCTION

The persistence and transport of bacteria and viruses in the subsurface is an area of major interest to those concerned with public health. Almost half of all waterborne diseases are caused by contaminated groundwater (1). In order to develop adequate guidelines for the placement of waste disposal sites and drinking-water wells, information is needed on the fate of pathogenic microorganisms in groundwater.

The fate of pathogenic bacteria and viruses in the subsurface is determined by their survival and retention by soil particles. Both survival and retention are largely determined by the three factors shown in Fig. 4.1. Climate controls two

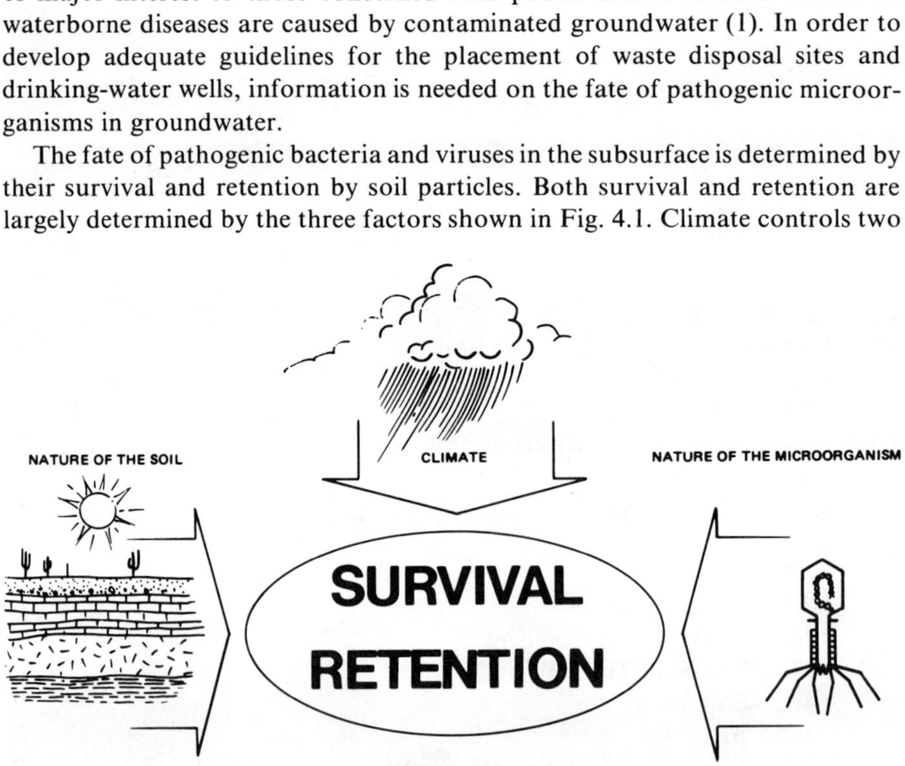

Figure 4.1. Factors affecting microbial fate in the subsurface.

important factors in determining viral and bacterial survival: temperature and rainfall. The survival of microorganisms is greatly prolonged at low temperatures; below 4°C they can survive for months or even years (2). At higher temperatures, inactivation or die-off is fairly rapid. In the case of bacteria (and probably viruses), the die-off rate is approximately doubled with each 10°C rise in temperature between 5°C and 30°C (3). Above 30°C temperature is probably the dominant factor determining virus survival time. Rainfall mobilizes previously retained bacteria and viruses and greatly promotes their transport to groundwater. Several studies have shown that the greatest degree of drinking-water well contamination occurs after periods of heavy rainfall (4–6).

The nature of the soil also plays a major role in determining survival and retention. Soil properties influence moisture-holding capacity, pH, and organic matter—all of which control the survival of bacteria and virus in the soil. Other soil properties such as particle size, cation-exchange capacity, and clay content influence retention. Microbial resistance to environmental factors varies among different species as well as strains. Bacteria are believed to be removed largely by filtration processes while adsorption is the major factor controlling virus retention (2). The following sections are a summary of the factors currently believed to influence microbial persistence and transport in the subsurface.

4.2. BACTERIAL MOVEMENT THROUGH SOILS

4.2.1. Filtration

The straining or filtration of bacteria at the soil surface is a major limitation in their travel through soils. When suspended particles, including bacteria, accumulate on the soil surface, these particles become the filter as water passes through the soil (7). Such a filter is capable of removing even finer particles by bridging or sedimentation before they reach and clog the original soil surface. This phenomenon will, in fact, be dominant if only a portion of the suspended particles are larger than the pore openings. As soon as a few such particles have accumulated, they become the straining surface for finer particles (7).

In studies in which *E. coli* suspended in distilled water were allowed to percolate into sand columns, Krone et al. (7) found that after the first arrival of bacteria, the concentration in column effluents continued to rise until a maximum was reached, after which it fell. This suggests that accumulating bacteria at the soil surface enhances removal by straining. When the capacity for removal by straining has been satisfied along the length of the column, sedimentation alone is operative and the concentration of organisms in the effluent becomes constant.

This same effect is seen during the land application of domestic sewage when repeated cycles of flooding and drying of infiltration basins are used (8). For example, at the Flushing Meadows Project near Phoenix, Arizona, treated sewage effluent is spread into basins underlaid with loamy sand. The greatest number of fecal coliforms are observed after the start of each new inundation period when nearly infiltrated water arrives at the bottom of sampling wells. Afterwards a general decrease in values occur (see Fig. 4.2). A similar phenomenon occurs when water containing microorganisms is pumped into recharge wells.

Mat formation may be more significant in retaining bacteria in some types of discharges, such as septic tank liquors. The high solids content of these discharges can act to retain significant numbers of bacteria. Mats adjunct to septic tank drain fields have been observed to retain as much as 99.9 percent of the original coliform population over a distance of less than a foot (9).

Studies using sandy soils of various effective porosities indicate removal of bacteria from a liquid percolating through a given depth of soil is inversely proportional to the particle size of the soil. The greatest removal of bacteria occurs on the surface mat (top 2–6 mm) that forms on the soil.

The size and shape of microorganisms also plays a role in their removal by filtration. Bacterial movement through saturated sandy soil columns can be treated in terms of gel permeation chromatography, where the pore volume available for cell movement is considered. If there is no adsorption, larger cells should move faster than smaller cells in a soil column or at a given distance from a subsurface injection point. This is due to the fact that soil capillaries are of various sizes and microorganisms move more freely through the larger pores. Bitton et al. (10) observed this effect, noting that a large encapsulated strain of

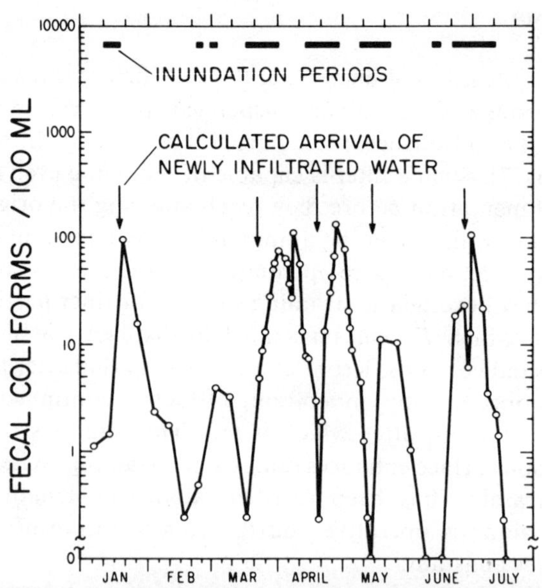

Figure 4.2. Fecal coliforms in well water during multiple flooding and drying cycles of treated wastewater (8). Courtesy of the Water Pollution Control Federation.

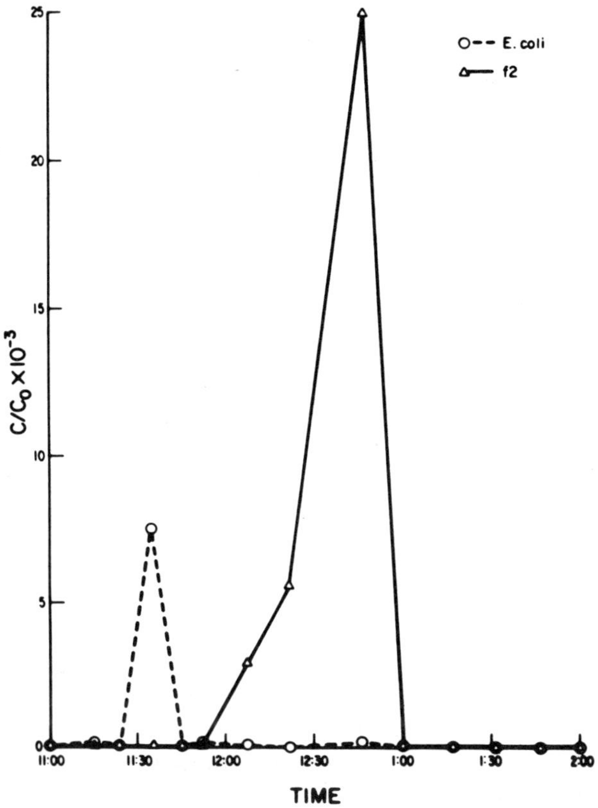

Figure 4.3. Breakthrough curves for *Escherichia coli* and coliphage f2 in a tracer test. Organisms were fed simultaneously into a 300-gpm flow of water that was being injected into a well. The figure shows the concentration of organisms collected from an observation well 500 feet away. The quantity C_0 is the concentration of the tracer in the injected water and C is the concentration measured in the observation well.

Klebsiella aerogenes moves faster than a smaller noncapsulated strain. The effect can also be seen in Fig. 4.3 which illustrates the arrival of the bacterium *E. coli* and coliphage f2 in an observation well 17 m from the point of injection. The larger *E. coli* arrived at the well first, followed by the much smaller coliphage f2.

4.2.2. Adsorption

Adsorption is also a factor in the removal of bacteria by soil. Factors that reduce the repulsive forces between two surfaces, such as the presence of cations, would be expected to allow closer interaction between them and to allow adsorption to proceed. A more detailed explanation of mechanisms involved in adsorption is found in the section on viruses. The very small size of clays, their generally platy shapes, and the occurrence of large surface area per given volume make them ideal adsorption sites for bacteria and viruses in soils. Thus, adsorption phenomena play a more important role in the removal of microorganisms in soils that contain clays (11).

Goldshmid et al. (12) found that filtration efficiency of coliforms through sand is higher when the medium is tap water rather than distilled water. No bacterial removal occurred when triple-distilled water was used. Increases in the cation concentration and valence resulted in increases in the retention efficiency. The retention efficiencies were similar to those predicted from the Shulze–Hardy laws, which describe the effect of the addition of electrolytes on the stability of lyophobic colloids. In the presence of oxidation pond effluents, more cations were found to be necessary to achieve the same retention efficiency as with tap water. In addition, a decrease of pH from 9.3 to 3.9 increased the retention efficiency, again in agreement with adsorption phenomena. The reversibility of this phenomenon was shown when the ionic concentration of column influent was reduced and a corresponding increase in the number of coliforms in the column effluent occurred.

Other studies have demonstrated that common cations in solution will affect bacteria adsorption to soils (13). Certain metallic cations (Fe^{3+}, Cu^{2+}, Zn^{2+}) at concentrations not uncommon to soils have been shown to enhance the removal of bacteria. The common fertilizer ion NH_4^+ also enhances the removal of bacteria. The anions Cl^-, SO_4^{2-}, and NO_3^- have little effect on adsorption (13). Low soil pH also enhances bacterial retention (10). Soluble organics may also compete with bacteria for adsorption sites onto the soil, thereby reducing bacterial adsorption (14).

Rainfall will effect bacterial retention by lowering ionic concentration and increasing infiltration rates. Several surveys have indicated that rainfall and well depth are related to microbial groundwater quality. Studies in Washington indicate that shallow drinking-water wells average median coliform values of 8 MPN/100 ml with an average depth of 9.4 m (31 ft) while deep wells with an average depth of 153.3 m (503 ft) average 4 MPN/100 ml (14). They also observed that virtually all bacterial contamination coincided with the periods of heaviest rainfall. Brooks and Cech (15) observed in rural eastern Texas that practically all dug wells with depths of 15 m (50 ft) or less were positive for either fecal coliforms or fecal streptococci. While the presence of fecal bacteria was much less common in deeper wells, wells as deep as 80 m (250 ft) were positive. Increased levels of bacterial contamination of drinking-well water after periods of rain have been noted in several studies (4,5,6,16). It was also noted that while an increase in coliform bacteria appears almost immediately after periods of heavy rainfall in shallow wells, the increase did not occur until two weeks later in deeper wells (17). Thus, any satisfactory study of well water quality should include sampling during periods of highest rainfall. Sandhu et al. (18) found that basic well design and construction had little effect on the extent of microbial pollution in their study area.

4.3. SURVIVAL OF ENTERIC BACTERIA IN SOIL

At the turn of the century, it was found that eating raw vegetables grown on soil fertilized with raw sewage resulted in outbreaks of typhoid fever. As a result, the survival of enteric bacteria in soil systems has been extensively studied. There are several major reviews on the survival of enteric bacteria in soil (17–21), but we only consider those factors that affect the length of survival of these bacteria. Most enteric bacterial pathogens die-off very rapidly outside of the human gut, whereas indicator bacteria such as *Escherichia coli* persist for longer periods of time. Survival times among different types of bacteria vary greatly and are difficult to assess without studying each type individually. In most cases, it appears that 2–3 months are sufficient for the reduction of pathogens to negligible numbers once they have been applied to the soil. However, survival times of as long as five years have been reported (19). Factors known to influence bacterial survival in the soil are listed in Table 4.1.

4.3.1. Moisture

A major factor in determining the survival of bacteria in the soil is moisture. Young and Greenfield (22) showed that moisture was a factor in the viability of *Escherichia coli* in soils. Beard (23) stated that moisture was the most important determining factor in the survival of *Salmonella typhosa.* Bacterial survival was determined in various types of soil exposed outdoors in clay flowerpots. Survival in all types of soil tested was found to be greatest during the rainy season. In sand, where drying was rapid due to its low moisture-retaining power, survival time was short—between 4 days and 7 days during dry weather. In soils

Table 4.1. Factors Affecting Survival of Enteric Bacteria in Soil[a]

Factor	Comments
Moisture content	Greater survival time in moist soils and during times of high rainfall
Moisture holding capacity	Survival time is less in sandy soils with lower water-holding capacity
Temperature	Longer survival at low temperatures; longer survival in winter than in summer
pH	Shorter survival time in acid soils (pH 3–5) than in alkaline soils
Sunlight	Shorter survival time at soil surface
Organic matter	Increased survival and possible regrowth when sufficient amounts of organic matter are present
Antagonism from soil microflora	Increased survival time in sterile soil

[a]Modified from Ref. 2.

that retain a high amount of moisture, such as loam and adobe peat, the organisms persisted longer than 42 days.

Bouma et al. (24) have suggested that survival data for fecal organisms could be compared with soil-moisture characteristic curves, and hence, the distance of soil filtration necessary for removal be defined as a function of moisture content. In studies on the survival of enteric organisms in septic tank discharges, Parker and Mee (25) concluded that under field conditions survival is not likely to be limited by moisture in coarse sands. They felt that enough moisture was available in the drainfields to prevent moisture from playing a significant role and thus could not be used to predict survival under normal field conditions.

4.3.2. pH

Beard (23) also found that the death of *S. typhosa* was very rapid in peat soil with a pH between 3 and 4. Kligler (26) found that moist, slightly alkaline soils were most favorable for the survival of *S. typhosa*. Cuthbert et al. (27) inoculated various peat (pH 2.9–4.5) and limestone (pH 5.8–7.8) soils held in the laboratory with *E. coli* and *Strep. faecalis*. They found that both organisms could persist for several weeks in the limestone soils, but would die out in a few days in acid peat soils. They felt that the low pH could not only adversely affect the viability of the organism but the availability of nutrients and the action of inhibiting agents.

4.3.3. Sunlight

Beard (23) found that sunlight exerted a definite lethal action on typhoid organisms at the soil surface. Van Donsel et al. (28) reported a greater die-away of both *E. coli* and *Strep. faecalis* when added to soil plots exposed to direct sunlight rather than a shaded area. These results are not surprising, because the ultraviolet light present in sunlight is known to be bactericidal.

4.3.4. Temperature

Cold temperatures favor the survival of most microorganisms and enteric bacteria are no exception (29). *S. typhosa* may survive as long as 24 months at freezing temperatures (23). Mirzoev (30) points out that in areas with prolonged winters, for example, the Russian Arctic, the processes of soil self-disinfection are slowed down or suspended. He showed that low temperatures (down to −45° C) were very favorable for the survival of dysentery bacilli, which he was able to detect 135 days after it had been added to the soil. Van Donsel et al. (28) found that a 90 percent reduction in the number of fecal coliforms took 3.3 days in the summer and 13.4 days in the winter in exposed soil plots.

In contrast to these observations, Kibbey et al. (31) found that any length of freezing conditions was highly lethal to *Streptococcus faecalis*, and that numerous freeze–thaw intervals were more destructive than one extended freeze. Organisms surviving freezing declined more rapidly because of greater cellular damage caused by the freeze–thaw cycles. Soil moisture, the nature of the organism, and other factors greatly influence the effects of freezing on bacteria in soils.

4.3.5. Organic Matter

The frequent addition of broth culture fluid to soil has been found to increase the survival of *S. typhosa* (19). Under field conditions, it has been found that some aftergrowth of *E. coli* and *Strep. faecalis* can occur, particularly after wet weather (28). The survival of fecal coliforms is greatly extended in organic soils over that observed in mineral soils (32) and regrowth of *Salmonella typhimurium* and *E. coli* has been observed in buried feces (33). The extended survival and growth in organic soils may be due not only to the presence of organics but to the high moisture-holding capacity of these soils (32).

4.3.6. Other Microorganisms

Soil moisture, temperature, pH, and the availability of organic matter can also indirectly influence the survival of enteric bacteria by regulating the growth of antagonistic organisms (28). Bryanskaya (34) showed that actinomycetes in soil were capable of suppressing the growth of salmonella and dysentery bacilli. In addition, the longer survival time of enteric organisms after inoculation into sterilized soil as compared to unsterilized soil found by a number of workers (19) indicates that antagonism is an important factor. Tate (32) observed that the protozoan population of a muck soil increased dramatically after the addition of *E. coli* and suggested that soil protozoa could play a significant role in the decline of these organisms in these soils. Since it is evident that enteric bacteria are capable of utilizing nutrients found in nature, it could be argued that competition by the natural soil microflora is in large part responsible for their eventual disappearance from the soil.

4.4. VIRUS MOVEMENT THROUGH SOILS

Concern over public health risks associated with land disposal of domestic wastes that contain human and animal pathogens has been raised. Among these risks is the potential for viruses to be transported through soils and thus to possibly contaminate groundwater supplies. On many occasions, epidemics of infectious diseases were connected to the consumption of contaminated groundwater (this topic is thoroughly reviewed in Chapter 7). This may be due to the fact that viruses may travel through fissures and fractures within the substratum. On many other occasions, viruses were not detected in groundwater beneath sites used for land application of wastewater effluents and residuals (see Chapter 9 for more details). Thus, the public health risks associated with land disposal of human and animal wastes can be evaluated only through a critical review of the various factors controlling the transport pattern of viruses through soils (2, 35–39). Knowledge concerning virus movement through soils has been gained through monitoring land application sites and through batch and column experiments under laboratory conditions.

Batch and column experiments have been widely used to understand the transport pattern of viruses through soils. These studies have been criticized because of the lack of standardization in experimental conditions and because column experiments do not always simulate the natural field conditions (37). However, valuable information has been gained through these studies since many of the factors controlling virus transport through soils have been identified. These factors include soil type, pH, organic matter, cations, flow rate, degree of pore saturation, and virus type.

4.4.1. Soil Type

Viruses are retained by soils mainly through adsorptive phenomena. The adsorption of viruses to soils generally conforms to Langmuir and Freundlich-type isotherms (40–44). Most often, the slope of the Freundlich isotherm is close to unity, which means that the percent virus removal by soils is theoretically independent of virus concentration. Soils offer a large number of sites for viruses to be adsorbed. Moore et al. (43), using Ottawa sand, showed that 2.2×10^6 viruses adsorbed/g sand. They also reported that only 1 percent of the sand surface was covered with viruses. This indicates that soils are potentially efficient in virus binding. Soils are complex environments, and qualitative changes in their textural components (sand, silt, clay, organic matter, iron oxides) undoubtedly influence their sorptive capacity toward viruses. Moore et al. (43) have examined 34 soils and minerals for their ability to retain poliovirus. Virus adsorption varied from 16–79 percent for a muck soil to 99.99 percent for a magnetic sand. The poor adsorption to the muck soil confirms earlier studies (45, 46).

It is generally agreed that fine-textured soils retain viruses more effectively than sandy soils, since the soil clay mineral fraction displays a high sorptive capacity toward viruses as a result of its high surface area and ion-exchange capacity. Following the examination of nine soils from Arkansas and California, it was shown that virus adsorption increased with the clay content and the specific surface area of the soil (41). Koya and Chaudhuri (47) reported that, among all the soils studied, a lateritic soil (32.5 percent clay) was the most efficient in retaining viruses. Iron oxides, particularly magnetite, also display a high affinity for viruses (48, 49). A magnetic sand and hematite proved to be the most effective virus adsorbents (43). It thus appears that soil iron oxides can increase the ability of soils to retain viruses. Finally, a recent investigation has shown a highly negative correlation between virus adsorption and the capacity of soils to bind a cationic polymer, PDADM (Polydiallyldimethyl ammonium chloride). It was suggested that the ability to bind the polymer could serve as an indicator of the extent of viral adsorption (43).

4.4.2. pH

Most viruses behave as proteins, the net charge of which is dependent on the pH of the suspending medium. As the pH increases, there is an increase in the ionization of carboxyl groups and a decrease in the ionization of amino groups. Consequently, the net charge of the virus particle is negative at pHs above neutrality. Major components of soils (sand, clay minerals, organic matter) are also negatively charged at pHs above 7. Electrophoretic mobility studies have shown that a common soil clay mineral, montmorillomite, is negatively charged at pH between 4.5 and 10.5. Muck soils also have a high negative charge (50). It is tempting to conclude that at alkaline pHs virus adsorption will be at a minimum whereas at acid pHs virus adsorption to soils will reach its maximum. The relationship between virus adsorption and pH is not clear-cut, however, because of many complicating factors. The pH of the soil, as conventionally measured, does not reflect necessarily the pH at the surface of soil colloidal particles such as clays. Various soil components (clay, sand, oxides of Al and Fe) display different isoelectric points. There is also a lack of information on the isoelectric points of more than 100 viruses that occur in wastewater or groundwater. So far, we know that the isoelectric point varies with the virus type and strain (50–53). Even more complicating is the fact that a given virus may display two isoelectric points. For example, poliovirus 1 has two isoelectric points at pH 4.5 and 7.0 (52) and poliovirus 2 displays two isoelectric points at pH 4.5 and 7.5 (50). The factors responsible for the passage from one form to another are not known. It is possible that some other soil factors (e.g., cations or humic acids) may also influence the net electric charge of viruses. The previous discussion can help us understand the confusing trends reported in the literature. Some researchers have reported an increase in adsorption as the pH decreases (40, 41, 51) while others found no correlation between pH level and adsorption (43).

4.4.3. Conductivity of the Percolating Water

Conductivity is a measure of the ionic strength of a solution and is expressed in mmhos/cm or μmhos/cm at 25° C. It is generally agreed that the concentration and species of cations in the percolating water affects the extent of virus adsorption to soils. Early experiments with bacterial phages showed that adsorption to soil increased with the cation concentration of the percolating solution (41). Divalent cations (e.g., Ca^{2+}, Mg^{2+}) were very efficient in virus adsorption to a sandy soil and most of the virions were detected in the top layer of the soil (54). Thus, cations are necessary for reducing the repulsive forces on both the virus and soil particles and allow adsorption to take place. Virus retention by soils is generally greater in the presence of sewage effluents than in distilled water (55). Wastewater effluents have higher conductivity (500–600 μmhos/cm) than distilled water (2–10 μmhos/cm) or rainwater (20–40 μmhos/cm). Rainwater, being of lower conductivity than sewage effluents, may lead to reduced virus adsorption or to virus desorption with the subsequent redistribution of viruses within the soil profile. This phenomenon was well demonstrated via soil core studies under controlled laboratory conditions (46, 55, 56). Landry et al. (57) showed that virus penetration was more extensive in rainwater-rinsed cores than in wastewater-rinsed cores. Moreover, the desorbed viruses may readsorb at greater depths. Heavy rainfall might then remobilize soil-bound viruses with the potential contamination of groundwater supplies (58). However, it now appears that the ability of rainwater to release viruses depends on the soil type, the release being more pronounced in sandy then in clay soils (46, 59). The elution pattern also depends on the virus type and strain. For example, poliovirus 3 and echovirus 6 were mobilized by artificial rainwater whereas echovirus 1 was not affected. The elution pattern of the reference strain of poliovirus 1 differed from that of field and mutant strains (60). The application of these concepts to virus management during land application of wastewater effluents will be discussed by Lance (Chapter 9)

4.4.4. Soluble Organic Materials

Soluble organic materials are known to compete with viruses and bacteria for adsorption sites. It may then be possible that organics present in sewage may interfere with virus sorption to soils. However, several studies have shown that viruses are well adsorbed to various types of soils in the presence of secondary and even primary wastewater effluents (42, 55, 61). As discussed above, wastewater effluents contain enough salts to overcome any interference by soluble organic matter.

Humic and fulvic acids constitute a category of soluble organic matter that may interfere with virus adsorption to soils. They may increase virus transport through organic soils and, in particular, situations where sewage effluents are discharged into wetlands (45, 62). This phenomenon was confirmed by Bixby and O'Brien (63) who reported that fulvic acids complex MS2 phage and prevent its adsorption to soil. More recently, muck soils were found to display a lower adsorption capacity than mineral soils (43, 46).

4.4.5. Virus Type

During the last two decades great attention has been focused upon the quest for a virus indicator that could adequately simulate the adsorption–elution pattern of all the other enteric viruses. First, researchers focused on bacterial phages since their assay is rapid and inexpensive. This approach was criticized and numerous studies were then undertaken on the adsorption–elution pattern of poliovirus type 1. It is now agreed that poliovirus type 1 adsorbs well to soils. More recently, Goyal and Gerba (64) studied the adsorptive behavior of 27 enteroviruses, a simian rotavirus (SA-11), and 5 bacterial phages. It was found that most of the viruses adsorbed well to soils except for echo 1, echo 12, echo 29, SA-11, and phage f2. Echo 1 and RNA phage f2 were the least adsorbed among all the viruses tested. Others have confirmed the low ability of echovirus 1 to adsorb to soils, organic flocs, and sludges (65, 66). However, others showed that echovirus 1 displayed a high affinity for soils (60). These apparently conflicting findings show that virus adsorption–elution patterns in soils vary not only with virus type but with isolates within the same type. Thus, differences in sorptive ability were noted among natural isolates of echovirus 1 and coxsachivirus B4 (64). On the other side of the scale, poliovirus 1 and T-even phages displayed the highest adsorption to soils (64). This confirms the data of Bitton et al. (62) who showed that poliovirus 1 and T2 phage were equally well adsorbed to soils. It was recently concluded that viruses may be grouped into three categories according to their adsorptive behavior (51). Category 1 contains the poorly adsorbed viruses (echovirus 1, echovirus 11, coxsackievirus B4, ϕX174, MS2). Category 2 includes the highly adsorbed viruses (poliovirus 1, echovirus 7, coxsackievirus B3, T2, and T4). Phage f2 was placed in a third category.

This demonstrates that poliovirus 1 cannot serve as an indicator to predict the adsorption–elution behavior of enteric viruses in soils. Virus movement through soils to the groundwater appears to be controlled by the interplay between virus type and strain, soil type, and other known and yet unknown factors. The quest for a suitable virus indicator still continues.

4.4.6. Saturated versus Unsaturated Flow of Water

The degree of saturation of pores within the soil matrix is an important parameter to consider in studies of virus retention by soils. Under saturated-flow conditions, water fills all the pores whereas under unsaturated conditions it flows only through the small pores or is retained as a film around soil particles (see Chapter 2). Thus, unsaturated-flow conditions allow viruses to get closer to particle surfaces. This increases the opportunity for virus sorption to soil. Virus movement through soils was examined mainly under saturated-flow conditions using soil columns. These studies probably underestimated the degree of virus retention by soils because they do not simulate the unsaturated-flow conditions often prevailing in the field. Therefore, soil columns should be long enough to allow the establishment of an unsaturated zone (see Chapter 2). Soil drying between sewage applications is a useful management practice that can prevent the desorption of virus following rainfall (56).

4.4.7. Downward Flow Rate

The lower the infiltration rate of sewage into soil, the longer the retention of viruses within the vadose zone. This results in more efficient virus removal. Green (67) showed that viruses suspended in septic tanks effluents were retained more efficiently by sand mounds as the flow rate of sewage decreased. More recent studies have shown that virus movement through soils was promoted by increasing the flow rate of the percolating sewage (121–123). Management practices for the control of sewage effluents may be important to the treatment process with respect to viruses and other microbial pathogens (see Chapter 9). The factors that may influence virus transport to groundwater are summarized in Table 4.2.

Table 4.2. Factors that may Influence Virus Movement to Groundwater

Factor	Comments
Soil type	Fine-textured soils retain viruses more effectively than light-textured soils. Iron oxides increase the adsorptive capacity of soils. Muck soils are generally poor adsorbents.
pH	Generally, adsorption increases when pH decreases. However, the reported trends are not clear-cut due to complicating factors.
Cations	Adsorption increases in the presence of cations (cations help reduce repulsive forces on both virus and soil particles). Rainwater may desorb viruses from soil due to its low conductivity.
Soluble organics	Generally compete with viruses for adsorption sites. No significant competition at concentrations found in wastewater effluents. Humic and fulvic acid reduce virus adsorption to soils.
Virus type	Adsorption to soils varies with virus type and strain. Viruses may have different isoelectric points.
Flow rate	The higher the flow rate, the lower virus adsorption to soils.
Saturated versus unsaturated flow	Virus movement is less under unsaturated flow conditions.

4.5. VIRUS SURVIVAL IN SOILS

We have reviewed the various factors controlling the transport of viruses through soils. It is now realized that some of the soil-associated viruses may be desorbed and thus redistributed within the soil matrix following changes in certain characteristics of the soil solution (e.g., ionic properties). This may result in groundwater contamination. Another important aspect of groundwater pollution is the persistence of virus particles in the soil environment. This subject has been extensively reviewed by various investigators (2, 35, 68, 69). We shall limit ourselves to the analysis of the most pertinent factors controlling virus survival in soils. It is worth mentioning at this point that significant contributions were made by phytopathologists as well as insect virologists. Phytopathogenic viruses (e.g., wheat yellow mosaic virus, barley yellow mosaic virus) are known to survive for extended periods in soils (70–72). Clay minerals appear to play an important role in the extended survival of plant viruses (71, 72). Similarly, insect viruses (e.g., nuclear polyhydrosis virus, granulosis virus) persist well in soils (73–75). The above findings have inspired investigators interested in the fate of human and animal viruses in soils. It was, however, essential to learn more about virus detection in soils. Soil-associated viruses can be recovered using a variety of eluents ranging from tryptose phosphate broth (55), minimal essential medium and antibiotics (76), glycine–EDTA (77), beef extract (78, 79), isoelectric casein to nonfat dry milk (78). Quantitative techniques have been developed for virus recovery from soils. These methods include an elution step followed by a concentration step (77, 79).

Soil is a complex heterogenous environment the properties of which may be altered in response to climatic changes and agricultural practices. This means that any alien microorganism (e.g., viruses) introduced into soils may be subjected to a variety of environmental factors or to a combination of factors that are dictated by the interplay between climate, vegetation, soil, and microorganisms. It is also expected that sewage and sludge application to land may alter some of the soil's physical and chemical properties with regard to virus survival (69). However, from an experimental viewpoint it is rather difficult to study the effect of combinations of factors on virus persistence. One can then limit our discussion to an overview of the various factors and speculate when possible on the impact of their interrelationship with respect to virus survival.

4.5.1. Temperature

Temperature is probably the most detrimental factor affecting virus survival in soils as well as in other habitats (68). Temperature obviously affects chemical and biological processes in soils and this may indirectly affect virus decline. Owing to the heterogeneity of viral populations, it is likely that some thermoresistant isolates may survive extreme temperatures in soils (80, 81).

Bagdasar'yan (82) observed that viruses could survive up to 170 days in soil at 3°–10°C and that survival was higher at 3°–10°C than at 18°–23°C. Similar observations were made by Lefler and Kott (54) with regard to poliovirus type 1 and bacteriophage f2 survival in a sandy soil in Israel. The heavy dependence of viral persistence on soil temperature was subsequently confirmed by other investigators (55, 67, 83–87).

Land application of municipal sludge offers many advantages but one should be concerned with the survival of viruses in the sludge–soil mixtures and with the possible contamination of groundwater. It appears that virus survival in sludge-amended soils is controlled primarily by soil temperature and moisture (83, 84, 88, 89). Larkin et al. (76) also observed that poliovirus survival was greater in winter than in the summer in Cincinnati, Ohio. In Denmark, at temperatures ranging from 0°C–10°C, coxsackievirus B3 was shown to survive 161 days in a sludge-amended soil (90). One would expect viruses to survive longer in sludge injected 10–15 cm below the soil surface (89) than in surface-spread sludge (84).

4.5.2. Soil Moisture

Soil desiccation is directly related to temperature and both factors synergetically influence virus survival in soils. Bagdasar'yan (82) was one of the first to report the marked effect of soil moisture on enteroviruses (P1, CB3, E7, E9). Viruses survived no more than 15–25 days in air-dried soil compared to 60–90 days in samples with 10 percent moisture. Similarly, DuBoise et al. (55) reported an increased reduction of poliovirus 1 in a dry sandy soil. Hurst et al. (91) observed that the drying period is important with regard to virus survival in rapid-infiltration basins and that the rate of inactivation depends on the rate of soil moisture loss. Poliovirus 1 could not be recovered from a brown-red sandy soil after 16 days when the moisture dropped from 15 percent to 3 percent (79). Use of labeled viruses has shown that they are truly inactivated rather than irreversibly bound to soil particles (87). Evaporation of soil water may be primarily responsible for virus loss during soil drying (87). However, in controlled laboratory experiments, a survey of the soil characteristics and environmental factors affecting virus survival did not reveal any clear trend regarding the effect of soil moisture (85).

This contradictory finding may be explained by the heterogenous nature of soils. Small variations in texture may lead to drastic changes in the water-holding capacity of soil, thus affecting virus survival. Soil desiccation was one of the most detrimental factors controlling virus persistence in sludge–soil mixtures (84, 88).

4.5.3. Sunlight

It is known that viruses may be inactivated by sunlight in the aquatic environment (92, 93). Sunlight at the soil surface is detrimental and its role is minor in comparison to other environmental factors.

4.5.4. Soil Characteristics

Soils vary in their chemical and physical properties and this, in turn, probably influences the fate of viruses. Early studies (82) did not reveal any clear trend as to the effect of soil type on virus survival. Later, Sadovsky et al. (79) compared poliovirus survival in two soils, a desert alluvial soil and a coastal brown-red sandy soil, but it was difficult to distinguish between environmental factors and soil characteristics. A more expanded study that included nine different soils was undertaken by Hurst et al. (85). It showed that virus survival correlated with the extent of virus adsorption to soil, soil saturation, pH, and exchangeable Al. Viruses are known to survive better in the sorbed state than in suspension (46, 67). The significant correlation observed between virus survival and exchangeable Al may be due to increased viral adsorption in the presence of Al (85). DuBoise and his co-workers (69) speculated that cations and pH affect adsorption, which in turn influences viral stabilization in soils. It has been established that certain cations (Ca, Mg) and pH may increase the thermal stabilization of enteroviruses (94–96). This phenomenon may be of some significance with regard to virus survival in soils.

The two most active components of soil are clay minerals and humic materials. Clay minerals influence the ecology of soil microbial populations as well as microbial pollutants (97, 98). One may speculate that clays increase viral adsorption to soil and thus indirectly increase their stabilization in soils. Viral genome can persist in soils and be protected from nucleases by clay minerals (99–101). In aquatic environments, clays protect viruses from light (92), heat (Schiffenbauer and Stotzky, unpublished), and biological degradation (102, 103). In soils, water may be tightly bound to clays, even though the soil is considered dry and this may influence the survival of microorganisms subjected to desiccation (104). Humic materials probably have an indirect influence on virus survival in soils since they affect the cation-exchange capacity, pH, and moisture retention of soils. No experimental evidence is available, however, to support or refute these speculations.

4.5.5. Biological Factors

As in the aquatic environment (68), it is likely that biological factors may play a role in virus inactivation in soils. There is no clear trend, however, regarding the contribution of soil microflora to virus decline. Some investigators did not observe any effect (76, 82) while others reported a greater virus decline in nonsterile than in sterile soils (46, 85). This reflects the complex and heterogenous nature of soil environment.

4.5.6. Virus Survival Under Field Conditions

Viruses may survive long enough to be detected in groundwater. Enteroviruses have been detected at the surface of soils irrigated with raw sewage in the

U.S.S.R. (105). A field study revealed virus survival for at least 28 days in soil following application of a package treatment plant effluent in a cypress dome in Gainesville, Florida (58). Other field studies confirm the important role played by temperature and soil moisture in virus persistence in soils (79, 91). Similarly, it appears that virus survival in sludge-amended soils is controlled primarily by desiccation and soil temperature (84, 88, 106). During surface application of digested sludge on soils in Pensacola, Florida, it was shown that indigenous enteroviruses were able to survive only nine days after sludge application (107).

4.5.7. Mechanism of Viral Inactivation in Soil

We have discussed the major factors that influence viral persistence in soil but we have yet to learn more about the mechanism(s) of inactivation. It is a fact that desiccation and temperature are the leading factors controlling virus survival. This has led to a study on the structural changes associated with poliovirus decline in soil (108). It was shown that damage to the enterovirus consisted of dissociation of viral components followed by degradation of viral genome. The above investigators also reported that viral RNA was degraded more rapidly in moist than in dry soils. The viral genome may persist in soil and could be protected from nuclease action by soil colloids (69). The factors that may influence virus persistence in soil are summarized in Table 4.3.

4.6. BACTERIAL AND VIRAL SURVIVAL IN GROUNDWATER

We have reviewed the various parameters that control the transport and survival of pathogenic microorganisms in soils. Comparatively little is known about the

Table 4.3. Factors that may Influence Virus Survival in Soils

Factor	Comments
Temperature	One of the most detrimental factors.
Desiccation	One of the most detrimental factors. Increased virus reduction in drying soils.
Sunlight	May be detrimental at the soil surface.
Soil pH	May indirectly affect virus survival by controlling their adsorption to soils.
Cations	Certain cations have a thermal stabilizing effect on viruses. May also indirectly influence virus survival by increasing their adsorption to soil (viruses appear to survive better in the sorbed state).
Soil texture	Clay minerals and humic substances increase water retention by soil and thus have an impact on viruses subjected to desiccation.
Biological factors	No clear trend with regard to the effect of soil microflora on viruses.

Table 4.4. Die-off Rate Constants (day^{-1}) for Viruses and Bacteria in Groundwater

Microorganism	Die-off[a] rate (day^{-1})	Reference
Poliovirus 1	0.046	110
	0.21	112
	0.77	114
Coxsackievirus	0.19	112
Rotavirus SA-11	0.36	112
Coliphage T7	0.15	115
Coliphage f2	1.42	110
	0.39	112
Escherichia coli	0.32	112
	0.36	111
	0.16	110
Fecal Streptococci	0.23	112
	0.24	111
	0.03	110
Salmonella	0.13	112
typhimurium	0.22	111

[a]As $\log_{10} N_T/N_0$, where N_T equals concentration of organisms after 24 hrs and N_0 equals the initial concentration of organisms.

survival of these pathogens in groundwater. Microbial persistence has been studied using flasks incubated under laboratory conditions (109, 110), McFeter-type chambers immersed in flowing groundwater (111, 112), or dialysis tubing suspended directly into the wells (113). Table 4.4 gives the decay rate constants for some viruses and bacteria in groundwater. Data available indicate that bacteria and viruses survive longer in groundwater than in surface waters. For example, the decay rate of poliovirus type 1 in groundwater is 0.0019 hr^{-1} (110) compared to 0.031 hr^{-1} in river water (114) or 0.020 hr^{-1} in seawater (116). Microcosm studies also show that viruses survive longer than indicator bacteria in groundwater (110, 112, 114, 117). There is the need to find suitable indicators of viral contamination since no correlation was found between indicator bacteria and viruses in groundwater (118). Indicator bacteria may grow in groundwater if sufficient nutrients are present. *E. coli* growth was observed during groundwater recharge operations in Israel (119). It was also found that coliform growth occurred during the period between recharge and pumping (120).

It appears, from the above investigation, that more research needs to be done on the persistence of microbial pathogens in groundwaters from various geological formations and depths. A suitable indicator of viral contamination in groundwater should also be found.

REFERENCES

1. G. F. Craun, *Ground Water* **17,** 183 (1979).
2. C. P. Gerba, C. Wallis, and J. L. Melnick, *J. Irrig. Drain. Div.* **101,** 157 (1975).
3. K. R. Reddy, R. Khaleel, and M. R. Overcash, *J. Environ. Qual.* **10,** 255 (1981).
4. F. B. DeWalle, R. M. Schaff, and J. B. Hatlen, *J. Am. Water Works Assoc.* **72,** 533 (1980).
5. W. J. Lewis, J. F. Farr, and S. S. D. Foster, *Proc. Instn. Div. Engrs.* **69,** 281 (1980).
6. R. A. E. Barrell and M. G. M. Rowland, *J. Hyg., Lamb.* **83,** 143 (1979).
7. R. B. Krone, G. T. Orlob, and C. Hodgkinson, *Sew. Indust. Wastes* **30,** 1 (1958).
8. H. Bouwer, R. C. Rice, and E. D. Escarcega, *J. Water Pollut. Contr. Fed.* **46,** 844 (1974).
9. E. McCoy and W. A. Ziebeil, In N. I. McClelland, Ed., *Individual Onsite Wastewater Systems,* Ann Arbor Science, Ann Arbor, 1977, p. 67.
10. G. Bitton, N. Lahav, and Y. Henis, *Plant Soil* **40,** 373 (1974).
11. R. B. Krone, In *Symposium on the Use of Municipal Sewage Effluent for Irrigation,* Louisiana Polytechnic Institute, Ruston, LA, 1960.
12. J. Goldshmid, D. Zohar, Y. Argaman, Y. Kott, In S. H. Jenkins, Ed., *Advances in Water Pollution Research,* Pergamon Press, New York, 1973, p. 147.
13. J. W. Boyd, T. Yoshida, L. E. Vereen, R. L. Cada, and S. M. Morrison, Sanitary Engineering Papers, No. 5, Colorado State University, Fort Collins, 1969.
14. D. W. Hendricks, F. J. Post, and D. R. Khairnar, *Water, Air, Soil Pollut.* **12,** 219 (1979).
15. D. Brooks and I. Cech, *Water Res.* **13,** 33 (1979).
16. K. G. Lamka, M. W. LeChevallier, and R. J. Seidler, *Appl. Environ. Microbiol.* **39,** 734 (1980).
17. E. P. Loehnert, In W. van Duijvenbooden, P. Glasbergen, and H. van Lelyveld, Eds., *Quality of Groundwater,* Elsevier, Amsterdam, 1981, p. 147.
18. S. S. Sandhu, W. J. Warren, and P. Nelson, *Appl. Environ. Microbiol.* **37,** 744 (1979).
19. W. Rudolfs, L. L. Frank, and R. A. Ragotzkie, *Sewage Indust. Waste* **22,** 1261, (1950).
20. D. H. Foster and R. S. Engelbrecht, In W. E. Sopper and L. T. Kardos, Eds., *Recycling Treated Municipal Wastewater and Sludge through Forest and Cropland,* The Pennsylvania State Press, University Park, PA, 1973, p. 247.
21. E. Sepp, "The Use of Sewage for Irrigation—A Literature Review," Bureau of Sanitary Engineering, California State Dept. of Public Health, Sacramento, 1971.
22. C. C. Young and H. Greenfield, *Am. J. Public Hlth.* **13,** 270 (1923).
23. P. J. Beard, *Am. J. Public Hlth.* **30,** 1077 (1940).
24. J. Bouma, W. A. Ziebell, W. G. Walker, P. G. Olcott, E. McCoy, and F. D. Hole, "Soil Adsorption of Septic Tank Effluent," University of Wisconsin—Extension Geological and Natural History Survey, Information Circular No. 20, 1972.
25. W. F. Parker and B. J. Mee, *Appl. Environ. Microbiol.* **43,** 981 (1982).
26. I. J. Kligler, "Investigations of Soil Pollution and the Relation of the Various

Types of Privies to the Spread of Intestinal Infections," International Health Board Monograph No. 15, Rockefeller Institute of Medical Research, New York, 1921.

27. W. A. Cuthbert, J. J. Panes, and E. C. Hill, *J. Appl. Bacteriol.* **18,** 408 (1950).
28. D. J. Van Donsel, E. E. Geldreich, and N. A. Clarke, *Appl. Microbiol.* **15,** 1362 (1967).
29. M. Ostrolenk, N. Kramer, and R. C. Cleverdon, *J. Bacteriol.* **53,** 197 (1947).
30. G. G. Mirzoev, *Hyg. Sanit.* **31,** 437 (1968).
31. H. J. Kibbey, C. Hagedorn, and E. L. McCoy, *Appl. Environ. Microbiol.* **35,** 711 (1978).
32. R. L. Tate, *Appl. Environ. Microbiol.* **35,** 925 (1978).
33. K. L. Temple, A. K. Camper, and G. A. McFeters, *Appl. Environ. Microbiol.* **40,** 794 (1980).
34. A. M. Bryanskaya, *Hyg. Sanit.* **31,** 123 (1966).
35. G. Bitton, *Water Res.* **9,** 473 (1975).
36. G. Bitton, In *Adsorption of Microorganisms to Surfaces*, G. Bitton, and K. C. Marshall, Eds., Wiley Interscience, New York, 1980.
37. G. Bitton, J. M. Davidson, and S. R. Farrah, *Water, Air, Soil Poll.* **12,** 449 (1979).
38. W. D. Burge and P. B. March, *J. Environ. Qual.* **7,** 1 (1978).
39. W. D. Burge, and J. F. Parr, In M. R. Overcash and J. M. Davidson, Eds., *Environmental Impact of Non-Point Source Pollution*, Ann Arbor Sci., Ann Arbor, Mich., 1980, p. 107–124.
40. W. D. Burge and N. K. Enkiri, *J. Environ. Qual.* **7,** 73 (1978).
41. W. A. Drewry and R. Eliassen, *J. Water Pollut. Control Fed.* **40,** R 257 (1968).
42. C. P. Gerba and J. C. Lance, *Appl. Environ. Microbiol.* **36,** 247 (1978).
43. R. S. Moore, D. H. Taylor, L. S. Sturman, M. M. Reddy, and G. W. Fuhs, *Appl. Environ. Microbiol.* **42,** 963, 1981.
44. R. F. Reece, "Virus Sorption on Natural Soils," M.S. Thesis, University of Arkansas, 1967.
45. P. R. Scheuerman, G. Bitton, A. R. Overman, and G. E. Gifford, *J. Environ. Eng. Div.* **105,** 629 (1979).
46. M. D. Sobsey, C. H. Dean, M. E. Knuckles, and R. A. Wagner, *Appl. Environ. Microbiol.* **40,** 92 (1980).
47. K. V. A. Koya and M. Chaudhuri, *Prog. Water Technol.* **9,** 43 (1977).
48. G. Bitton, O. Pancorbo, and G. E. Gifford, *Water Res.* **10,** 973 (1976).
49. V. C. Rao, S. V. Waghme, and S. B. Lakhe, *Ann. Meeting Am. Soc. Microbiol.*, Miami Beach, FL, Abstract No. Q136 (1980).
50. D. H. Taylor, R. S. Moore, and L. S. Sturman, *Appl. Environ. Microbiol.* **42,** 976 (1981).
51. C. P. Gerba, S. M. Goyal, I. Chech, and G. F. Bogdan, *Environ. Sci. Technol.* **15,** 940 (1980).
52. B. Mandel, *Virology* **44,** 554 (1971).
53. D. H. Taylor, In *Chemistry in Water Reuse*, Vol. 2, W. J. Cooper, Ed., Ann Arbor Science, Ann Arbor, Michigan, 1981, pp. 595–612.

54. E. Lefler and Y. Kott, In *Virus Survival in Water and Wastewater Systems*, J. F. Malina, Jr., and B. P. Sagik, Eds., Center for Research in Water Research, Austin, Texas.
55. S. M. DuBoise, B. E. Moore, and B. P. Sagik, *Appl. Environ. Microbiol.* **31,** 536 (1976).
56. J. C. Lance, C. P. Gerba, and J. L. Melnick, *Appl. Environ. Microbiol.* **32,** 520 (1976).
57. E. F. Landry, J. M. Vaughn, and W. F. Penello, *Appl. Environ. Microbiol.* **40,** 1032 (1980).
58. F. M. Wellings, A. L. Lewis, C. W. Mountain, and L. V. Pierce, *Appl. Microbiol.* **29,** 751 (1975).
59. G. Bitton, P. R. Scheuerman, G. E. Gifford and A. R. Overman, In H. T. Odum and K. C. Ewel, Eds., *Cypress Wetlands* (in press).
60. E. F. Landry, J. M. Vaughn, M. Z. Thomas, and C. A. Beckwith, *Appl. Environ. Microbiol.* **38,** 680 (1979).
61. C. P. Gerba and S. M. Goyal, In H. L. McKim, Ed., *State of Knowledge in Land Treatment of Wastewater*, U.S. Army Corps of Engineers, Cold Regions Research Engineering Laboratory, Hanover, NH, 1978.
62. G. Bitton, N. Masterson, and G. E. Gifford, *J. Environ. Qual.* **5,** 370 (1976).
63. R. L. Bixby and D. J. O'Brien, *Appl. Environ. Microbiol.* **38,** 840 (1979).
64. S. M. Goyal and C. P. Gerba, *Appl. Environ. Microbiol.* **38,** 241 (1979).
65. G. Bitton, B. N. Feldberg, and S. R. Farrah, *Water, Air, Soil Pollut.* **12,** 187 (1979).
66. S. A. Schaub, K. F. Kenyon, B. Bledsoe, and R. E. Thomas, In H. L. Kim, Ed., *State of Knowledge in Land Treatment of Wastewater*, U.S. Corps of Engineers, Cold Regions Research Engineering Laboratory, Hanover, NH, pp. 245–252, 1978.
67. K. M. Green, "Sand Filtration for Virus Purification of Septic Tank Effluents," Ph.D. thesis, Univ. of Wisconsin, Madison, WI (1976).
68. G. Bitton, In R. Mitchell, Ed., *Water Pollution Microbiology*, Vol. 2, Wiley, New York, 1978.
69. S. M. DuBoise, B. E. Moore, C. A. Sorber, and B. P. Sagik, *Crit. Rev. Microbiol.* **7,** 245 (1979).
70. H. H. McKinney, *Soil Sci.* **61,** 93 (1946).
71. Y. Miyamoto, *Virology* **7,** 250 (1958).
72. Y. Miyamoto, *Virology* **9,** 290 (1959).
73. W. A. L. David and B. O. C. Gardiner, *J. Invertebr. Pathol.* **9,** 342 (1967).
74. T. Hukuhara and H. Namura, *J. Invertebr. Pathol.* **19,** 308 (1972).
75. R. P. Jaques, *J. Invertebr. Pathol.* **6,** 251 (1964).
76. E. P. Larkin, J. T. Tierney, and R. Sullivan, In L. B. Baldwin, J. M. Davidson, and J. Gerber, Eds., *Virus Aspects of Applying Municipal Waste to Land*, Univ. of Fla., Gainesville, FL, 1976.
77. C. P. Gerba, E. M. Smith, and J. L. Melnick, *Appl. Environ. Microbiol.* **34,** 158 (1977).
78. G. Bitton, M. J. Charles, and S. R. Farrah, *Can. J. Microbiol.* **25,** 874 (1979).

79. A. Y. Sadovski, B. Fattal, D. Goldberg, E. Katzenelson, and H. I. Shuval, *Appl. Environ. Microbiol.* **36,** 824 (1978).

80. D. N. Medearis, Jr., J. H. Arnold, and J. F. Enders, *Proc. Soc. Exp. Biol. Med.* **104,** 419 (1960).

81. N. F. Stanley, D. C. Dorman, J. Poncford, and M. Larkin, *Aust. J. Exp. Biol.* **34,** 297 1956).

82. G. A. Bagdasar'yan, *J. Hyg. Epidemiol. Microbiol. Immunol.***7,** 497 (1964).

83. G. Bitton and S. R. Farrah, *Am. Soc. Microbiol. News* **46,** 622 (1980).

84. G. Bitton, S. R. Farrah, O. C. Pancorbo, and J. M. Davidson, in M. Goddard and M. Butler, Eds., *Viruses and Wastewater Treatment*, Pergamon Press, Oxford, 1981.

85. C. J. Hurst, C. P. Gerba, and I. Cech., *Appl. Environ. Microbiol.* **40,** 1067 (1980).

86. J. T. Tierney, R. Sullivan, and E. P. Larkin, *Appl. Environ. Microbiol.* **33,** 109 (1977).

87. J. G. Yeager and R. T. O'Brien, *Appl. Environ. Microbiol.* **38,** 694 (1979).

88. B. E. Moore, B. P. Sagik, and C. A. Sorber, in B. P. Sagik and C. A. Sorber, Eds., *Risk Assessment and Health Effects of Land Application of Municipal Wastewater and Sludges*, Center for Applied Research Technology, Univ. Texas, San Antonio, TX. (1978).

89. B. P. Sagik, in J. L. Smith and E. H. Bryan, Eds., *Williamsburg Conference on Management of Wastewater Residuals*, Williamsburg, Va. (1975).

90. S. Damgaard-Larsen, K. O. Jensen, E. Lund, and B. Nissen, *Water Res.* **11,** 503 (1977).

91. C. J. Hurst, C. P. Gerba, J. C. Lance, and R. C. Rice, *Appl. Environ. Microbiol.* **40,** 192 (1980).

92. G. Bitton, R. Fraxedas, and G. E. Gifford, *Water Res.* **13,** 225 (1979).

93. Kapuscinski, R. B. and R. Mitchell, *Am. Soc. Microbiol. Meeting,* Abstract No. QT3, Miami Beach, FL (1981).

94. C. Wallis and J. L. Melnick, *Tex. Rep. Biol. Med.*, **19,** 683 (1961).

95. C. Wallis and J. L. Melnick, *Virology* **16,** 504 (1962).

96. J. L. Melnick and C. Wallis, *Proc. Soc. Exp. Biol. Med.* **112,** 894 (1963).

97. K. C. Marshall, In A. D. McLaren and J. J. Skujins, Eds., *Soil Biochemistry*, Vol. 2, Marcel Dekker, New York, 1971, pp. 409–445.

98. G. Stotzky, *Trans. N.Y. Acad. Sci. Ser. II* **30,** 11 (1967).

99. H. Fraenkel-Conrat, B. Singer, and A. Tsugita, *Virology* **14,** 54 (1961).

100. M. P. Greaves and M. J. Wilson, *Soil Biol. Biochem.* **2,** 257 (1970).

101. B. Singer and H. Fraenkel-Conrat, *Virology* **14,** 59 (1961).

102. G. Bitton and R. Mitchell, *Water Res.* **8,** 227 (1974).

103. C. P. Gerba and G. E. Schaiberger, *J. Water Poll. Contr. Fed.* **47,** 93 (1975).

104. G. Bitton, Y. Henis, and N. Lahav, *Plant and Soil* **45,** 65 (1976).

105. L. V. Grigor'eva, G. J. Korchek, V. I. Bondarenko, and T. V. Bei, *Hyg. Sanit.* **33,** 360 (1966).

106. E. P. Larkin, J. T. Tierney, and R. Sullivan, *J. Environ. Eng. Div.* **102,** 29 (1976).

107. S. R. Farrah, G. Bitton, E. M. Hoffman, O. Lanni, O. C. Pancorbo, M. C. Lutrick, and J. E. Bertrand, *Appl. Environ. Microbiol.* **41,** 459 (1981).

108. J. G. Yaeger and R. T. O'Brien, *Appl. Environ. Microbiol.* **38,** 702 (1979).
109. L. W. Sinton, *J. Hydrol.* (*N.Z.*) **19,** 119 (1980).
110. G. Bitton, S. R. Farrah, R. H. Ruskin, J. Butner, and Y. J. Chou, *Ground Water* **21,** 405 (1983).
111. G. A. McFeters and D. G. Stuart, *Appl. Environ.* **27,** 823 (1974).
112. B. H. Keswick, C. P. Gerba, S. L. Secor, and I. Cech, *J. Environ. Sci. Health* **A17,** 903 (1982).
113. G. N. Martin, and M. J. Noonan, In *Water and Soil Tech. Publ. #7*, Water and Soil Div., Ministry of Work and Dev., New Zealand, 1977, 25 pp.
114. R. T. O'Brien and J. S. Newman, *Appl. Environ. Microbiol.* **33,** 334 (1977).
115. M. Niemi, *Water Res.* **10,** 751 (1976).
116. A. M. Matossan and G. A. Garabedian, *Am. J. Epidemiol.* **85,** 1 (1967).
117. J. G. Yeager and R. T. O'Brien, *Am. Soc. Microbiol. Meeting*, Abs. #N51, New Orleans, LA. (1977).
118. Y. Marzouk, S. M. Goyal, and C. P. Gerba, *Water Res.* **14,** 1585 (1980).
119. J. Goldshmid, *J. Am. Water Works Assoc.* **66,** 163 (1974).
120. M. Rebhun and J. Schwarz, *Water Resources Res.* **4,** 1207 (1968).
121. J. M. Vaughn, and E. F. Landry, In H. L. McKim, Ed., *State of Knowledge in Land Treatment of Wastewater*, U.S. Corps of Engineers, Cold Regions Research Engineering Laboratory, Hanover, N.H., 1978, pp. 233–245.
122. J. C. Lance and C. P. Gerba, In *Proceedings of the Water Reuse Symposium*, Vol. 3, Am. Water Wks. Assoc. Res. Found., Denver, Colorado, 1979.
123. D. S. Wang, C. P. Gerba, and J. C. Lance, *Appl. Environ. Microbiol.* **42,** 83 (1981).

5

MICROBIOLOGICAL PROCESSES AFFECTING CHEMICAL TRANSFORMATIONS IN GROUNDWATER

P. L. McCarty

Department of Civil Engineering
Stanford University,
Stanford, California

B. E. Rittmann

Department of Civil Engineering
University of Illinois at Urbana-Champaign
Urbana, Illinois

and

E. J. Bouwer

Department of Civil Engineering
University of Houston
Houston, Texas

5.1. INTRODUCTION

Both inorganic and organic chemicals that enter the groundwater environment can be transformed by microbiological processes. Nonphotosynthetic microorganisms obtain energy for growth by oxidation of organic compounds, hydrogen, or reduced inorganic compounds of iron, nitrogen, or sulfur. Electron acceptors are required for these oxidations; electron acceptors used by microorganisms include oxygen under aerobic conditions, and nitrate, sulfate and carbon dioxide under anaerobic conditions. The energy obtained from the oxidation/reduction reactions is used for cell growth and maintenance, which bring about direct chemical transformations of inorganic and organic materials. Besides transforming chemical compounds directly, the growth of microorganisms can also alter the permeability of aquifer material and affect the chemical environment surrounding the microorganisms, resulting in change in pH and the oxidation–reduction potential of the system. Such environmental alterations can lead to other strictly chemical phenomena, such as precipitation or dissolution of phosphates and heavy metals, or chemical oxidation or reduction of iron and sulfur salts. Thus, many physical and chemical changes can be brought about in association with the decomposition or transformation of organic and inorganic materials by microorganisms living in the subsurface environment.

This chapter considers only one aspect of the many changes brought about by microbial activity in groundwater, that is, the kinetics of oxidation of inorganic and organic compounds when used for energy. Such oxidations represent the major route for biotransformation of organic contaminants in groundwater. Chemical oxidation of organic material is generally too slow for significant transformations to occur, and photolysis is virtually absent in groundwaters because of the absence of sunlight.

The nature of the groundwater environment dictates that the predominant type of microbiological activity will be bacteria attached to solid surfaces in the form of biofilms. This results from the usually low substrate concentration and the high specific surface area, both of which strongly favor the predominance of biofilm bacteria (1, 2). The attached bacteria remain generally fixed in one place and obtain energy and nutrients from the groundwater that flows by. A third microbiologically important characteristic of the groundwater environment is that heterotrophic bacteria are usually required to utilize numerous different organic compounds to obtain sufficient energy to sustain themselves. Furthermore, a significant portion of the organic matter, especially in groundwater containing recharged wastewaters, leachates, or spilled chemicals, consists of anthropogenic compounds which are resistant to bacterial degradation (3–5).

5.2. BIOFILM MODELS

5.2.1. Introduction

To adequately understand the kinetics of microbiological processes in groundwater, an understanding of biofilm kinetics and their application to groundwater systems is necessary. The following sections review the fundamentals of biofilm kinetics and the special advancements necessary for their application in groundwater systems.

5.2.2. Biofilm Kinetics

Biofilm kinetics are best understood in terms of an idealized biofilm that is composed of a homogeneous matrix of bacteria and the extracellular polymers that bind the bacteria together and to the surface (6–8). Figure 5.1 is an illustration of an idealized biofilm having a uniform cell density of X_f (M_xL^{-3}) and a locally uniform thickness of L_f. Groundwater flows past the biofilm in the x direction, while substrates are transported from the water to the biofilm in the z direction. The distance L represents the thickness of a mass-transport diffusion layer through which substrate must pass in order to go from the bulk liquid into the biofilm, where utilization occurs.

Since details of the development of the kinetic models for an idealized biofilm have been presented elsewhere (9–12), only a summary is given here. Within the biofilm, two processes occur simultaneously: namely, utilization of the substrate by the bacteria, assumed to follow a Monod-type relation (13), and diffusion of the substrate through the biofilm according to Fick's Law (14). Equation 5.1 is the differential equation that equates steady-state substrate utilization (the right-hand term) with molecular diffusion (the left-hand term) at all points in the biofilm.

$$D_f \frac{d^2S_f}{dz^2} = \frac{kX_f S_f}{K_s + S_f} \qquad [5.1]$$

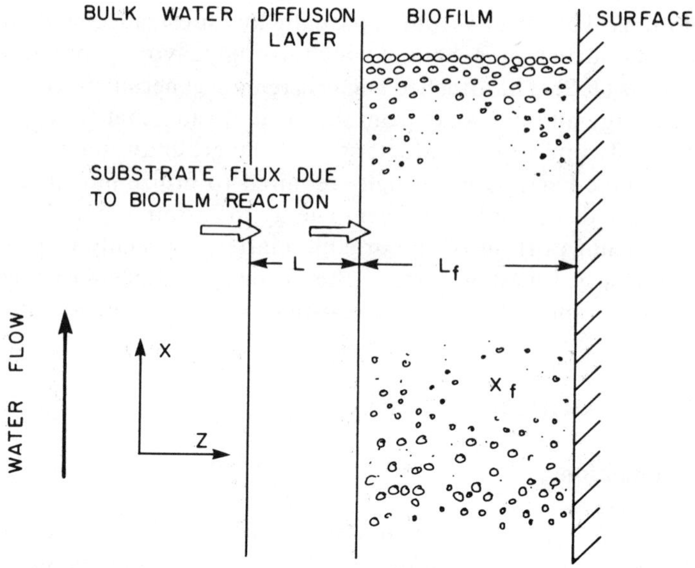

Figure 5.1. Idealized biofilm illustrating uniform cell density (X_f), thickness (L_f), water flow, and substrate flux into biofilm.

where S_f = concentration of rate-limiting substrate at a point within the biofilm (M_sL^{-3}); D_f = molecular diffusivity of the substrate within the biofilm (L^2T^{-1}); k = maximum specific rate of substrate utilization by the bacteria ($M_sM_x^{-1}T^{-1}$); K_s = Monod half-maximum rate concentration (M_sL^{-3}); z = distance normal to the surface of the biofilm (L). In order for substrate to reach the biofilm surface, mass transport across the diffusion layer occurs according to Equation 5.2.

$$J = -D\frac{dS}{dz} = D\frac{S - S_s}{L} \qquad [5.2]$$

where J = flux of rate-limiting substrate across the diffusion layer and into the biofilm ($M_sL^{-2}T^{-1}$); D = molecular diffusivity of the rate-limiting substrate in water; S and S_s = substrate concentrations in the bulk liquid and at the biofilm surface, respectively. Figure 5.2 shows how the three processes—substrate utilization, molecular diffusion within the biofilm and mass transport across the diffusion layer—interact to give substrate concentration profiles. For a thick biofilm (case A), the substrate concentration approaches zero and the biofilm is called deep. If the biofilm is very thin (case C), almost no substrate utilization occurs and the biofilm is essentially fully penetrated at the surface concentration S_s. Cases intermediate between deep and fully penetrated are termed shallow (case B).

Biofilm modeling has had numerous applications in the chemical industry and in wastewater treatment. For some of these applications, the models developed allow somewhat different assumptions than given here. Table 5.1 lists

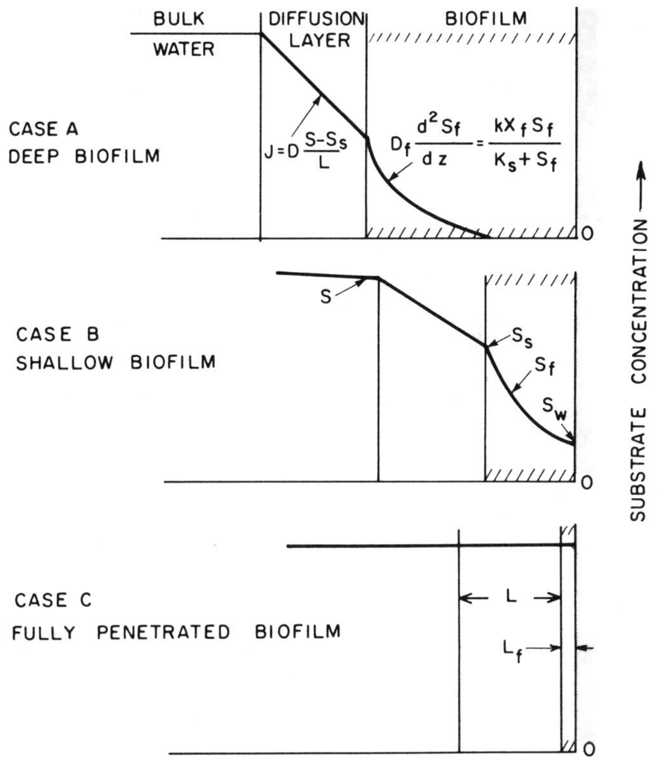

Figure 5.2. Interactions of substrate utilization, molecular diffusion within biofilm, and mass transport across diffusion layer.

the leading alternate formulations and their differences from the model presented above. In most cases, the alternate forms result from simplifications, such as ignoring mass-transport resistance from the bulk liquid to the biofilm surface or reducing the Monod substrate-utilization expression to a zero-order approximation. Although the alternate formulations may be adequate for the original applications, the simplifications used are generally not appropriate when applied to a low-concentration groundwater situation.

Table 5.1. Summary of Alternate Formulations of Biofilm Models

Workers	Differences
Harremoës (15, 16)	Ignores diffusion layer Assumes zero-order utilization
LaMotta (17)	Assumes zero-order utilization
Atkinson (18, 19)	Ignores diffusion layer
Jennings (20)	Uses spherical coordinates
Harris and Hensford (21)	Ignores diffusion layer

5.2.3. Steady-State Biofilm Model

Because organic substrate is normally present at low concentrations, the mass of biofilm that can be grown and sustained is limited and of critical importance for determining the rate and extent of reactions. Rittmann and McCarty (11) specifically addressed biofilm growth for low-concentration systems. A key to the analytical development was a definition of a steady-state biofilm, which is one undergoing no net growth or decay. In other words, biofilm growth through substrate utilization is just balanced by biofilm loss through cell decay or maintenance. The steady-state biofilm concept is particularly applicable to the groundwater environment, in which low substrate concentration severely limits the growth potential and, hence, the available biomass.

The net growth of biofilm in a small segment of biofilm is represented by Equation 5.3 (11).

$$\underset{\text{net growth}}{\frac{\partial(X_f A dz)}{\partial t}} = Yk \underset{\text{growth}}{\frac{S_f}{K_s + S_f}} (X_f A dz) - \underset{\text{Losses}}{b(X_f A dz)} \qquad [5.3]$$

where Y = true yield of bacterial mass per unit of substrate utilized ($M_x M_s^{-1}$), b = specific biofilm-loss coefficient (T^{-1}), A = cross sectional area of biofilm in a small biofilm segment (L^2), and dz = thickness of small biofilm segment (L). When growth exceeds losses, the right-hand side of Equation 5.3 is positive and the biofilm segment has net growth; when it is negative, the segment is decaying.

A steady-state biofilm undergoes no net decay or growth for the whole biofilm, and the integral of Equation 5.3 throughout the biofilm is zero. In other words, the total biofilm mass remains constant with time. If the bulk substrate concentration remains relatively steady, a biofilm should approach a steady-state thickness. In this case, the growing sections near the surface are continually taking up space lost by the decaying sections near the solid surface. Thus, the biofilm steady-state is dynamic within the biofilm.

The mathematical solution of the steady-state model (11), which requires simultaneous solution to Equations 5.1, 5.2, and 5.3, gives the substrate flux and biofilm thickness for a steady-state biofilm. The maximum thickness of a biofilm at steady-state is determined from the substrate flux and biomass losses. The biomass produced per unit time and per unit cross-sectional area can likewise be expressed as $-bX_f L_f$ (11). Conservation of mass (i.e., growth = losses) yields the following relationship for L_f:

$$L_f = \frac{YJ}{bX_f} \qquad [5.4]$$

Figure 5.3 is a sample solution for the steady-state model and shows how L_f (proportional to J) varies with substrate concentration. The most dramatic aspect of Fig. 5.3 is the rapid decrease to zero of biofilm thickness as the substrate concentration approaches a limiting value, S_{min}. At concentrations below S_{min}, the entire biofilm would be in net decay and would eventually cease to exist. Hence, in order to sustain a biofilm reaction for the long term, the rate-limiting substrate must be present at a concentration greater than S_{min}, which is equal to S_f in Equation 5.3 when net growth equals zero.

$$S_{min} = K_s \frac{b}{Yk - b} \quad [5.5]$$

Experiments have been performed to evaluate the concepts of a steady-state biofilm and the predictive capability of the model (22). The rate-limiting substrate, acetate, was continuously fed to 12-cm long by 2.5-cm diameter columns filled with 3-mm glass beads and initially seeded with bacteria. After being fed continuously for three to four weeks, a steady-state biofilm formed. Then the substrate utilization and the biofilm thickness on the glass beads were analyzed.

Figure 5.4 shows that S_{min} existed in each experiment and that the observed substrate utilizations and S_{min} values were very close to those predicted by theory from independently determined parameters (11, 12). Measurements of biofilm thickness along the reactor also confirmed that the biofilm mass declined to an insignificant level as the substrate concentration approached S_{min} (22). Similar results, having the substrate concentrations level off at a distinct S_{min} value, were obtained for dichloromethane, alanine, galactose, phenol, and thymine (23, 24).

Application of the steady-state biofilm concepts to groundwater systems suggests that a residual amound of biodegradable substrate should remain in the

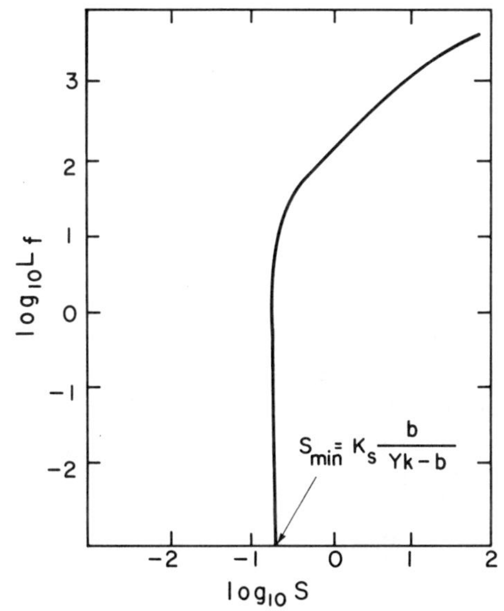

Figure 5.3. Illustration of how biofilm thickness L_f varies with substrate concentration S for a steady-state biofilm (11). Note that $J = L_f X_f / bY$.

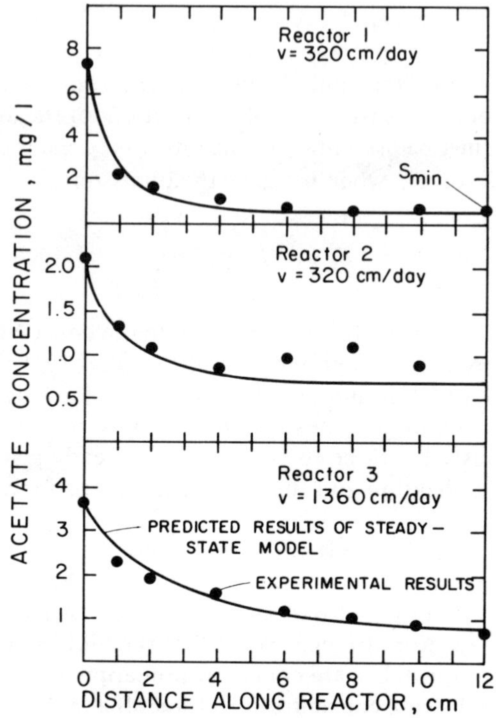

Figure 5.4. Steady-state results and model predictions for three biofilm reactors fed acetate (22). $T = 20°C$.

water and that no removal will occur if substrate conditions are less than S_{min}, which is typically in the range of 0.1–1.0 mg/l for aerobic utilization of organic metabolites. Since the organic matter in groundwater is often comprised of many individual compounds generally present at sub-S_{min} concentrations, the steady-state model would seem to indicate that they should remain nonbiodegraded. This is not necessarily true, as non-steady-state and secondary utilization can partially mitigate the biofilm growth limitations represented by S_{min}.

5.2.4. Non-Steady-State Biofilm Model

A biofilm having a given thickness will undergo net growth when the bulk substrate concentration is greater than the steady-state concentration necessary to sustain that thickness. A substrate concentration less than the steady-state concentration causes the biofilm to decay to a smaller thickness. While undergoing net decay, however, an established biofilm can still carry out substrate utilization. In the non-steady-state situation, the biofilm grows when substrate concentration is greater than S_{min}. When substrate concentration becomes less than S_{min}, the biofilm begins to decay, but it also utilizes the substrate, reducing the substrate concentration to less than S_{min}.

Non-steady-state conditions thus represent one way in which biofilm reactions can reduce substrate concentrations to well below S_{min}, even when the feed concentration is less than S_{min}. Non-steady-state biofilms work because the steady-state biofilm thickness is uncoupled from the substrate flux, or utilization rate, at a particular time. Once a biofilm is established, it will persist for some time thereafter, and for this time period it will be capable of removing substrate no matter how low the substrate concentration may be. Thus, biofilm thickness at any time is not directly dependent on the substrate concentration at that time, but is a result of the history of its development.

A model has been developed and evaluated for substrate flux into biofilms that are of any thickness (12), steady-state or non-steady-state, but presentation of the details are beyond the scope of this chapter. The non-steady-state concepts and models, however, were evaluated in a series of experiments in which steady-state biofilms, having a known biofilm thickness, were subjected to short-term non-steady-state conditions (12). To effect non-steady-state conditions, the influent feed concentration or flow velocity was changed from its steady-state values. Figure 5.5 presents typical results that show that concentra-

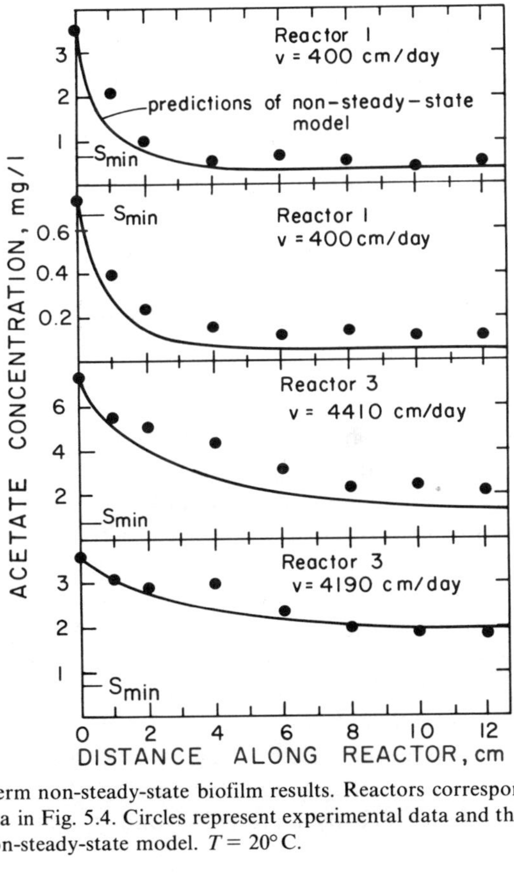

Figure 5.5. Short-term non-steady-state biofilm results. Reactors correspond to the steady-state reactors used for data in Fig. 5.4. Circles represent experimental data and the solid lines represent predictions of the non-steady-state model. $T = 20°C$.

tions less than S_{min} are achieved and that the model can predict the short-term effects of non-steady-state conditions. When the input concentrations were reduced to Reactor 1, the effluent concentration went below S_{min}; on the other hand, increases in feed concentration or flow rate (Reactor 3) resulted in effluent concentrations higher than S_{min}.

The non-steady-state biofilm concept shows that very low concentrations are achievable, but that the biofilm must have already been grown to a significant thickness to achieve those removals when the feed concentration is less than S_{min}. Since the growth rate of a biofilm is usually greater than the decay rate, short-term periods of super-S_{min} concentration can allow enough biofilm growth to sustain substrate removal for longer periods during which the substrate concentration is low.

5.3. SECONDARY UTILIZATION

When a substrate is always present at a concentration less than its S_{min}, the non-steady-state situation of alternating growth and decay cannot work. Nonetheless, biofilm thickness can be uncoupled from substrate utilization through secondary utilization. In secondary utilization, a substrate that does not provide significant energy or carbon for bacterial growth can be utilized as a secondary substrate by a biofilm that is grown and sustained through utilization of a primary substrate that is present at a concentration greater than its S_{min} (26, 28). A compound always present at a sub-S_{min} concentration is an excellent example of a secondary substrate, since present alone it could not supply enough energy to meet the maintenance requirements of the bacteria. The primary substrate can be a single compound or an aggregate of many compounds. In fact, the primary substrate can be composed of many secondary substrates, none of which is able to support a steady-state biofilm if present alone, but collectively provide enough energy to support a biofilm.

Modeling of secondary utilization requires a combination of the steady-state and non-steady-state modeling discussed previously. Since the primary substrate supports a steady-state biofilm (for steady input conditions), utilization of the primary substrate and growth of the biofilm are modeled with the steady-state model (11). The steady-state utilization of the secondary substrate is then modeled with the non-steady-state model (12), for which the steady-state model using the primary substrate provides the biofilm thicknesses and for which kinetic parameters are determined individually for each secondary substrate. For the situation in which the primary substrate is an aggregate of several compounds, an aggregate parameter such as chemical oxygen demand (COD) or total organic carbon (TOC) may be used for the primary substrate, providing that only the biodegradable materials are considered.

An example has been presented (26) showing how to apply steady-state and non-steady-state concepts to a groundwater recharge site. Using COD as the primary substrate, relatively easily degraded secondary substrates were reduced

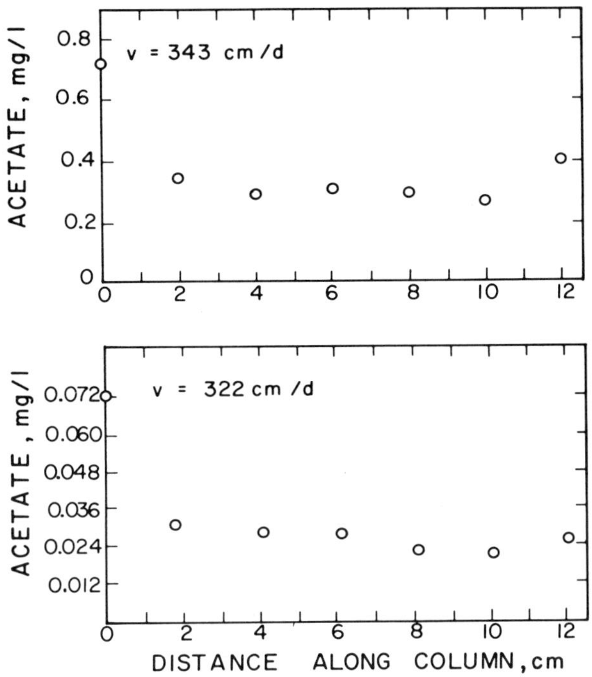

Figure 5.6. Secondary utilization of acetate by biofilm culture grown on residual organics in laboratory-deionized water (27).

to microgram per liter levels; however, a slowly degraded secondary substrate could pass through the biologically active zone with little removal. Thus, substrate removal kinetics are important in determining whether or not a biodegradable secondary substrate can be biologically removed from groundwater. If the removal kinetics of a particular compound are too slow, it may pass through the biologically active zone with little or no removal.

Both laboratory and field studies have demonstrated secondary utilization. A particularly good example is illustrated in Fig. 5.6 (27). A biofilm was grown in a once-through biofilm reactor through utilization of 0.17 mg/liter of the original 0.59 mg/liter of residual TOC in laboratory deionized water plus mineral nutrients. When 0.72 and 0.072 mg/liter of acetate were added to the feed, the visible biofilm in the first two cm of the reactor removed acetate to values well below its S_{min} of 0.66 mg/liter. Since acetate was not added to the feed solution on which the biofilm was grown, acetate did not contribute to growth of the biofilm and was a secondary substrate.

Stratton et al. (24) grew five different steady-state aerobic biofilms on galactose, acetate, alanine, phenol, or thymine. Secondary utilization of the compounds other than the primary substrate was then tested. Figure 5.7 shows the secondary-utilization response of the galactose-grown biofilm for phenol, acetate, thymine, and alanine. Acetate and alanine were rapidly and easily removed to low levels, but phenol and thymine showed little utilization. This

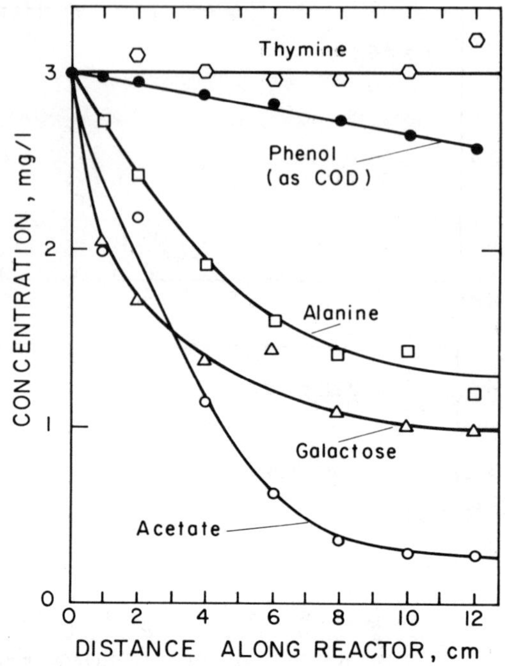

Figure 5.7. Secondary utilization of alanine, acetate, phenol, and thymine by biofilm culture grown with galactose as a primary substrate (24).

indicates that successful secondary utilization requires a primary substrate capable of supporting bacteria that also utilize the secondary substrate of interest at a fast enough rate.

In other studies (28), an aerobic biofilm was fed 1.0 mg/liter acetate and about 10μg/liter each of several aromatic and aliphatic organic compounds as secondary substrates. Table 5.2, which lists the results for primary and secondary utilization, points out that secondary utilization was possible for several nonchlorinated and chlorinated aromatics. Some compounds such as 1,3-dichlorobenzene and the chlorinated aliphatics were not degraded by the aerobic biofilm. These demonstrations of secondary utilization of at least some chlorinated organics are especially encouraging because of their anthropogenic and potentially hazardous nature.

In field results from Palo Alto, California recharge project, evidence for secondary utilization of naphthalene and heptaldehyde was found (4, 26). Once a biofilm was formed, naphthalene was nearly completely degraded, while heptaldehyde was partially removed. However, the same studies showed no evidence for secondary utilization of chloroform, bromoform, bromodichloromethane, and 1,1,1-trichloroethane.

Secondary utilization is a promising concept for explaining and enhancing biodegradation in groundwater. Knowledge of the processes involved is still rather limited, and more research is necessary to understand the response of

Table 5.2. Average Utilization of Substrates Fed Continuously to Aerobic Biofilm Reactor (28)

Substrate	Influent Concentration (μg/liter)	Effluent Concentration (μg/liter)	Percent Removal
Primary			
Acetate	1000	50	95
Secondary			
Chlorinated aromatics			
Chlorobenzene	11	1	91
1,2-Dichlorobenzene	9.6	0.4	96
1,3-Dichlorobenzene	9.8	8.3	15
1,4-Dichlorobenzene	11	0.2	98
1,2,4-Trichlorobenzene	9.2	0.9	90
Nonchlorinated organics			
Ethylbenzene	8.7	<0.05	>99
Styrene	7.2	<0.05	>99
Naphthalene	15.1	<0.05	>99
Chlorinated aliphatics			
Chloroform	29	29	0
1,1,1-Trichloroethane	16	16	0
Tetrachloroethylene	9.8	9.9	−1

bacteria to very low concentrations and the interactions of primary and secondary substrates.

5.4. SEQUENTIAL AND COMPETING REACTIONS

5.4.1. Introduction

Contaminated groundwaters may contain a variety of inorganic or organic compounds that could serve as energy substrates for bacteria under aerobic conditions, anaerobic conditions, or both. Certain compounds, such as ammonia, are oxidized sequentially by different bacterial strains. Ammonia is oxidized first to nitrite by *Nitrosomonas*, and then the nitrite is oxidized to nitrate by *Nitrobacter*. If nitrification were to proceed in a groundwater system, then questions may be asked as to whether the two strains would exist side by side within a biofilm, would they occupy different depth locations within the biofilm, or would they reside at different biofilm locations along the flow path of water? Little information is available to answer such questions. Reactions that occur in sequence are important, however, and occur in heterotrophic as well as autotrophic reactions. For example, in the complete anaerobic conversion of a substrate as simple as ethanol to methane, at least three different bacterial species are involved (29). Studies to clarify how biofilms act with sequential reactions are needed.

Perhaps more important than sequential reactions are transformations in which there is a competition for a common resource. Oxygen is generally quite limited, if present at all, in groundwaters, and different organisms may compete for it when available. For example, if insufficient oxygen for oxidation of both ammonia and an organic substrate such as glucose were present in a given groundwater, would only one of the compounds be oxidized, and if so, which one and why? Might they both be oxidized, but incompletely? The same questions would arise if two organic compounds were present, but used by different species. Unfortunately, little experience is available to give definitive answers to such questions, but some basis for judgment might be developed from our current understanding of biofilm processes. First the question of the limiting nutrient for a given organism will be addressed and then the effects of sequential and competing reactions will be discussed.

5.4.2. Limiting Nutrients

An assumption made in developing the models already discussed for kinetics of biofilms was that only one substrate limits the rate of the reaction. The electron acceptor, oxygen, rather than an organic electron donor may frequently be rate-limiting for aerobic reactions because of the limited oxygen concentrations normally found in groundwater. Whether the electron acceptor or the electron donor would be rate limiting in a given situation would depend upon the relative demand for both by a given organism, the relative kinetic coefficients for utilization of each nutrient, and their relative concentration in the bulk solution. Consider the general overall stoichiometric reaction by a given species within a biofilm:

$$\nu_d C_d + \nu_a C_a + \text{Growth Requirements} = \text{Cells} + \text{End Products} \quad [5.6]$$

where ν_d and ν_a are the stoichiometric reaction coefficients for the electron donor (C_d) and the electron acceptor (C_a), respectively. Based upon a relationship developed by Frank-Kamenetskii (30), it was shown (9) that if the following relationship holds, then the rate-limiting substrate is the electron donor.

$$\frac{S_d D_d}{\nu_d M_d} < \frac{S_a D_a}{\nu_a M_a} \quad [5.7]$$

Otherwise, the electron acceptor limits. M_d and M_a refer to the molecular weights of the donor and acceptor, and D_d and D_a to their molecular diffusion coefficients, respectively.

As an example, a typical overall stoichiometric equation for aerobic oxidation of glucose by bacteria can be formulated (32) as follows:

$$C_6H_{12}O_6 + 2O_2 + 0.8HCO_3^- + 0.8NH_4^+ \rightarrow$$
(glucose)
$$0.8C_5H_7O_2N + 2.8CO_2 + 5.2H_2O \quad [5.8]$$
(cells)

For complete oxidation of one mole of glucose to carbon dioxide and water, 6 moles of oxygen would be required, but in the bacterial reaction some of the glucose is converted into bacterial cells and, hence, only 2 moles of oxygen are required. For this example, $\nu_d = 1$, $\nu_a = 2$, $M_d = 180$, $M_a = 32$, $D_d = 0.6\ cm^2/day$ and $D_a = 2\ cm^2/day$ at 20° C. If the concentration of glucose at a given point in the aquifier were 20 mg/liter, then from Equation 5.7 the glucose would be rate-limiting when the dissolved oxygen concentration was greater than 2.1 mg/liter. However, according to Equation 5.8, the 20 mg/liter of glucose would require 20 $\times$ (64/180) = 7 mg/liter of oxygen for complete oxidation; thus, further downstream oxygen would be depleted before glucose if the concentration were less than this value.

Since at ambient temperatures the saturation concentration for oxygen is 8–10 mg/liter and reaeration is not possible in subsurface systems, the capacity for aerobic oxidation of substrates in groundwater is rather limited. If different organisms are competing for the same oxygen for oxidation of different electron donors, then the oxygen demands for all the reactions need to be considered. A complete model for biofilm reactions would need to handle these many competing needs. Once oxygen was depleted, then nitrates, sulfates, and carbon dioxide would become the electron acceptors for bacterial reactions in the order stated. In a complete evaluation of transformations in a groundwater system, these various factors should be included.

5.4.3. Competing Reactions

When several bacterial species within a biofilm are competing for a common resource such as oxygen, which one will win out? The example of competition between autotrophic oxidation of ammonia and heterotrophic oxidation of glucose can serve for illustration. It is generally recognized from empirical observations of fixed-film biological treatment processes that the organisms using the organic substrate generally win out in the competition with the result that organic oxidation proceeds first, followed by nitrification, providing sufficient oxygen still remains. A commonly held view is that ammonia cannot be utilized in the presence of organic material, but a more correct explanation is that the autotrophic growth rate is lower than the heterotrophic growth rate. For this reason, the autotrophs lose out in the competition. This has been illustrated previously (25). Organisms that grow faster come to dominate at the biofilm/water interface where they have access to the oxygen as it begins to diffuse into the biofilm. Slower growing organisms tend to reside deeper within

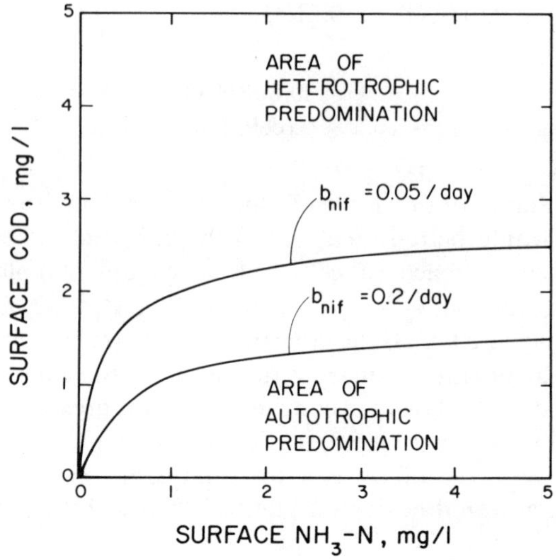

Figure 5.8. Lines of equal growth rate and area of predominance for biofilms with competing heterotrophic and nitrification reactions [modified from McCarty et al. (32)].

the biofilm; if oxygen concentration were insufficient to reach these depths, then the slower growing organisms could not act.

Being based upon typical growth kinetic coefficients for heterotrophic growth and autotrophic growth with ammonia, the concentration boundaries where growth rates for the different organisms would be equal appear in Fig. 5.8. At organic concentrations below this boundary, as measured by biodegradable COD, nitrification would dominate, but if above, the heterotrophic reactions would dominate. These concepts fit in with general empirical observations, but have not been experimentally verified. However, they are useful for gaining an understanding of why certain reactions tend to dominate others in fixed-film bacterial systems present in aquifers.

5.5. ENVIRONMENTAL FACTORS AFFECTING TRANSFORMATIONS

5.5.1. Introduction

Many of the synthetic organic contaminants found in groundwaters are not readily transformed by biological processes in the soil. The persistence of an organic compound may be attributable to physical, chemical, or environmental causes. In addition to low concentration, reasons for recalcitrance include inaccessibility of the substrate, absence of essential growth factors, toxic conditions, structural characteristics of the molecule that prevent enzymatic attack, and inability of the community of microorganisms to metabolize the compound because of some physiological inadequacy (33). Some organic contaminants are transformed to a small degree, whereas others can be completely mineralized to inorganic end products.

5.5.2. Transformations Under Aerobic and Anaerobic Conditions

One of the most important environmental factors affecting transformation, as already discussed, is the electron acceptor. Some compounds appear to degrade only under aerobic conditions and others only under anaerobic conditions. Some compounds are transformed under either condition, while others are not transformed at all.

Batch and continuous-flow laboratory studies under aerobic and anaerobic conditions were performed to determine the potential for degradation of some anthropogenic organics found in contaminated groundwaters, often at relatively low concentrations (27, 35). The organic compounds used were the halogenated one- and two-carbon aliphatic compounds and aromatic compounds listed in Table 5.3. In order to evaluate their potential for biodegradation, several series of batch biodegradation studies were conducted for aerobic and anaerobic conditions with different initial concentrations of each compound.

Table 5.4 demonstrates that the selected aromatic compounds were biodegraded aerobically. The nonhalogenated compounds, naphthalene and styrene, were removed below the detection limit without a noticeable lag period. Of the chlorinated benzenes, 1,4-dichlorobenzene showed the fastest biodegradation, while 1,3-dichlorobenzene was not significantly removed at any of the concentrations over the 11-week period. Table 5.4 also shows that no anaerobic biodegradation was obtained during 11 weeks of batch incubation with an active methanogenic inoculum. Thus, the batch studies indicate that the aromatic compounds studied were susceptible to aerobic, but not anaerobic, biodegradation.

In strong contrast, the halogenated aliphatic compounds evaluated were

Table 5.3. Organic Compounds Used in Biodegradability Studies at Low Concentrations (27, 35)

Chloroform	Ethylbenzene
Bromodichloromethane	Styrene
Dibromochloromethane	Naphthalene
Bromoform	Chlorobenzene
1,1,1-Trichloroethane	1,2-Dichlorobenzene
Trichloroethylene	1,3-Dichlorobenzene
Tetrachloroethylene	1,4-Dichlorobenzene
	1,2,4-Trichlorobenzene

Table 5.4. Degradability of Aromatic Compounds Present at Low Concentration (27)

	Aerobic			Anaerobic	
	Concen-tration	Percent Degraded		Concen-tration	Percent Degraded,
Compound	(μg/liter)	1 week	11 weeks	(μg/liter)	12 weeks
Naphthalene	190	>99	>99	93	0
	56	>99	>99	32	0
	24	>99	>99	10	0
	7.4	>99	>99		
Styrene	215	>99	>99	29	0
	51	>99	>99	7.8	0
	26	>99	>99	2.4	0
	5	>99	>99		
Chlorobenzene	280	23	>99	72	0
	81	0	99	31	0
	21	19	25	8	0
	8	22	90		
1,4-Dichlorobenzene	200	26	>99	74	0
	72	0	>99	29	6
	21	26	>99	7.4	18
	6	13	>99		
1,3-Dichlorobenzene	220	17	83	59	0
	79	0	36	23	0
	22	27	0	6	0
	7	19	0		

degradable only under anaerobic conditions and not aerobic conditions (Table 5.5). Chloroform was degraded under anaerobic conditions with the lowest concentration degraded to a value of 0.2 μg/liter within 16 weeks. The two other trihalomethanes, bromodichloromethane and dibromochloromethane, were degraded with no detectable amounts remaining at the end of two weeks. Also, these compounds decreased significantly in concentration even in the sterile controls, suggesting that degradation may occur chemically. However, degradation was much more rapid in the seeded cultures. Anaerobic degradation of trichloroethylene and tetrachloroethylene appears to have occurred, but at much slower rate than for the trihalomethanes.

The continuous-flow fixed-film studies on the aerobic transformations of these compounds at low concentrations reported in the section on secondary utilization are in agreement with the batch results. Some time was required for acclimation of the fixed-film cultures to the chlorinated aromatics as shown in Fig. 5.9. 1,4-dichlorobenzene (1,4-DCB) required the smallest acclimation period, 10 days, after which biodegradation resulted in 98 percent removal. Biodegradation of 1,2-dichlorobenzene (1,2-DCB) (96 percent removal) and

Table 5.5. Degradability of Halogenated Aliphatic Compounds Present at Low Concentration (4)

	Aerobic		Anaerobic		
				Percent Degraded	
Compound	Concentration (μg/liter)	Percent Degraded, 24 weeks	Concentration (μg/liter)	4 weeks	16 weeks
Chloroform	80	0	118	24	71
	29	0	38	97	99
	13	0	14	86	99
Bromodichloromethane	85	0	161	99	99
	31	0	35	99	99
	12	0	16	99	99
Dibromochloromethane	84	0	205	99	99
	30	0	40	99	99
	12	0	19	99	99
Trichloroethylene	81	0	124	15	44
	31	0	38	0	21
	11	0	13	8	31
Tetrachloroethylene	74	0	118	21	53
	33	0	36	3	44
	9	0	12	8	42

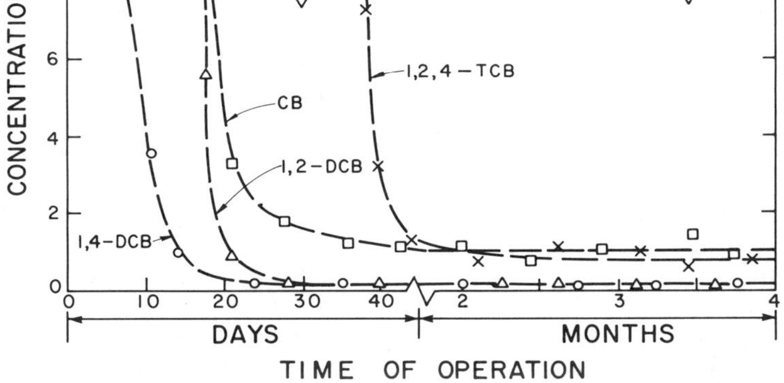

Figure 5.9. Chlorinated benzene degradation in a continuous-flow laboratory scale column (28). Influent concentrations remained at about 10μg/liter ± 20 percent.

chlorobenzene (CB) (91 percent removal) began after 20 days, and 1,2,4-trichlorobenzene (1,2,4-TCB) took 40 days to achieve significant removal of 90 percent. The effluent concentration of 1,3-dichlorobenzene (1,3-DCB) appeared to be lower than the influent concentration, but the difference was small and positive conclusions about its biodegradability at low concentration cannot be made either for the batch or the continuous flow studies.

A continuous-flow fixed-film system was also used to confirm transformations at low concentrations under anaerobic conditions. Procedures and apparatus were similar to the aerobic system except a higher concentration, 500 mg/liter, of acetate was used as the primary substrate. Halogenated organic compounds were added to the feed at about 30 μg/liter. A syringe pump maintained a feed rate of 20 ml/day (hydraulic loading of 4 cm/day), resulting in an actual detention time in the anaerobic column of 2 days.

High removal efficiencies of all the halogenated aliphatics were obtained (Table 5.6). Degradation of the brominated methanes to below detection limits occurred with almost no acclimation period required, while significant removals of the chlorinated aliphatics occurred after longer lag periods. Aromatic compounds were not significantly degraded in the column, confirming the batch results. Again, the continuous-flow results reinforce that halogenated aliphatics are susceptible to anaerobic decomposition, but the aromatic compounds are resistant.

In the laboratory, continuous-flow systems simulate the processes occurring during groundwater movement in the subsurface environment better than batch tests, and the results were more conclusive than batch results. Parallel columns were observed under sterile conditions, and it was determined that no sorption or volatilization removals were occurring in the system. Hence, the removals observed with the seeded columns were the result of biological or chemical transformations.

The degradation of chlorinated aromatics only under aerobic conditions

Table 5.6. Average Organic Concentrations in Anaerobic Column Influent and Effluent

Compound	Acclimation period in weeks	Column influent (μg/liter)	Column effluent (μg/liter)	Removal percent
Chloroform	4	33.3 ± 7	1.2 ± 0.6	96
Bromodichloromethane	0	30.3 ± 4.0	<0.1	99
Dibromochloromethane	0	34.1 ± 4.5	<0.1	99
Bromoform	0	33.8 ± 3.6	<0.1	99
1,1,1-Trichloroethane	8	25.4 ± 3.4	0.55 ± 0.3	98
Trichloroethylene	6	26.2 ± 4.0	1.6 ± 0.6	94
Tetrachloroethylene	16	18.3 ± 2.8	2.6 ± 1.3	86
Acetate	0	604 mg/l	42 mg/l	93

suggests the need for mixed-function oxidase systems to bring about dehalogenation and ring cleavage of these compounds. Many other biodegradability studies of substituted aromatics have demonstrated the requirement of aerobic conditions for ring cleavage (37–39). Recently, Marinucci and Bartha (45) found that O_2 was needed for biodegradation of 1,2,3- and 1,2,4-trichlorobenzene in soil samples. Reductive pathways, however, do exist for anaerobic decomposition of a range of ligno-aromatic compounds to methane under strict anaerobic conditions. The anaerobic transformations of ferulic acid and benzoic acid appeared to be carried out by a consortium of bacteria (43). The basic requirement for cleavage of the aromatic ring anaerobically was the presence of oxygen on the ring or ring substituent.

The degradation of halogenated 1- and 2-carbon aliphatics under anaerobic conditions is a new finding and has been confirmed by Lang et al. (5). Hydrolysis and photochemical oxidation were the only important removal mechanisms for trihalomethanes in natural waters cited in a review by the National Academy of Sciences (48). The estimated half-lives with such reactions in aqueous solutions are on the order of years or longer, while degradation obtained here occurred within days.

The removal of the brominated trihalomethanes in both the batch anaerobic controls and the seeded cultures indicates degradation by a chemical mechanism. The absence of an acclimation period prior to removal in the continuous-flow anaerobic column is also consistent with a chemical mechanism. A possible chemical degradation reaction is nucleophilic substitution of sulfhydryl for bromine to produce a thiol, for example, $CHBr_3 + HS^- \rightarrow CHBr_2SH + Br^-$ (47). Sodium sulfide was used to maintain reducing conditions in the anaerobic medium and may have reacted with the brominated organics to produce thiols. Another likely explanation is that the brominated aliphatics were unstable in the reduced environment and rapid hydrolysis occurred. The removal rate was increased in the presence of microbial activity relative to sterile controls. It is likely that the microbial activity maintained the environment in a more reduced state than in sterile controls, thereby increasing the rate of thiol formation and/or hydrolysis reactions. This behavior suggests a combination of chemical and biological mechanisms.

Anaerobic degradation of chlorinated aliphatics only occurred in the presence of anaerobic bacteria and not in sterile controls, suggesting that biological transformations were responsible for their removal. Lang et al. (5) similarly reported that almost all chlorinated methane, ethane, and ethene compounds were susceptible to biodegradation in anaerobic environments. They have tentatively isolated organisms capable of degrading trichloroethylene and tetrachloroethylene. The degradation pathway appears to be by way of a series of reductive dechlorination reactions. The final products were not identified. Reductive dehalogenation reactions are known to be mediated by microorganisms capable of degrading insecticidal chemicals (46). The most widely studied example is the reductive conversion of DDT to TDE under anaerobic, but not aerobic, conditions (38). Lindane and heptachlor can also be transformed by reductive dechlorination, which appears to involve discrete enzymes and electron-carrying proteins.

Alkyl chlorides are more stable and have reported reaction rates 25–50 times slower than corresponding alkyl bromides (47). This is qualitatively consistent with the transformation rates observed for anaerobic bacterial cultures. Hence, a combination of chemical and biological mechanisms for the observed chlorinated aliphatics degradation cannot be ruled out with certainty.

5.5.3. Sorption

Sorption is another process that can affect biodegradation in groundwater. Both stimulation and retardation of biodegradation has been observed for sorbed substrates. Since removal by sorption decreases the concentration in the groundwater, the biodegradation by suspended microorganisms is usually slowed. However, sorption can increase biodegradation in organic poor environments by concentrating organic compounds at a surface/water interface for utilization by attached bacteria. Heukelekian and Heller (44) and others (6, 31) have demonstrated that increasing the surface area to liquid–volume ratio improved substrate utilization and bacterial growth rates when the concentration of organic substrate was low, and organic sorption was implicated as the rate-enhancing step.

Sorption can alter the molecular character of a compound and favor enzymatic attack. With aromatic compounds, humic polymers can act as stabilizing agents, making the compounds less resistant to biodegradation (50). The opposite effect has also been observed, as compounds that are sorbed become sequestered and inaccessible to microbial attack (36). Steen et al. (49) found that sorption of toxic organic compounds to suspended sediments rendered them unavailable for biodegradation. They conducted batch tests and observed as the sediment:water ratio increased, the rate of biodegradation of the compound decreased. In another study using *n*-alkylamines bound to clay, it was found that CO_2 production from sorbed decylamine was lower than from the free amine (51). Some portion or all of the bound amine on the clay surfaces was poorly accessible to the microbes. In a study of bioregeneration of activated carbon, Benedek (34) reported that only 50 percent of phenol previously adsorbed on activated carbon was biodegraded. The remainder of bound phenol could not be utilized by bacteria. Additional research leading to a better resolution of sorption/biodecomposition interactions is desirable.

5.6. FIELD EVIDENCE FOR BIODEGRADATION

Evidence for biodegradation of trace organic compounds in groundwater was obtained from studies of the injection of reclaimed municipal wastewater at Palo

Alto, California (4). The movement and fate of organic compounds in the shallow confined aquifer was followed through several monitoring wells. After about a 10-day acclimation period, both heptaldehyde and naphthalene were found to be readily biodegraded with movement through 8 m of subsoil, which required about 12 hrs of passage time for the injected water. However, no reduction in concentration through biodegradation was obtained in these initial experiments for chlorinated benzenes or trihalomethanes, even after a three-month period of injection. The concentrations of all the compounds were in the microgram/liter range. During this period the aquifer was largely anaerobic because of the presence of a relatively large concentration of biodegradable materials in the injected water. Such organics were sufficient to insure the growth of a good biofilm.

Once the above injection was stopped, monitoring of the concentrations of trace organics remaining in the aquifer was continued (4). The concentrations of chloroform and other trihalomethanes began to decrease at a first-order rate of about 0.03/day, and for trichloroethylene, tetrachloroethylene, and 1,1,1-trichloroethane at a rate of 0.003/day. The laboratory studies reported in Section 5.5.2 confirm that biodegradation was the likely cause of these decreases. Anaerobic degradation of these compounds has also been confirmed through field and laboratory studies by Lang et al. (5). Their results suggest that anaerobic transformations of the halogenated aliphatic compounds results from reductive dehalogenation. Thus, under strictly anaerobic conditions, degradation of many aliphatic compounds is likely.

No evidence for degradation of chlorinated benzenes was obtained at the Palo Alto injection site, even though aquifer conditions during injection at one well appeared to be aerobic (4, 52), as apparently required. Two explanations seem possible. One is that the low concentration of oxygen found in the aquifer was too low for acclimation to the chlorinated benzenes. The other is that the concentration of biodegradable organics was too low in the injection water to support a sufficiently active biofilm. Schneider et al. (53) did find evidence for 1,4-dichlorobenzene degradation in an aquifer system, confirming the possibility shown through laboratory studies.

Field and soil column evidence was also obtained for the biodegradation of nonhalogenated aliphatic and aromatic hydrocarbons during rapid infiltration of secondary wastewater from spreading basins down through an unsaturated zone to groundwater (54, 55). However, the reduction in concentration within the groundwater underlying the basin was not sufficient to confirm biodegradation of any of the chlorinated organics. The anoxic waters did contain nitrate, and under such an environment, both chlorinated aromatic and aliphatic transformation may be retarded.

A better understanding of the conditions required for biodegradation in groundwaters is still needed. At least laboratory evidence clearly indicates that under proper conditions, many potentially hazardous organic compounds can be biodegraded in the ground, even when present at very low concentrations. Also, some compounds appear biodegradable only under aerobic conditions and others only under anaerobic conditions. Such biodegradation requires an active biofilm that is supported by other organic materials occurring in perhaps low concentration, but still greater than trace levels. Unless all the proper conditions are met for a given compound, biodegradation is not likely to occur. Better methods for evaluating when these conditions are met and how they affect the rate of degradation are clearly needed.

5.7. ACKNOWLEDGMENTS

This work was supported in part by Grant No. EPA-R-804431, Robert S. Kerr Environmental Research Laboratory; Grant No. EPA-R-808034010, Office of Research and Development, U.S. Environmental Protection Agency; Grant No. CME-7905707, National Science Foundation; National Science Foundation Graduate Fellowship.

5.8. SYMBOLS

Fundamental Quantities

L	Length
M_s	Mass of substrate
M_x	Mass of bacteria
T	Time

Latin Symbols

b	Specific bacterial loss or decay coefficient (T^{-1})
D	Molecular diffusivity in water (L^2T^{-1})
D_f	Molecular diffusivity in biofilm (L^2T^{-1})
J	Flux of rate-limiting substrate into biofilm surface ($M_sL^{-2}T^{-1}$)
k	Maximum specific rate of substrate utilization ($M_sM_x^{-1}T^{-1}$)
K_s	Monod half-maximum-velocity concentration (M_sL^{-3})
L	Depth of diffusion layer (L)
L_f	Thickness of biofilm (L)
S	Concentration of rate-limiting substrate in the bulk liquid (M_sL^{-3})
S_f	Concentration of rate-limiting substrate in the biofilm (M_sL^{-3})
S_{min}	Minimum bulk concentration of rate-limiting substrate able to sustain a steady-state biofilm (M_sL^{-3})
S_s	Concentration of rate-limiting substrate at the biofilm/liquid interface (M_sL^{-3})

t	Time (T)
v	Superficial flow velocity (LT^{-1})
x	Length along flow streamlines parallel to biofilm surface (L)
X_f	Bacterial density in biofilm (M_xL^{-3})
Y	True yield of bacterial mass per unit of substrate mass utilized ($M_xM_s^{-1}$)
z	Distance normal to biofilm surface (L)

REFERENCES

1. C. A. ZoBell, *J. Bacteriol.* **33,** 86 (1937).
2. K. Wuhrmann, "Steam Purification," in R. Mitchell, Ed., *Water Pollution Microbiology*, Wiley-Interscience, New York, 1972.
3. P. V. Roberts, P. L. McCarty, and W. M. Roman, *J. Environ. Engr. Div. ASCE* **104**(EE5), 933 (1978).
4. P. V. Roberts, P. L. McCarty, M. Reinhard, and J. Schreiner, *J. Water Pollut. Contr. Fed.* **52,** 161 (1980).
5. R. F. Lang, P. R. Wood, F. A. Parson, J. DeMarco, H. J. Harween, I. L. Payan, L. M. Meyer, M. C. Ruiz, and E. D. Ravelo, "Introductory Study of the Biodegradation of the Chlorinated Methane, Ethane, and Ethene Compounds," presented at the Annual Conference of the American Water Works Assn., 1981.
6. W. G. Characklis, *Water Res.* **7,** 1113 (1973).
7. B. A. Friedman, P. R. Dugan, R. M. Pfister, and C. C. Remsen, *J. Bacteriol.* **98,** 1329 (1969).
8. H. C. Jones, I. L. Roth, and W. M. Sanders, III, *J. Bacteriol.* **99,** 316 (1969).
9. K. Williamson and P. L. McCarty, *J. Water Pollut. Contr. Fed.* **48,** 9 (1976).
10. B. E. Rittmann and P. L. McCarty, *J. Environ. Engr. Div. ASCE* **104**(EE5), 889 (1978).
11. B. E. Rittmann and P. L. McCarty, *Biotech. Bioeng.* **22,** 2343 (1980).
12. B. E. Rittmann and P. L. McCarty, *J. Environ. Engr. Div. ASCE* **107**(EE4), 831 (1981).
13. A. W. Lawrence and P. L. McCarty, *J. Sanitary Engr. Div. ASCE* **96**(SA3), 757 (1970).
14. O. Levenspiel, *Chemical Reaction Engineering*, 2nd ed., John Wiley & Sons, Inc, New York, 1972
15. P. Harremoës, *J. Water Pollut. Contr. Fed.* **48,** 377 (1976).
16. P. Harremoës, *VATTEN* **2,** 122 (1977).
17. E. J. LaMotta, *Biotech. Bioeng.* **18,** 1359 (1976).
18. B. Atkinson and I. S. Daoud, *Trans. Inst. Chem. Eng.* **46,** 19 (1968).
19. B. Atkinson, and I. J. Davies, *Trans. Inst. Chem. Eng.* **52,** 248 (1974).
20. P. A. Jennings, V. L. Snoeyink, and E. S. K. Chian, *Biotech. Bioeng.* **18,** 1249 (1976).
21. N. P. Harris and G. S. Hansford, *Water Res.* **10,** 935 (1976).

22. B. E. Rittmann and P. L. McCarty, *Biotech. Bioeng.* **22,** 2359 (1980).
23. B. E. Rittmann and P. L. McCarty, *Appl. Environ. Microbiol.* **39,** 1225 (1980).
24. R. Stratton, E. Namkung, and B. E. Rittmann, "Biodegradation of Trace Organics by Biofilms on Porous Media,"*Proceedings of the 1981 Annual Conference of the American Water Works Assn.*, **877,** (1981).
25. P. L. McCarty, B. E. Rittmann, and M. Reinhard, *Environ. Sci. Tech.* **15,** 40 (1981).
26. B. E. Rittmann, P. L. McCarty, and P. V. Roberts, *Ground Water* **18,** 236 (1980).
27. B. E. Rittmann, E. J. Bouwer, J. E. Schreiner, and P. L. McCarty, "Biodegradation of Trace Organic Compounds in Groundwater Systems, Technical Report No. 255, Stanford University, Department of Civil Engineering, (1980).
28. E. J. Bouwer and P. L. McCarty, "Biofilm Degradation of Trace Chlorinated Organics," *Proceedings of the ASCE Environmental Engineering Division Specialty Conference* **196** (July 8–10, 1981).
29. P. L. McCarty, "One Hundred Years of Anaerobic Treatment," in Hughes et. al., Eds., *Anaerobic Digestion*, Travemunde, Germany Elsevier Scientific Publishing Company, The Netherlands, **3** (1982).
30. D. A. Frank–Kamenetskii, *Diffusion and Heat Exchange in Chemical Kinetics*, Princeton Univ. Press, Princeton, N.J. 1955.
31. C. E. ZoBell, *J. Bacteriol.* **46,** 39 (1943).
32. P. L. McCarty, *Progress in Water Technology* **7,** 157, Pergamon Press, London (1975).
33. M. Alexander, *Science* **211,** 9, (1981), p. 132.
34. A. Benedek, *Activated Carbon Adsorption of Organics from the Aqueous Phase, Vol. 2*, M. J. McGuire, and I. H. Suffet, Eds. Ann Arbor Science Publishers, Inc., Ann Arbor, MI, **273,** 1980.
35. E. J. Bouwer, B. E. Rittmann, and P. L. McCarty, *Environ. Science Tech.* **15,** 5, (1981), p. 596.
36. R. R. Colwell and G. S. Sagler, *Water Pollution Microbiology, Vol. 2,* R. Mitchell, Ed., John Wiley and Sons, New York, 111 1978.
37. S. Dagley, *Degradation of Synthetic Organic Molecules in the Biosphere*, National Academy of Sciences, Washington, D.C., 1971, p. 1.
38. E. G. Essaac and F. Matsumura, *Pharmac. Ther.* **9,** 1 (1980).
39. W. C. Evans, *J. Gen. Microbiol.* **32,** 177 (1963).
40. W. C. Evans, *Nature* **270,** 17 (1977).
41. J. B. Healy and L. Y. Young, *Appl. Env. Microb.* **35,** 1, (1978), p. 216.
42. J. B. Healy and L. Y. Young, *Appl. Env. Microb.* **38,** 1, (1979), p. 84.
43. J. B. Healy, L. Y. Young, and M. Reinhard, *Appl. Env. Microb.* **39,** 2, (1980), p. 436.
44. H. Heukelekian and A. Heller, *J. Bacteriol.* **40,** 547 (1940).
45. A. C. Marinucci and R. Bartha, *Appl. Env. Microb.* **38,** 811 (1979).
46. F. Matsumura and H. J. Benezet, *Pesticide Microbiology*, Academic Press, New York, (1981), p. 623.
47. R. T. Morrison and R. N. Boyd, *Organic Chemistry*, Allyn and Bacon, Inc., Boston, MA 1974, p. 452.
48. National Academy of Sciences, "Nonfluorinated Halomethanes in the Environment," Washington, D.C., (1978), p. 36.

49. W. C. Steen, D. F. Paris, and G. L. Baughman, "Effects of Sediment Sorption On Microbial Degradation of Toxic Substances," *Contaminants and Sediments*, Vol. 1, R. A. Baker, Ed., Ann Arbor Science Publishers, Inc., Ann Arbor, MI, (1980), p. 477.

50. L. Verma, J. P. Martin, and K. Haider, *Soil Sci. Soc. Am. Proc.* **39,** 2, (1975), p. 279.

51. P. C. Wszolek and M. Alexander, *J. Agr. Food Chem.* **27,** 2, (1979), p. 410.

52. P. V. Roberts, J. E. Schreiner, and G. D. Hopkins, *Wastewater Reuse for Groundwater Recharge*, T. Asano and P. V. Roberts, Eds., Calif. State Water Resources Control Board, 283, 1980.

53. J. K. Schneider, R. P. Schwarzenbach, E. Hoehn, W. Giger, and H. R. Wasmer, "The Behavior of Organic Pollutants in a Natural River—Groundwater Infiltration System," In W. Van Duijvenbooden, P. Glasbergen, and H. Van Lelyveld, Eds., *Quality of Groundwater*, Studies in Environmental Science, Vol. 17, Elsevier Scientific Publishing Company, The Netherlands, 565, 1981.

54. E. J. Bouwer, P. L. McCarty, and J. C. Lance, *Water Res* **18,** 236 (1981).

55. E. J. Bouwer and P. L. McCarty, "Organic Contaminant Behavior During Rapid Infiltration of Secondary Wastewater," *Proceedings of the Research Symposia*, Water Pollution Control Federation Annual Conf., Detroit, Michigan (1981).

6

MICROBIAL ACTIVITY IN MODEL AQUIFER SYSTEMS

John Wilson

United States Environmental Protection Agency
Robert S. Kerr Environmental Research Laboratory, Ada, Oklahoma

Michael James Noonan

Department of Agricultural Microbiology
Lincoln College, Canterbury, New Zealand

6.1. NATURE AND PROPERTIES OF MICROCOSMS

The word *microcosm* comes from the Greek for "little world." We shall use the term very broadly to refer to any experimental system in which various components of the subsurface environment are isolated and then studied as the parameters of interest change with time. The approach contrasts with repeated sampling of the original environment to monitor its behavior.

Ideally, a natural system should be studied directly and in its entirety, so that the components, processes, and interactions that comprise its structure and function can be described in a holistic and integrated manner. Unfortunately, constraints on physical and human resources make such an approach infeasible in many natural systems. This is particularly true for the terrestrial subsurface. The cost of drilling sampling wells or bore holes can be inordinately expensive. Fortunately, the use of microcosms allows detailed study of the processes occurring in the subsurface environment without the need for a large number of wells or bore holes. A second advantage of microcosms is their flexibility. The components of a microcosm can be tailored to focus on particular processes or interactions. As a result, the design of a microcosm is usually keyed to the purpose of a particular study.

Microcosms fall into two broad classes. Those in the first class are designed to reveal the effects of an environmental perturbation on the composition and functioning of a biological community. The focus is on the biota. Little work has been done that is relevant to regions deeper than root zones; however, the subsurface microflora may affect the quality of groundwater in ways that are not appreciated at present, and their importance in the overall functioning of the biosphere should be investigated.

Microcosms in the second class are used to define the processes that affect the transport and fate of specific environmental pollutants, that is, synthetic organic chemicals, heavy metals, pathogens, radionuclides, salts, and heat. The focus is on the pollutant.

There are two approaches to the second class of microcosms. The first is empirical. The behavior of the microcosm is considered a simulation of the larger system. Ideally, the microcosm contains all the components of the larger system and exhibits processes and interactions that parallel the larger system. This approach is commonly used when the goal is to define the processes that occur in the subsurface environment, but not necessarily to describe them precisely.

The second approach uses the microcosm (here more properly called a physical model) to obtain values required to solve mathematical equations that describe the microcosm and that are used to simulate the larger natural system. If the equations are considered hypotheses that are tested by their ability to predict the behavior of other microcosms operated under new conditions, or more preferably the larger natural system, the approach has the structure of science. If, however, the equations are tested by their ability to explain the behavior of the same physical models that were used to supply the necessary values, the process reduces to curve fitting, and the validity of the assumptions in the equation has not been subjected to a proper scientific test.

The first approach is the quickest and cheapest way to get started on a problem. However, the results depend on the faithfulness of the simulation and have limited generality. The second approach requires more time and is more expensive, particularly if the mathematical models must be validated against the larger system, but the predictions are much more general.

6.2. VARIABILITY IN THE SUBSURFACE ENVIRONMENT AS IT AFFECTS INVESTIGATIONS USING MICROCOSMS

There are several kinds of variation within the subsurface environment that can complicate the extrapolation of data gained in microcosms to larger natural systems. This variation can be either temporal or spatial. Temporal variation may reflect recent precipitation, seasonal changes, or shifts in climate, depending upon the rate of recharge. Spatial variation reflects both the distribution of point sources of pollution about the landscape and the variations in the geological composition of the subsurface. In comparison to natural systems on the surface, the subsurface is strongly three-dimensional. Processes in the subsurface may vary profoundly with depth.

If the information needed is very site-specific and the size is relatively uniform, samples can be taken from the whole site, mixed and used to fill replicate microcosms. In this way variability between microcosms will be reduced and, if enough subsamples are taken initially to represent the whole site, the data will be representative of the area. If, on the other hand, the structural integrity of the sample must be maintained or if minimum contamination of the sample by foreign microorganisms is desired, undisturbed cores may need to be taken from the smallest possible area to minimize variability.

Organisms sorbed onto solid material will be subjected to a changing environment as groundwater moves through the matrix. A recent study has been chosen as an example of the changes in conditions experienced in a small area throughout a period of one year. Spalding and Exner (1) monitored a network of 53 wells ten times within one year. The principal aquifer consisted of Pleistocene sand and gravel. The groundwater was unconfined, was recharged by river water, and moved at approximately 0.35 m/day. A secondary aquifer was confined by a clay deposit and consisted of lime-cemented silty sandstone of Tertiary age. Most of the study area was used for the production of irrigated corn, so that pollutants were probably moving down from the surface. Coefficients of variation from polled variances for two parameters were calculated from their data for each set of wells. Results are shown in Table 6.1. There was appreciable variation between wells in a relatively homogeneous area and within the same wells at different seasons of the year. Shallow groundwater showed higher variability of temperature since it is the immediate sink for vertically transported water originating in the vadose zone. Deep wells in the principal aquifer some distance away from the river showed less variation in chloride content than water from the secondary aquifer. Water temperature in both aquifers was considerably less variable than that taken from the river, although chloride levels were at least as variable. In contrast to this natural variability, most groundwater microcosms kept in the laboratory to date have been maintained under conditions that were as constant as possible. (11, 14, 17, 18)

If a microcosm is used as a physical model and a wide range of environmental conditions studied, extrapolation can be carried out over a large area. If, for example, one hopes to apply the results to one rock type throughout the world, then the data given by Freeze and Cherry (2) for some parameters give some indication of the range of conditions that would need to be tested to make the results generally applicable to one rock type. (Table 6.2)

As with most environments in the world, conditions in groundwater range from those that permit the growth of microorganisms to those that do not. For example, Dunlap and McNabb (3) have discussed changes of temperature with depth. Groundwater temperatures range from values that roughly correspond to

Table 6.1. Variability of Conservative Groundwater Quality Parameters in the Valley of the Platt River in Nebraska (1)

	Coefficient of Variation	
Depth of well[a]	Temperature	Chloride
Terrace Wells Slotted in Principal Aquifer		
Shallow	8.3%	26%
Terrace Bottomland Well Slotted in Secondary Aquifer		
Deep	4.6%	26%
Bottomland Wells Slotted in Principal Aquifer and Downgradient from Irrigated Cropland		
Shallow	28.2%	13%
Deep	6.3%	12%
Bottomland Wells Slotted in Principal Aquifer and Influenced by River Seepage and Near Pristine Conditions		
Shallow	43.3%	10%
Deep	5.6%	9.6%
Platte River		
No well	72.6%	12.9%

[a]Shallow and deep refer to wells which were slotted 2 and 18 m or more below the static water level, respectively.

Table 6.2. Variation in Solutes in Groundwater from Igneous Rocks Throughout the World (2)

Parameter	Range of Means (Data taken from many parts of the world) (pH units and g/m^3)
pH	5.5–7.1
HCO_3^-	2.0–59.2
Cl^-	0.6–22.0
SO_4^{2-}	0.4–10.9
SiO_2	5.9–85
Na^+	0.2–16.5
K^+	0.6–3.4
Ca^{2+}	04.–11.9
Mg^{2+}	0.1–4.9

the mean annual air temperature of the given region at the water table to temperatures which would preclude microorganisms at a depth below about 2500 m in areas of low geothermal activity.

6.3. MICROCOSMS TO STUDY EFFECTS ON THE BIOTA

Several simple microcosms have been constructed to study the rate of decline of populations of microcosms that have been introduced into aquifers as a result of waste disposal practices. Sinton (4) used a very simple microcosm to identify microorganisms that could survive in an aquifer long enough to be useful as a tracer. These organisms could then be used to determine the probable movement of bacteria of sanitary significance.

Water was taken from wells and stored in the dark at 12.5°C in 5-liter flasks. Cultures of three strains of *Escherichia coli* were centrifuged, resuspended, and then added to well water at densities between 10^8 and 10^9 cells/100 ml. The flasks were sampled 11 times over a period of 22 days. This flask microcosm has several deficiencies. The incubation temperature was the mean annual groundwater temperature of the area and did not duplicate fluctuations present in the natural groundwater. Additionally, washing the cultures may not have removed all the nutrients in the culture medium. Autolysis of dead cells at the original high densities may have provided nutrients to the survivors, which would have increased their longevity relative to the natural aquifer. Finally, solid surfaces may enhance survival in the natural system, but the only surface available in this microcosm was the walls of the container. On the positive side, a flask microcosm is the ultimate in simplicity. It is easy to construct, easy to replicate, and easy to sample.

Martin and Noonan (5) designed a more complex microcosm to study the survival of natural populations of coliform bacteria in well water. Dialysis tubing was filled with groundwater that had been contaminated by percolate from a municipal waste-disposal site. The tubing was hung inside a slotted PVC frame and the assembly was lowered into the same well that provided the contaminated water. The assembly was raised periodically and sampled for fecal coliform bacteria.

A possible drawback of this microcosm is a tendency for microorganisms to sorb to the walls of the tubing. However, the survival of the microbes is studied under more natural conditions. Using this microcosm, Martin and Noonan (5) determined that cell death and dilution rather than sorption were primarily responsible for the reduction in numbers of fecal coliforms in groundwater downgradient from the disposal site.

McFeters et al. (6) used a more elaborate microcosm to compare survival of strains of pathogenic bacteria or the other fecal bacteria used as indicators of the pathogenic forms. Suspensions of bacteria were placed in chambers sealed with membrane filters (Fig. 6.1). The chambers were placed in a 40-liter tank that was replenished with well water at a rate of 5 liter/min. The contents of the tank were

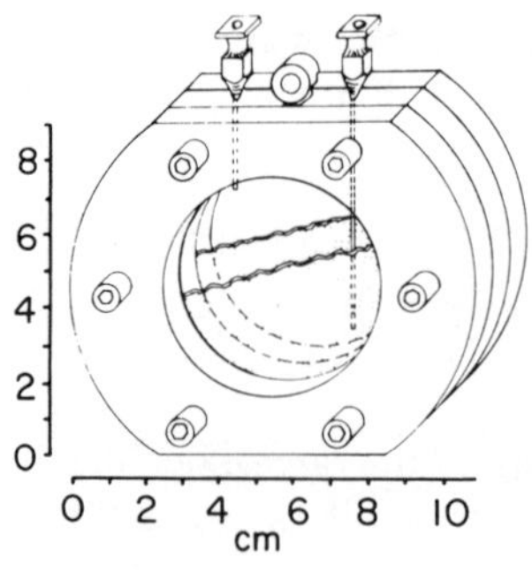

Figure 6.1. A microcosm to study survival of microorganisms in ground water (6) Used with the permission of the American Society for Microbiology.

mixed to ensure homogeneous conditions. The membrane filters probably allowed more rapid exchange between the microcosm and its surroundings than the dialysis tubing. Additionally, the membrane would not exclude large organic polymers that might serve as food for the microorganisms.

McFeters et al. (6) showed that the survival of pathogens and their indicator organisms in groundwater at their study site were similar. A comparison of

Table 6.3. Survival of Bacteria in Groundwater Microcosms

Microcosm	Organism	T_{90}[a] (days)	Reference
Flask	Antibiotic Resistant		4
	E. coli		
	(J 6-2)	2.6	
	(PB 922)	3.0	
	Neotype *E. coli*	3.2	
Dialysis Tubing	Faecal coliforms	6.2	5
Membrane Filter	**From Culture**		6
	Coliform bacteria	2.4	
	Enterococci	3.0	
	Streptococcus equinus	1.4	
	S. bovis	0.6	
	Shigella dysenteriae	3.1	
	S. sonnei	3.4	
	S. flexneri	3.7	
	Salmonella enteritidis		
	(ser. paratyphi A)	2.2	
	S. enteritidis		
	(ser. typhimurium)	2.2	
	(ser. paratyphi B)	0.3	
	S. typhi	0.8	
	Vibrio cholerae	1.0	
	From Raw Sewage		
	Coliform bacteria	2.4	
	Streptococci	2.7	

[a]Time required for loss of 90 percent of the original density.

survival times determined in the three microcosms discussed above reveals a good deal of consistency between microcosms. (Table 6.3).

6.4. MICROCOSMS THAT REVEAL THE FATE OF POLLUTANTS

The impetus for the development of these microcosms was the need for information about the specific pathways whereby wastes enter groundwater, in order to facilitate intelligent environmental management and regulation. Attention has been directed toward deep-injection wells for industrial wastes, toward materials leaching from landfills, and toward contaminant plumes from spills or other accidental releases of synthetic organic chemicals to the surface of the land.

Horvath and Elkan (7) designed a simple microcosm to study biological activity in a particular industrial waste under conditions pertaining to a deep saline aquifer. The microcosm was a glass column 8.5-cm I.D. × 95-cm long, filled with sand and closed with a black rubber stopper (Fig. 6.2). The microcosm was autoclaved and then inoculated by pumping water from an uncontaminated well in the aquifer being simulated. The inoculum was protected from atmo-

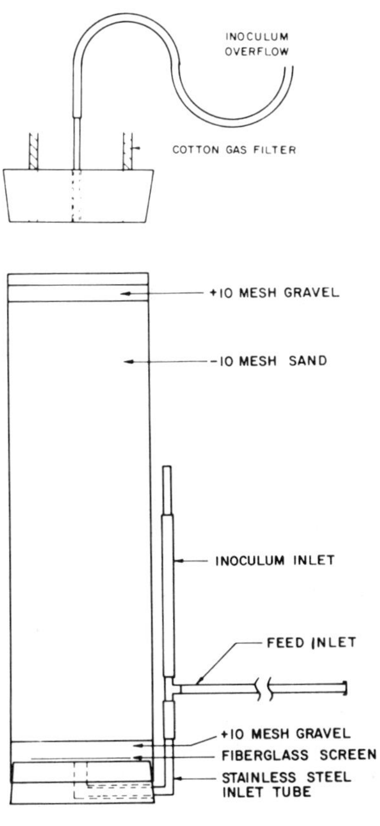

Figure 6.2. A microcosm to study biotransformations of concentrated industrial wastes after deep-well disposal (7) Used with the permission of the National Well Water Association.

spheric oxygen and drip tubes were used to prevent contamination of the microcosm by laboratory microorganisms. The feed solution was a synthetic chloride-brine dosed with acetic, formic, *para*-toluic, terephthalic, phthalic and benzoic acids, methanol, and formaldehyde to simulate a particular industrial waste. The concentration of individual solutes varied from 52 to 9359 mg/liter. Feed solution under gravity flow was introduced into the bottom of the column. Any gas produced within the column was separated from the column effluent at the top of the column. Flow was controlled by metering the column effluent through a pump. Only formic acid was degraded in the microcosm. This result is in good agreement with the behavior of the genuine industrial waste after injection into a deep saline aquifer. Leenheer et al. (8) found that formate was completely degraded in the aquifer, acetate was degraded only slightly, and the other organic constituents were not degraded (methanol was not part of the injected waste).

This microcosm is well suited for the study of anaerobic processes. Its components are cheap and easily obtainable. Because it does not require actual core material from the aquifer, the microcosm is inexpensive to set up and easily standardized. However, the microcosm has two limitations. The microflora within a well may be very different from the microflora of the aquifer *per se* (see Chapter 11). Additionally, aquifer solids are rarely entirely sand. Low concentrations of natural organic matter present in almost all aquifers may be necessary for the maintenance of the indigenous microflora. Organic matter and clays in the aquifer matrix may also sorb waste components and retard their movement.

This microcosm is well suited to define the biological fate of organic wastes present in concentrations great enough to enrich an active population (>10 mg/liter, aerobically; >500 mg/liter, anaerobically). Under these conditions, constraints on biodegradation imposed by a very low density or activity of the native microflora are not important. Use of rubber stoppers may present problems with certain reactive or very hydrophobic organic compounds, but in such a case, a Teflon closure can be substituted.

Horvath and Elkan (9) also described a simple microcosm that can be used to determine the effects of pressure on biodegradation. A screw-cap test tube is filled with a liquid medium inoculated with the test organism. The tube is closed with a flexible diaphragm held in place by a perforated screw cap. The tube is inserted into a steel pressure chamber and the remaining space is filled with water. Then a piston fitted to a screw is bolted to the chamber and the screw turned until the required pressure is reached. When a flora from a deep well (pressure at the well screen 32 atm) was examined in this microcosm for the ability to reduce nitrate and degrade formate, the attack on these substrates was greater in the range of 34–100 atm. Because 100 atm corresponds to a depth near 1000 m and few disposal wells exceed this depth, it is unlikely that pressure will limit biodegradation in most aquifers.

Horvath and Elkan's (9) device can be adapted to any liquid or slurry that can be forced into a test tube. It would be particularly useful to test the barotolerance of laboratory cultures that are selected or genetically engineered for the degradation of waste in disposal wells or the biologically enhanced recovery of oil from oil fields.

Grula and Grula (10) constructed a similar microcosm that was pressured with compressed gas. When helium was used, their results were similar to those of Horvath and Elkan (9). When air was used, oxygen toxicity destroyed the microorganisms at relatively low pressures (16 atm).

Biodegradation in the subsurface may be limited by the concentration of electron acceptors (O_2, NO_3^-, SO_4^{-2}). One remedy is to pump air into the aquifer. Grula and Grula's (10) device could be used to determine the maximum depth at which such a remedy could work or to select tolerant strains that could be used at greater depth.

Dunlap et al. (11) designed a unique microcosm to determine the fate of nitrilotriacetic acid (NTA) at environmentally relevant concentrations under conditions simulating as nearly as possible those prevailing in a natural aquifer (Fig. 6.3). Authentic aquifer material was collected from a shallow sandy aquifer in the flood plain of a river. Then 3.0-liter of aquifer matrix material was added to a 4.0-liter aspirator jar. A glass tube penetrated the mass of sand. Connected to the tube was a length of Tygon tubing that passed through a peristaltic pump and returned to the microcosm. The apparatus allowed internal circulation of water to facilitate sampling and to simulate groundwater flow. As far as possible, aseptic technique was employed in the collection of the aquifer matrix material and in the construction of the microcosm. Biodegradation was probably

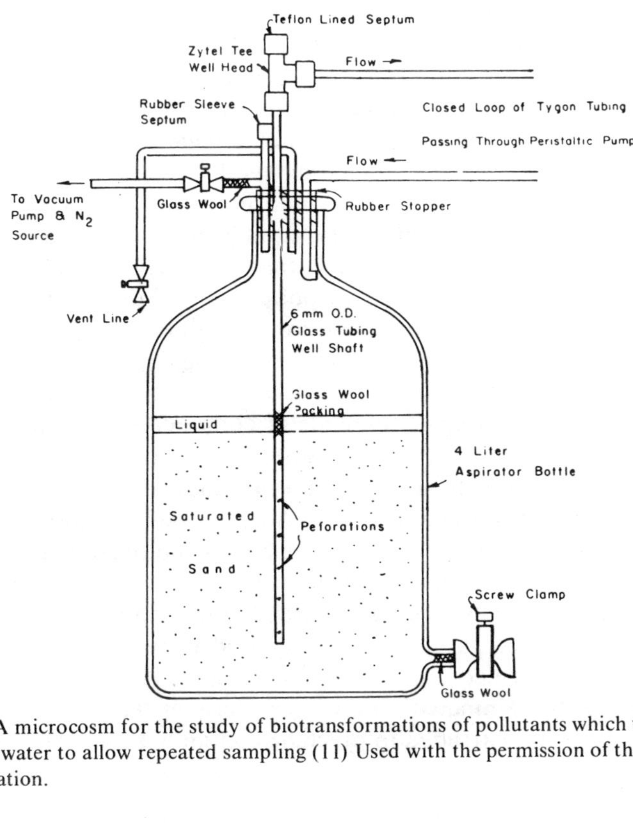

Figure 6.3. A microcosm for the study of biotransformations of pollutants which uses internal recirculation of water to allow repeated sampling (11) Used with the permission of the National Well Water Association.

restricted to organisms originally present in the aquifer. Most of the native pore water was replaced with sterile solutions of Na_3NTA at an initial concentration of 30–55 mg/liter. The headspace of the microcosm was filled with N_2; any gases evolved within the microcosm were sampled from the headspace. Microcosms were prepared in duplicate; there was also an autoclaved control.

Degradation of NTA was slow (0.5–1.0 mg liter^{-2} day^{-1}); there was no degradation in the autoclaved microcosm. Degradation was roughly zero order with concentration over the range of 50–5 mg/liter; although there were some shifts in the rate of degradation that were probably due to changes in the density of active organisms. There was good agreement between the duplicate microcosms, even after 260 days of incubation and three additions of NTA.

This microcosm configuration has a number of advantages. It can be sampled repeatedly (within reason), because sampling is not destructive. As a result, the microcosm can be sampled intensively at the beginning of an assay, and then preliminary results can be used to schedule the remaining samples. This way, a wide range of biodegradation rates can be adequately sampled. A second advantage is that the entire sample of aquifer material has the same history and is subjected to the same concentration of test substrate. As a result, plots of concentration against time can be used to infer the rate law for degradation. This is often not the case with a column configuration.

The recirculating loop in the microcosm of Dunlap et al. (11) was made of Tygon, which might present problems of sorption of various organic compounds. However, Teflon-lined peristaltic tubing is now available, this microcosm could be constructed entirely of nonreactive materials. The large headspace in the microcosm presents formidable difficulties in the study of volatile organic compounds and would allow escape of carbon dioxide and methane produced by the microbes. Escape of carbon dioxide tends to buffer the water in the microcosm from changes in pH that would occur in an aquifer. Because this microcosm is constructed with authentic aquifer matrix material, it is useful at lower concentrations of test compound that, although environmentally relevant, are too low to enrich an active population of degraders. It would be particularly useful in cases where the rate law and values for the rate constants for degradation are needed for mathematical modeling.

Bengtsson (12) designed a microcosm in a column configuration that can be used to study highly volatile compounds or compounds that strongly sorb to rubber or plastic (Fig. 6.4). A modification of the procedures described by Dunlap et al. (13) was used to acquire samples of the deeper subsurface that were not contaminated with surface microbes. A bore hole of the desired depth was drilled with an auger. Then the auger was removed and the sample taken with an autoclaved thin-wall core barrel 9.5-cm I.D. × 45-cm long. The core was extruded 3–5 cm and then broken off to produce an aseptic face. The remainder of the core was extended past a device that pared away the outer 1.2 cm of core material. As a result, the core material that was in contact with the core barrel, and thus might be contaminated with surface microbes, was discarded. The last 4 cm of the core was not extruded to preclude contamination from the extruder.

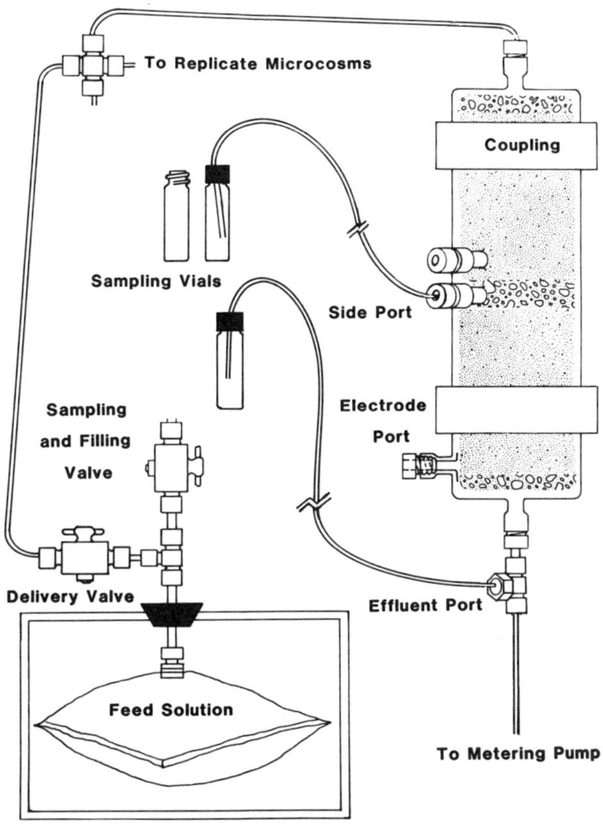

Figure 6.4. A microcosm in column configuration to study biotransformations of pollutants which are highly volatile or sorb strongly to rubber or plastic (12).

The aseptic sample was extruded directly into a glass column 7.0 cm I.D. × 40 cm long. The column had a sampling port at 20 cm and at the effluent end. A special Teflon well was constructed to provide water from the same lens of sand that was sampled by the core barrel. The water was filter sterilized, dosed with a suite of organic chemicals, and stirred in a sealed container until all compounds were dissolved. The resulting feed solution was stored in flexible Teflon bags supported in a Plexiglass box filled with water. The bags had virtually no headspace, which prevented loss of volatile organic compounds from the feed solution. Pressure from a reservoir forced the feed solution onto three replicate columns. Flow through the columns was maintained at 2 cm/day by a peristalic pump downstream of the column effluent port.

Samples for volatile constituents were drawn by connecting a sterile gas-tight syringe to the sample port. Samples for nonvolatile constituents were drawn by opening the port and allowing the sample to drip into a vial. All materials touching the feed solution, aquifer matrix material, or samples, were either Teflon or borosilicate glass. All experimental manipulations were performed aseptically. The microcosm was incubated at 17°C, the *in situ* temperature.

Using this microcosm, Wilson et al. (14) found no evidence of degradation of chloroform, 1,1-dichloroethane, 1,1,1-trichloroethane, trichloroethene, tetrachloroethene, toluene, chlorobenzene, or 2-chloroethyl vinyl ether in sandy aquifer material collected just below the water table at a depth of 6.5m. This is somewhat surprising because chlorobenzene and toluene, at least, are easily degraded in surface soil (15), and the aquifer material contained an appreciable density of bacteria (10^6/g dry material) as determined by both plate count and direct enumeration using epifluorescence microscopy (16).

This microcosm is very flexible in terms of the nature of the compounds that may be studied. It is also a very faithful simulation of the subsurface environment; it is built with authentic and uncontaminated subsurface material that is exposed to dose water prepared with well water from the same subsurface location and then percolated through the column at hydrologically realistic flow rates. As is the case with most column configurations, the relative breakthrough times of the compounds of interest and some conservative tracer like chloride or tritiated water reveal the extent of sorption of the compound to subsurface solids. After the concentration of compounds in the column effluent comes to a steady state, a comparison of the concentration of the compounds in the feed solution and column effluent reveals the extent of degradation during passage through the column.

This microcosm is, however, expensive to build and laborious to operate. Most of the labor involves aseptic procedures for preparing feed solutions and acquiring samples. Additionally, it is only suitable for subsurface materials that are relatively permeable, because water must pass through the microcosm. Also the aquifer matrix material must be sufficiently fluid to seal against the walls of the column and prevent channeling or short circuiting of the feed solution. This microcosm would be appropriate for a site study where very trustworthy information is needed about the fate of specific pollutants in a particular subsurface environment. It is also an excellent research tool to evaluate the behavior of simpler microcosms.

Wilson et al. (14) developed a simple microcosm that can also be used to study volatile or reactive compounds (Fig. 6.5). Authentic uncontaminated samples of the subsurface were obtained as described above (12). Sterile water was mixed into the sample until it became a slurry, then the slurry was transferred to 35-ml screw-cap test tubes. Space was left for approximately 1 ml of dose solution. Sterile dose solution was prepared as described above. The remaining space in each of the microcosms was filled with dose solution, the tip of the tube was wiped clean with a piece of sterile paper, the tube sealed with a 25-mm Teflon-lined screw cap, and then the dose solution was dispersed through the slurry with a vortex mixer. A portion of the subsurface material was autoclaved prior to dispensing into the test tubes to provide an abiotic control. The microcosms were sampled by diluting the pore water. The cap was removed and the microcosm was screwed into one end of a large Teflon stopcock with a 10-mm bore. A 35-ml test tube filled with water was screwed into the other end. Then the cock was opened and the contents of the two tubes were mixed together with a vortex mixer. Finally the cock was closed, the tubes were separated and sealed with Teflon-

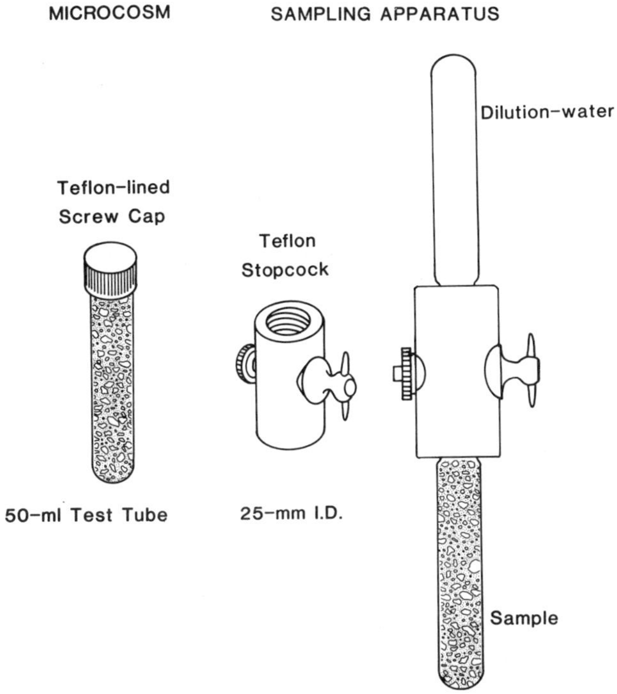

Figure 6.5. A simple microcosm to study biotransformations of pollutants which are highly volatile or sorb strongly to rubber or plastic (14).

lined screw caps, and centrifuged gently until the diluted pore water was clear. Microcosms were analyzed after 0, 1, 3, and 9 weeks of incubation. At each time period, at least two microcosms plus an autoclaved microcosm were analyzed.

In samples from immediately above and below the water table from a site at Fort Polk, Louisiana and a site near Pickett, Oklahoma, there was no detectable degradation of chloroform, 1,1-dichloroethane, 1,1,1-trichloroethane, trichloroethene, tetrachloroethene, and chlorobenzene. Styrene and 2-chloroethyl vinyl ether disappeared in all four samples. Toluene disappeared slowly in both samples from Fort Polk, but not in the samples from Pickett. In the autoclaved samples, toluene and styrene did not disappear, suggesting that biodegradation had occurred in the unautoclaved samples. 2-Chloroethyl vinyl ether disappeared in the autoclaved samples, indicating an abiotic process.

This microcosm is well-suited for screening compounds or subsurface materials. It faithfully simulates the subsurface environment with the exception of the flow of water. It can be used with highly volatile or hydrophobic compounds. The microcosms are easy to set up and require no maintenance. They accept materials with a wide range of textures. The experimental manipulations are so simple that the entire process from extruding the core to completion of the microcosm can be carried out in a glove box under a nitrogen atmosphere. As a result, biological processes can be examined in samples from anaerobic subsurface environments that might harbor oxygen-sensitive microbes. The recovery of added compounds from the autoclaved microcosms provides information about aerobic degradation or sorption; although the quality of the data is generally not as good as column-breakthrough data, it is usually adequate for screening purposes.

The microcosm has several limitations. The volume of water available for analysis is small, generally 15–30 ml, and the compounds of interest are diluted by a factor of 4–8 in the process of sampling. As a result, a sensitive analytical scheme is required to examine the fate of compounds at environmentally relevant concentrations. Additionally, sampling is destructive; thus, a fairly large amount of core material is required to set up a properly replicated and adequately controlled assay. Finally, the subsurface material is mixed to a slurry before being loaded into the test tubes. This destroys much of the spatial heterogeneity present in the sample and may destroy microhabitats required by certain groups of microbes. Work is underway at the Robert S. Kerr Environmental Research Laboratory that compares the behavior of several organic compounds in both the test tube microcosm and the glass column microcosm in order to determine whether disruption of potential microhabitats is a serious difficulty.

6.5. MICROCOSMS AS PHYSICAL MODELS

Rees and King examined the degradation of phenol in aquifer matrix material under aerobic (17) and anaerobic (18) conditions. Glass columns 4.5 cm I.D. × 60 cm long were packed to a depth of 40 cm with material from the Lower Greensand formation that was collected from a depth of 1.5 m. They were dosed with sterile solutions of phenol. In the case of the aerobic columns, no mineral nutrients were added; in the anaerobic columns, acetate, formate, phosphate, ammonia, and a number of other salts were added to simulate leachate from a domestic refuse landfill. Apparently aseptic precautions were not taken in collection of the aquifer matrix material or preparation of the columns. Water was collected from positions within the aerobic columns through porous ceramic thimbles. The columns were operated at various flow rates and concentrations of phenol until they came to a steady state. Starting at 50 mg/liter, degradation of phenol under aerobic conditions was first order up to approximately 10,000 mg/liter. At higher concentrations, the phenol was inhibitory. These results are surprising, because Yang and Humphry (19) found that aerobic degradation of phenol by *Pseudomona putida* and the yeast *Trichosporum cutaneum* in culture medium had a K_s in the range of 2 mg/liter. Degradation would be essentially zero order at concentrations above 10 mg/liter. This is three orders of magnitude lower than the results of Rees and King (17, 18).

Almost all of the phenol-degrading bacteria that could be enumerated by the MPN technique were found in the first 4 cm of the column; the great majority were found in the first 2 cm. However, the first sampling thimble was placed at 6 cm. This situation illustrates one of the primary difficulties with a column configuration. The microcosm is often spatially heterogeneous with respect to degradative activity, although calculation of a rate constant assumes homogeneous conditions over the entire microcosm or at least over the interval between sampling ports. As a result, the value of the rate constant determined in a column microcosm can depend on the length of the column.

Degradation of phenol under anaerobic conditions was studied by purging oxygen from the feed solution and the column headspace with N_2. The columns were inoculated with leachate from a laboratory-scale compost pile of municipal waste. At steady state, degradation was generally first order at concentrations from 200–1500 mg/liter, and zero order above 1500 mg/liter.

Rees and King (17, 18) used the Michaelis–Menten equation to extrapolate their laboratory measurements to hypothetical field situations. Such an approach will probably be necessary in order for the science of environmental projection to reach maturity. However, it must be established that the assumptions inherent in the mathematics are met by the subsurface environment being described. For example, the Monod equation assumes that the concentration of organic compound is the limiting factor for its degradation. Frequently, however, degradation in the deeper subsurface will be limited by other requirements for metabolism, particularly, oxygen needed for respiration and phosphorous and nitrogen needed for growth.

6.6. RATIONAL DESIGN OF MICROCOSMS

Microcosm design involves a number of tradeoffs between convenience of operation and/or cost and faithful simulation of the environmental conditions that significantly impact the process of interest. The judgment involved in identifying these important environmental conditions can only come from experience with the larger physical system. Therefore, microcosms should be designed by specialists in the environment being described. If the microcosm is to be used as a physical model, additional constraints are imposed on the experimental design to ensure that the data required by the equations are properly collected. The designer must either understand the mathematics or work closely with someone who does, prior to designing and constructing a microcosm.

Microcosms obey a law of diminishing returns. As microcosms become larger and more complex, the amount of time spent designing, constructing, and maintaining the system seems to increase exponentially. Microcosms should be as simple as possible. In many cases, it is cheaper to build two systems modeling two unrelated phenomena and to optimize the design of each system for its particular phenomenon than to build one common system that does a poor job on both. One useful approach to scaling a microcosm is to first determine a concentration of pollutant to be studied that is environmentally realistic. Then determine the extent of sorption or degradation the microcosm should be able to detect. A change in concentration of at least two orders of magnitude is needed to determine a rate law. Next, determine the sensitivity of the analytical instrument. Then calculate the quality of sample that must be extracted and concentrated to produce a detectable response on the instrument at the lowest concentration the microcosm is expected to detect. Finally, scale the microcosm to produce that quantity of sample.

If at all possible, microcosms should be replicated. Variability in many environmental parameters increases with a decrease in the size of the sample, due in part to stochastic processes. Unless microcosms are replicated, it is impossible to tell if an unexpected result is a statistical aberration or if it represents the usual behavior of the system.

Additionally, the container and plumbing of the microcosm must not affect the processes under study. The experimenter should demonstrate that such is the case prior to the initiation of a long and complex study. If the system is built of nonreactive materials (e.g., glass, Teflon, or stainless steel rather than PVC and Tygon), it is less likely that problems of this nature will occur. The major cost in most microcosm studies is chemical analysis of the samples, particularly analysis of organic compounds at trace concentrations. It is foolish to try to save a relatively small amount of money on the microcosms *per se* and risk compromising the results of expensive analytical procedures.

REFERENCES

1. R. F. Spalding and M. E. Exner, *J. Environ. Qual.* **9,** 466 (1980).
2. R. A. Freeze and J. A. Cherry, *Groundwater*, Prentice–Hall, Inc., Englewood Cliffs, N.J., 1979, p. 276
3. W. J. Dunlap and J. F. McNabb, EPA 600/2-73-014, Environmental Protection Technology Series, U.S. Environmental Protection Agency, Washington, D.C. 1973.
4. L. W. Sinton, *J. Hydro.* (N.Z.) **19,** 119 (1980).
5. G. N. Martin and M. J. Noonan, Water and Soil Technical Publication No. 7, National Water and Soil Conservation Organization, Wellington, New Zealand (1977).
6. G. A. McFeters, G. K. Bissonette, J. J. Jezeski, C. A. Thomson, and D. G. Stuart, *Appl. Microbiol.* **27,** 823 (1974).
7. E. Horvath and G. H. Elkan, *Ground Water* **16,** 174 (1978)
8. J. A. Leenheer, R. L. Malcolm, and W. R. White, *Environ. Sci. Technol.* **10,** 445 (1976).
9. E. Horvath and G. H. Elkan, *Appl. Environ. Microbiol.* **35,** 1221 (1978).
10. M. M. Grula and E. A. Grula, *Feasibility of Microbial Decomposition of Organic Wastes under Conditions Existing in Deep Wells*, BERC/R1-76-6, United States Energy Research and Development Administration, Office of Public Affairs, Technical Information Center, 1976.
11. W. J. Dunlap, R. L. Cosby, J. F. McNabb, B. E. Bledsoe, and M. R. Scalf, *Ground Water* **10,** 107 (1972).

12. S. G. L. Bengtsson, Presented at the First International Conference on Ground Water Quality Research, October 7–10, 1981, Rice University, Houston, Texas.

13. W. J. Dunlap, J. F. McNabb, M. R. Scalf, and R. L. Cosby, EPA-600/2-77-176, Environmental Protection Technology Series, U.S. Environmental Protection Agency, Washington, D.C., 1977.

14. J. T. Wilson, M. J. Noonan, and J. F. McNabb, Presented at The First International Conference on Ground Water Quality Research, October 7–10, 1981, Rice University, Houston, Texas.

15. J. T. Wilson, C. G. Enfield, W. J. Dunlap, R. L. Cosby, D. A. Foster, and L. B. Baskin, *J. Environ. Qual.* **10,** 501 (1981).

16. W. G. Ghiorse, Presented at The First International Conference on Ground Water Quality Research, October 7–10, 1981, Rice University, Houston, Texas.

17. J. F. Rees and J. W. King, *J. Chem. Tech. Biotechnol.* **30,** 296 (1980).

18. J. F. Rees and J. W. King, *J. Chem. Tech. Biotechnol.* **31,** 306 (1981).

19. R. D. Yang and A. E. Humphrey, *Biotech. Bioeng.* **17,** 1211 (1975).

7

HEALTH ASPECTS OF GROUNDWATER POLLUTION

G. F. Craun

Health Effects Research Lab
United States Environmental Protection Agency
Cincinnati, Ohio

Waterborne illness can be contracted in a number of ways: (1) ingestion of contaminated water or ice; (2) contact through bathing, swimming or wading; (3) inhalation of aerosols, gases, or vapors from contaminated water or wastewater;

This chapter was written by Gunther F. Craun in his private capacity. No official support or endorsement by the Environmental Protection Agency or any other agency of the Federal Government is intended or should be inferred.

(4) insufficient supply of water, which makes water generally unavailable for bathing and washing and contributes to the person-to-person transmission of infectious diseases. The emphasis of this chapter is on illness and the human health effects associated with ingestion of contaminants in drinking water.

The transmission of infectious disease through contaminated drinking water has been a frequent and well-documented occurrence in the past. Chemical poisonings caused by contamination of source waters and cross-connections in water distribution systems have also occurred. However, with improvements in source water protection, treatment, and distribution of drinking water, outbreaks of waterborne disease have been infrequently reported. The chronic ingestion of low levels of chemical contaminants in drinking water can be responsible for human health effects; however, these effects are generally less well-documented and with some exceptions are less known and understood.

Among the etiologic agents of waterborne disease, some are found recurrently while others are not. This varies by geographic area and depends upon the general level of sanitation, endemic levels of disease, degree of industrialization, cultural and socioeconomic characteristics of the population, and climate. Any disease acquired by ingestion of contaminated drinking water can be transmitted through groundwater if the etiologic agent reaches the water source in sufficient concentration to cause illness. Some contaminants, however, are more likely than others to be present in groundwater, resulting in some waterborne diseases that are more frequently transmitted by groundwater sources. The degree to which groundwater is developed for drinking water supplies, geological and hydrological conditions, sewage disposal practices, and general technological development all influence the importance of groundwater in the transmission of a particular disease in any country.

In the United States, groundwater is an important source of drinking water. Approximately 25 percent of all water withdrawn in the United States is groundwater and about 50 percent of the U.S. population depends on groundwater as its principal source of drinking water (1). Many smaller communities rely totally on groundwater and most noncommunity water supplies are obtained from wells or springs. Many states are heavily dependent upon groundwater: 86 percent of all water withdrawn in Kansas, 68 percent in Nebraska, and 61 percent in Arizona is groundwater (1).

Relatively few infectious diseases are likely to be transmitted through groundwater. Contamination generally occurs in local areas by seepage of sewage into aquifers and improperly developed wells or the entry of sewage-contaminated surface water into poorly protected wells. The list of possible chemical contaminants of groundwater is much larger. Chemical contaminants originating from natural geological deposits or as a result of man's activities can contaminate entire aquifers as well as local areas (2–4). Groundwater contaminants such as arsenic, barium, fluoride, selenium, and radionuclides are often related to the natural composition of soils and rock. Contamination with synthetic organic chemicals and heavy metals such as cadmium, chromium, and lead is generally associated with the manufacture, disposal, and transportation of chemicals. Once groundwaters are contaminated with chemicals, they tend to remain contaminated because of insufficient dilution and the slow movement of groundwater in most aquifers. Most aquifers lack any natural cleansing capabilities (2, 3).

Agricultural practices, use of chemicals, leaks and spills, disposal of chemicals in surface impoundments, landfills and underground injection wells, disposal of wastewater on land and through septic tanks, mining, petroleum and natural gas production contribute to the contamination of groundwater (2–4). (See Chapter 3 for more details.) The most important source of groundwater contamination in the United States is felt to be the disposal of industrial wastes in impoundments and landfills (2, 3). There are approximately 76,000 active industrial landfills in the United States; some 50,000 active and inactive sites are felt to contain potentially hazardous chemical wastes (4). Groundwater contaminants such as benzene, polychlorinated biphenyls, chlorinated phenols, organic solvents, heavy metals, selenium, and arsenic have originated from industrial landfills and in many instances have migrated some distance from the original contamination site (2–4). Secondary sources of chemical contamination of groundwater in the United States are septic tanks, mining, and petroleum exploration. Approximately 20 million homes in the United States use septic tank and cesspool disposal systems, discharging more than 1 trillion gallons of waste into the ground each year (4). Homeowners can buy septic tank cleaning fluids that contain trichloroethylene, benzene, or methylene chloride and can cause significant chemical contamination. On Long Island alone some 400,000 gallons of septic tank cleaners are used by homeowners (4). Mining causes groundwater contamination primarily through waste disposal practices such as slurry and tailings ponds, slag piles, and the failure to reclaim mined land. Petroleum production activities have caused groundwater contamination in 17 south-central and western states and contamination from brine pits to dispose of saline byproducts of drilling has caused groundwater to become unusable in many areas (4). Pesticides such as DBCP, Aldicarb, and Carbonfuran and nitrates have been found in farm wells. In many instances, the contamination was related to current agricultural practices (3).

Inorganic contamination of groundwater can be the result of mining, manufacturing, industrial waste disposal practices, and natural processes such as chemical weathering, soil leaching, and decaying vegetation. Depending upon factors such as solubility, pH, adsorption characteristics, and formation of complexes, the penetration and movement of rainwater through the soil results in the release of various trace and bulk inorganic constituents into the groundwater (5). Trace metals may also be contributed to drinking water through corrosion products or the solution of metals from water distribution system piping and materials (6). This is generally not a problem in water systems using groundwater, but could be if groundwater is corrosive or softened.

To prevent the occurrence of adverse health effects from contaminants in drinking water, the Environmental Protection Agency (EPA) has established drinking water standards for public water supplies. However, only a limited number of chemical contaminants are included in these standards. National Interim Primary Drinking Water Regulations for bacteria, turbidity, 10 in-

Table 7.1. Maximum Contaminant Levels for Inorganic Constituents in Community Water Systems in the United States (7)

Contaminant	Level (mg/liter)
Arsenic	0.05
Barium	1.
Cadmium	0.010
Chromium	0.05
Lead	0.05
Mercury	0.002
Nitrate (as N)	10.
Selenium	0.01
Silver	0.05
Fluoride	1.4–2.4[a]

[a]Depending upon the annual average of maximum daily air temperature.

organic constituents (Table 7.1), 6 pesticides (Table 7.2), and radionuclides became effective on June 24, 1977 (7). Amendments to the regulations (8) include the promulgation of a maximum contaminant level of 100 μg/liter, for total trihalomethanes in community water systems of over 10,000 population and requirements for monitoring and reporting sodium levels and corrosivity characteristics (pH, alkalinity, hardness, temperature, total dissolved solids, and Langelier Index). With the exceptions of bacteria, turbidity, and nitrate, the maximum contaminant levels are based on effects associated with chronic exposure. No maximum contaminant levels have been established for most

Table 7.2. Maximum Contaminant Levels for Organic Chemicals in Community Water Systems in the United States (7)

Compound	Level (mg/liter)
1. Chlorinated hydrocarbons	
Endrin	0.0002
Lindane	0.004
Methoxychlor	0.1
Toxaphene	0.005
2. Chlorophenoxys	
2,4-D	0.1
2,4,5-TP Silvex	0.01

Table 7.3. Organic Compounds for Which Suggested No-Adverse-Response Levels Have Been Established (10)

Benzene	Methylene chloride
Benzene hexachloride	Polychlorinated biphenyl
Carbon tetrachloride	Resorcinol
Catechol	Tetrachloroethylene
Chloroform	1,1,1-Trichloroethane
Dibromochloromethane	Trichloroethylene
Dichlorodifluoromethane	Trichlorofluoromethane
2,4-Dichlorophenol	Toluene
Epichlorohydrin	Xylenes
Hexachlorobenzene	

Table 7.4. Secondary Maximum Contaminant Levels for Public Water Supplies in the United States (11)

Contaminant	Level
Chloride	250 mg/liter
Color	15 color units
Copper	1 mg/liter
Corrosivity	Noncorrosive
Foaming Agents	0.5 mg/liter
Iron	0.3 mg/liter
Manganese	0.05 mg/liter
Odor	3 Threshold Odor Number
pH	6.5–8.5
Sulfate	250 mg/liter
Total Dissolved Solids (TDS)	500 mg/liter
Zinc	5 mg/liter

synthetic organic compounds that have been identified in groundwater, but in some instances, states have adopted standards for a few organic chemicals (4). Several states have also established guidelines for notification of customers and physicians in areas where water supplies have high sodium levels (9). Suggested no-adverse-response levels (snarls) based on expected human health risks and extrapolations from experimental animal data have been prepared by the National Research Council (10) for some 19 organic compounds (Table 7.3). National Secondary Drinking Water Regulations (11) for contaminants that may adversely affect the aesthetic quality of drinking water such as taste, odor, color, and appearance have also been established by the EPA (Table 7.4).

7.1. BIOLOGICAL CONTAMINANTS

Infectious diseases that have been transmitted by contaminated drinking water include:

1. Amebiasis
2. Balantidiasis
3. Cholera
4. Dracontiasis
5. Gastroenteritis
   a. Undetermined etiology
   b. Campylobacter
   c. Toxigenic *E. coli*
   d. Viral
6. Giardiasis
7. Hepatitis A
8. Paratyphoid Fever
9. Salmonellosis
10. Shigellosis
11. Typhoid Fever
12. Tularemia

The only disease that is transmitted exclusively through drinking water is dracontiasis or guinea worm disease, an infection of the subcutaneous and deeper tissues with a large nematode, *Dracunculus medinensis* (12). This disease is widely distributed in India and west Africa, but also occurs in northeastern Africa, the Middle East, the West Indies, and northeastern South America. Dracontiasis is primarily associated with poverty, particularly with rural communities with inadequate water supplies. The intermediate host, a crustacea of the genus *Cyclops*, is required for transmission. This disease is manifested by a vesicle that usually appears on a lower extremity, especially the foot. Within the vesicle lies the female worm which can discharge larvae when the vesicle is immersed in water. Larvae discharged into the water are ingested by *Cyclops* and develop into the infective stage in about two weeks. Man swallows the infected copepods by drinking water from contaminated step-wells and ponds. Larvae are liberated in man's stomach or duodenum, migrate through the viscera, become adults, and reach the subcutaneous tissues. The cycle is continued when infected individuals wade in step-wells and ponds used for water supplies and *Cyclops* are present.

Tularemia, an infectious disease of rodents and rabbits that also occurs in domestic animals and man, is occasionally transmitted to man through contaminated drinking water (13–15). This disease is not directly transmitted from man to man. Other routes of transmission include (1) contact through wounds, unbroken skin, or conjunctiva; (2) bites of arthropods; (3) inhalation of dust from contaminated areas; (4) ingestion of insufficiently cooked rabbit. Two small outbreaks caused by drinking contaminated water from surface sources have been documented in the United States.

Other infectious diseases are transmitted by the fecal–oral route from man to man or animal to man; drinking water is one of many possible sources of exposure. A cycle of human-sewage–water–humans becomes possible because sewage disposal systems and drinking water supplies are often in close proximity and enteric organisms can remain viable in untreated water. For some organisms such as *Giardia*, wild and domestic animals are included in the cycle of contamination. Although organisms excreted into sewage or water encounter an environment frequently hostile to their survival, organisms may still reach a drinking water supply in a concentration sufficient to cause infection unless the water is disinfected or in some instances filtered.

Cholera is a classic waterborne disease which fortunately rarely occurs today in the United States. Recently, however, several cases of nonlaboratory-acquired infection with *V. cholerae* have occurred in the United States where the source of the infection could not be identified (16, 17).* Contaminated, untreated water, especially from shallow wells subject to surface runoff, is an important means of transmission of cholera in underdeveloped countries where the disease is endemic. Commercially bottled water has also been shown to be a mode of transmission (18). In the 1974 cholera epidemic in Portugal, outbreaks in Faro and Lisbon were traced to a bottling plant using two untreated springs for the source of water; *V. cholerae* was identified in the spring water. Although bottled, uncarbonated mineral water from the bottling plant was epidemiologically implicated as the vehicle of transmission, carbonated water from the same bottling plant and water source was not. *V. cholerae* was found to survive less than a day in carbonated water, but 13–22 days in well water.

Two enteric diseases that have commonly been transmitted by drinking water, particularly in the United States (Table 7.5), are *Shigella* and *Salmonella* gastroenteritis. The symptoms caused by these diseases are relatively similar—diarrhea, abdominal cramps, fever, chills, and vomiting. Clinically, these enteric infections are usually indistinguishable; however, *Shigella* infection produces bloody stools and rectal pain more frequently (12).

Typhoid fever (*Salmonella typhi*) is an enteric infection often transmitted in past years by drinking water. The diagnosis of typhoid fever is confirmed by blood, urine, or stool culture. Patients usually experience fever, headache, and abdominal pain (12).

Contaminated water has also been responsible for outbreaks of paratyphoid fever. In 1972, a waterborne outbreak caused by *S. paratyphi B* occurred in Britain in a group of villages at the same time as an outbreak of the same type in cattle and humans at a farm several miles away (19). The water supply to the villages was obtained from several springs subject to pollution from the septic tank of a cottage occupied by workers at the infected farm. The infections at the farm originated from a stream that received sewage effluent from a village where a chronic carrier resided.

In recent years, giardiasis has been a frequent cause of waterborne-disease outbreaks in the United States and in travelers in the Soviet Union (20). The

*The largest waterborne outbreak of cholera during this century in the United States occurred in 1981; *V. cholera* 01 was responsible for 17 cases of severe diarrhea caused by sewage contamination of an oil rig's potable water system.

Table 7.5. Diseases Generally Transmitted by Contaminated Drinking Water in the United States

Disease	Incubation Period	Symptoms
Bacterial		
Shigellosis	1–7 days	Diarrhea, fever, vomiting, cramps tenesumus, blood in stools occasionally
Salmonellosis	6–72 hours	Abdominal pain, diarrhea, nausea, vomiting, fever
Typhoid Fever	1–3 days	Abdominal pain, fever, chills, diarrhea or constipation, intestinal hemorrhage or perforation
Parasitic		
Giardiasis	7–14 days	Chronic diarrhea, abdominal cramps, flatulence, malodorous stools, fatigue, weight loss
Viral		
Hepatitis A	15–45 days	Fever, malaise, anorexia, nausea, jaundice

etiologic agent is the protozoan *Giardia lamblia* which exists in a cyst form that is resistant to the usual disinfection provided for the treatment of drinking water. The organism produces a chronic (more than 1 week) diarrhea syndrome that is accompanied by weight loss, epigastric pain, and flatulence. Diagnosis is established by identifying cysts or trophozoites in fecal material or in fluid from the upper small bowel; however, laboratory confirmation may be difficult. Outbreaks are generally caused by the ingestion of untreated stream water in mountainous areas and treated surface waters that have been disinfected but not properly filtered. Beavers have been implicated as the source of infection where there has been no obvious source of human fecal contamination in surface water.

Amebiasis (*Entamoeba histolytica*) and balantidiasis (*Balantidium coli*) are two additional protozoan infections that are sometimes transmitted by contaminated drinking water. These organisms also exist in a cyst form that is resistant to the usual disinfection dosage. There have been only five waterborne outbreaks involving 75 cases of amebiasis in the United States since 1946 (15). An outbreak of 110 cases of balantidiasis occurred in Micronesia in 1971 when a typhoon destroyed water catchment facilities and people were forced to utilize water supplies contaminated by pig feces (21). Pigs are the principal reservoir of *B. coli*.

Hepatitis A is a viral disease that has long been recognized to be transmitted by drinking water. In the United States, less than 1 percent of the cases of hepatitis A are associated with drinking contaminated water. The virus produces symptoms of nausea, vomiting, muscle aches, and jaundice that occur approximately 15–45 days after exposure. The diagnosis can be made by history, physi-

Table 7.6. Recently Identified or Suspected Biologic Agents Transmitted by Drinking Water in the United States

Agent	Incubation Period	Symptoms
Enterotoxigenic *E. coli*	12–72 hours	Diarrhea, fever, abdominal cramps, vomiting, body aches
Norwalk-like agent	12–48 hours	Vomiting, abdominal cramps, headache, fever
Campylobacter fetus ssp. *jejuni*	1–7 days	Diarrhea, abdominal cramps, headache, fever, vomiting, occasional bloody stool
Yersinia enterocolitica	1–7 days	Abdominal pain suggesting acute appendicitis, fever, headache, malaise, diarrhea, vomiting

cal examination, tests of liver function, and serologic studies. Hepatitis A outbreaks are often not recognized because the long incubation period obscures the relationship of illness and consumption of water.

Waterborne outbreaks of gastroenteritis caused by enterotoxigenic *E. coli*, Norwalk-like agent, and *Campylobacter fetus* ssp. *jejuni* have also been documented in the United States (Table 7.6) Although gastroenteritis caused by *Yersinia enterocolitica* is suspected to be acquired by drinking contaminated water, sufficient documentation of this etiology from waterborne outbreaks is lacking at the present time.

7.1.1 Waterborne Outbreaks in the United States

During 1971–79, 267 outbreaks of waterborne disease affecting 57,974 persons were reported (22–26) by 45 states and Puerto Rico (Table 7.7). Most of the outbreaks (54 percent) occurred in noncommunity water systems, which serve a large transitory population (Table 7.8). A large number (78 percent) of the outbreaks in noncommunity water systems affect travelers, campers, restaurant patrons, and visitors to recreational areas; of the outbreaks involving this transitory population, 74 percent occurred from May to August. There are approximately 150,000 noncommunity water systems in the United States, most of which (96 percent) use groundwater sources.

Most of the illness (69 percent) resulted from outbreaks in community water

Table 7.7. Waterborne Disease Outbreaks in the United States (1971–1979)

	1971	1972	1973	1974	1975	1976	1977	1978	1979	Total
Outbreaks	20	30	25	25	24	35	34	32	42	267
Cases of Illness	5184	1650	1762	8356	10,879	5068	3860	11,435	9780	57,974

Table 7.8. Waterborne Disease Outbreaks in the United States, 1971–1979, by Type of Water System

	Outbreaks (Percent)	Cases of Illness (Percent)
Community systems	34	69
Noncommunity systems	54	30
Individual systems	12	1
	100	100

systems, which number about 64,000 and serve some 200 million people. Approximately 82 percent of the community water systems and 30 percent of the population served by community systems use groundwater. The larger community water supply systems generally rely upon surface water sources to meet the demands of large populations and industry. Twelve percent of the reported waterborne outbreaks and 1 percent of the reported illness occurred in individual water systems, which depend primarily on untreated groundwater; however, outbreaks in individual water systems are unlikely to be recognized and reported.

The use of untreated or inadequately treated groundwater was responsible for 52 percent of all waterborne outbreaks and 45 percent of all waterborne disease in the United States. In noncommunity water systems, the use of untreated, contaminated groundwater accounted for many of the outbreaks (44 percent) and much of the illness (41 percent); only 11 percent of the outbreaks and 2 percent of the illness in community water systems were caused by the use of untreated, contaminated groundwater (Tables 7.9 and 7.10).

An etiologic agent was identified in 45 percent of the outbreaks (Table 7.11). The remaining outbreaks were classified as acute gastrointestinal illness, characterized by symptoms such as abdominal cramps, nausea, vomiting, and diarrhea occurring 24–48 hrs after consumption of water. With some exceptions, outbreaks in groundwater systems exhibited similar etiologies as outbreaks in surface water systems. Five of the six outbreaks of acute gastroenteritis where a

Table 7.9. Deficiencies Responsible for Waterborne Disease Outbreaks in Noncommunity Water Systems in the United States (1971–1979)

	Outbreaks (Percent)	Illness (Percent)
Contaminated surface water, no treatment[a]	8	2
Contaminated ground water, no treatment	44	41
Deficiencies in treatment facilities	36	39
Deficiencies in distribution system	7	17
Miscellaneous	5	1
	100	100

[a]Includes outbreaks of giardiasis in which surface water was chlorinated, but not filtered.

Table 7.10. Deficiencies Responsible for Waterborne Disease Outbreaks in Community Water Systems in the United States (1971–1979)

	Outbreaks (Percent)	Illness (Percent)
Contaminated surface water, no treatment[a]	13	23
Contaminated ground water, no treatment	11	2
Deficiencies in treatment facilities	36	50
Deficiencies in distribution system	34	24
Miscellaneous	6	1
	100	100

[a]Includes outbreaks of giardiasis in which surface water was chlorinated, but not filtered.

viral etiology was identified, all outbreaks of typhoid fever, and the single outbreak of toxigenic *E. coli* gastroenteritis occurred in groundwater systems where either no disinfection was provided or disinfection was interrupted or inadequate. The single outbreak of *Campylobacter* gastroenteritis and the majority of waterborne outbreaks of giardiasis occurred in surface water systems. Only three giardiasis outbreaks occurred in groundwater systems: in one, the groundwater in the well was recharged from an adjacent stream; in the second, the well was heavily contaminated by human sewage; the third involved a groundwater system where contamination entered the distribution system through a cross-connection.

Outbreaks in groundwater systems were examined to determine the source of contamination or treatment deficiency responsible for the outbreak (Tables 7.12 and 7.13). Outbreaks in groundwater systems caused by contamination of the distribution system were excluded from this analysis, as the occurrence of these outbreaks is unrelated to the type of water source. The overflow or seepage of

Table 7.11. Etiology of Waterborne Disease in the United States (1971–1979)

	Outbreaks (Percent)
Acute gastrointestinal illness	55
Chemical poisoning	11
Giardiasis	11
Shigellosis	8
Hepatitis A	6
Salmonellosis	3
Viral gastroenteritis	2
Typhoid	2
Toxigenic *E. coli* gastroenteritis	<1
Campylobacter gastroenteritis	<1
	100

Table 7.12. Waterborne Disease Outbreaks Caused by Source Contamination of Groundwater Systems Used Without Treatment in the United States (1971–1979)

	Outbreaks	Illness
Overflow or seepage of sewage	33	4167
Data insufficient to classify	24	934
Contaminated springs	9	940
Chemical contamination	9	127
Contamination through limestone or fissured rock	5	880
Contamination by surface runoff	6	396
Flooding	2	588
	88	8032

sewage, primarily from septic tanks or cesspools, was responsible for 43 percent of the outbreaks and 63 percent of the illness caused by the use of untreated, contaminated groundwater. Chemical contamination of groundwater (arsenic, ethyl acrylate, leaded gasoline, nitrate, phenol, polychlorinated biphenyl, selenium, and waste oil) and contamination by surface runoff or flooding caused 19 percent of the outbreaks in systems using untreated groundwater. Ten percent of the outbreaks involved contamination of springs. A source of contamination for the remaining outbreaks in systems using untreated groundwater could not be readily established either because of incomplete investigation or reporting.

Groundwater systems usually depend on a source water of relatively good bacteriologic quality where disinfection is used to protect against possible contamination of the distribution system. In these situations, unexpected contamination of the source could overwhelm the disinfectant. For groundwater systems that use a source known to be frequently or intermittently contaminated with bacteria, continuous disinfection is necessary to ensure potability until the sources of contamination are located and removed, improvements are made in source protection, or alternative water sources are developed. Outbreaks were caused by both inadequate disinfection and interruption of the disinfection

Table 7.13. Waterborne Disease Outbreaks Caused by Treatment Deficiencies in Groundwater Systems in the United States (1971–1979)

	Outbreaks	Illness
Interruption of disinfection		
Chlorine	31	9721
Iodine	3	71
Inadequate disinfection		
Chlorine	10	1469
Iodine	1	72
Problems with addition of other chemicals	5	434
	50	11,767

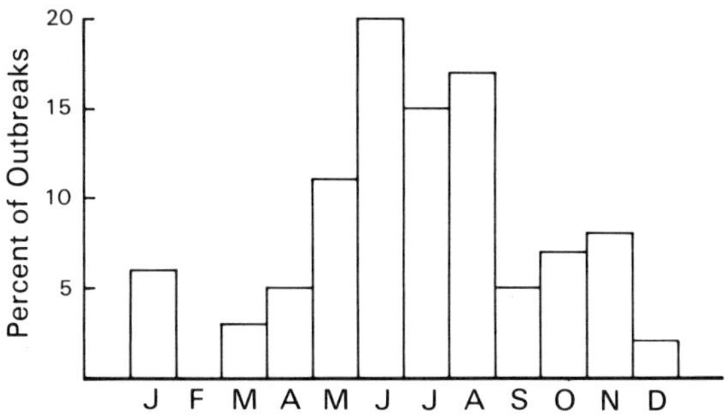

Figure 7.1. Seasonal distribution of waterborne disease outbreaks in water systems using untreated groundwater in the United States (1971–1979).

process (equipment malfunction or insufficient disinfectant). Interrruption of disinfection was responsible for most of the outbreaks (62 percent) and illnesses (73 percent) caused by a deficiency in treatment of groundwater. Most of the outbreaks in treated groundwater systems were the result of improper chlorination, but four outbreaks were caused by interruption or inadequate iodination in small noncommunity water systems serving a primarily transitory population.

An increase in the number of outbreaks resulting from the use of untreated, contaminated groundwater was noted during the summer months (Figure 7.1), implying increased contamination of groundwater during the summer or greater use of untreated groundwater by larger numbers of susceptible individuals during this period. Outbreaks caused by the use of untreated surface water showed a similar increase, but the increase was more evident during late summer and early fall.

7.1.2. Case Histories of Groundwater Outbreaks

7.1.2.1. Large Outbreaks

The four largest outbreaks caused by contaminated groundwater in the United States during 1971–79 occurred in Pico Rivera, California; Comerio, Puerto Rico; Richmond Heights, Florida; and Crater Lake, Oregon (Tables 7.14 and 7.15).

Between July 20, 1971 and August 7, 1971, approximately 62 percent of the people living within a 1.25-square-mile area of Pico Rivera became ill with gastroenteritis (22). The outbreak was confined to that part of Pico Rivera served by the Pico County Water District and was most severe in the western portion, close to a reservoir. A chlorinator provided disinfection to water entering the reservoir via a gravity line from a well. It was discovered that the chlorinator had broken on July 20 and had not been repaired for approximately one week. Water samples taken from the reservoir, from a gravity water line feeding the reservoir, and from a trailer park within the area revealed heavy contamination by fecal coliforms but no pathogens. The most likely source of the contamination was

Table 7.14. Large Waterborne Outbreaks Caused by Use of Contaminated Groundwater by Noncommunity Systems in the United States (1971–1979)

Year	Location	Cases	Etiology	Deficiency
1975	Crater Lake, Oregon	1000	Toxigenic *E. coli*	Inadequate disinfection, spring
1971	Restaurant, New York	500–1000	Gastroenteritis	Contaminated groundwater
1978	West Plains, Missouri	750	Gastroenteritis	Contamination of groundwater, limestone area
1974	Resort, Montana	615	Gastroenteritis	Contaminated groundwater
1974	Country Club, Pennsylvania	600	*Shigella sonnei*	Contaminated groundwater
1975	Camp, Arkansas	500	Gastroenteritis	Interruption of disinfection, wells
1977	Restaurant, Pennsylvania	500	Gastroenteritis	Interruption of disinfection, wells

along the gravity line serving the reservoir, since samples collected from the well supplying the gravity line yielded no fecal coliforms.

The second largest outbreak in a groundwater supply, an estimated 2105 cases of shigellosis, occurred in Comerio, Puerto Rico, in 1976 (22). *Shigella sonnei* was isolated from clinical specimens but could not be isolated from water samples collected two weeks after the epidemic had peaked. High coliform counts were found in the water distribution system during the outbreak, and one of seven wells supplying the system was found to be contaminated with total coliforms of >4900/100 ml and fecal coliforms of 23/100 ml. Although the wells were chlorinated, insufficient chlorine contact time was provided prior to the water's distribution and the facilities were not maintained to provide continuous, effective disinfection. Echovirus 8 was also found in the water from the contaminated well; however, its significance was not evaluated by epidemiologic studies.

Between January 17, 1974 and March 15, 1974, approximately 1200 cases of acute gastrointestinal illness occurred in Richmond Heights, Florida, a residential community of 6500 persons (22, 27). *Shigella sonnei* was considered the most likely etiologic agent. Epidemiological investigation showed that tap water consumption was associated with illness. It was found that one of the two wells providing water to the community had been continuously contaminated with excessive levels of fecal coliforms. The source of the contamination was traced by dye studies to the septic tank of a church and a day-care center located approximately 150 ft from the well. A breakdown in chlorination allowed approximately 1 million gallons of unchlorinated or insufficiently chlorinated water from the contaminated well to be distributed to the community 48 hours before the epidemic began. This outbreak illustrates the need for disease

Table 7.15. Large Waterborne Disease Outbreaks Caused by Use of Contaminated Groundwater by Community Systems in the United States (1971–1979)

Year	Location	Cases	Etiology	Deficiency
1971	Pico Rivera, California	3500	Gastroenteritis	Interruption of disinfection, wells
1976	Camerio, Puerto Rico	2150	*Shigella sonnei*	Inadequate disinfection, wells
1974	Richmond Hts., Florida	1200	*Shigella sonnei*	Interruption of disinfection, wells
1979	Templeton, Maine	500	Gastroenteritis	Flooding of well

surveillance to detect outbreaks, as only 10 cases of shigellosis were initially reported to health authorities. If local health authorities had not been conducting shigellosis surveillance, the initial 10 cases might never have been recognized as an unusual occurrence, and a waterborne outbreak as large as this might not have been detected.

At Crater Lake National Park, more than 1000 cases of diarrhea occurred in 1975 following contamination of park water by an obstructed sewer line that caused sewage to overflow from a manhole into a spring (22, 28). The spring was chlorinated, but facilities were improperly designed and did not provide adequate disinfection prior to distribution. Spring water was chlorinated in a holding reservoir and either gravity-fed to the headquarters area or pumped to the rim area of the park. No detectable chlorine residual could be found in the headquarters area, even when high levels (1.0–2.0 mg/liter) were present in the rim area. Coliform contamination was documented before the outbreak and fluorescent dye studies confirmed the source of the contamination. Enterotoxigenic *E. coli* serotype 06:H16 was isolated from ill persons and from a water sample from the spring. This is the first well-documented report of an outbreak of waterborne disease caused by an enterotoxigenic strain of *E. coli*. However, four waterborne outbreaks caused by enteropathogenic *E. coli* serotypes 0111:B4 and 0124:B27 were reported in 1967 and 1969 before tests for toxin production were commonly done (15). Waterborne outbreaks of toxigenic *E. coli* have also occurred outside the United States. For example, an epidemiologic investigation of an outbreak of acute gastroenteritis in September 1973 at Nagoya airport in Japan showed the source of infection to be one of two wells supplying drinking water to a restaurant; 956 persons were affected, with diarrhea being the primary symptom. Enterotoxigenic *E. coli* serotypes 06:K:H16, 027:K:H7, 0148:K:H28 were isolated from human fecal samples and the well water (29).

7.1.2.2. Contamination Through Limestone and Rock

An outbreak of 98 cases of viral hepatitis in Polk County, Arkansas, in 1971 illustrates the problems associated with wells developed in limestone and rock formations (30). The outbreak was traced to commercial pellet ice made from well water at a general store. Both the ice and the well water showed heavy coliform contamination. Dye studies revealed the sedimentary rock strata in the area permitted lateral drainage of a septic tank effluent from a nearby home occupied by residents who had had infectious hepatitis six weeks previously.

Large-scale outbreaks can also occur in areas where groundwater is obtained from limestone or rock aquifers (22, 31). During May 7–26, 1978, several outbreaks of gastroenteritis were reported in high school teams that had visited West Plains, Missouri. An investigation revealed that the municipal-sewage oxidation-lagoon had leaked approximately 18 million gallons of its contents into the ground on May 5–6. The surrounding rural terrain is composed of porous limestone with numerous subterranean streams, which permitted rapid movement of the contaminated groundwater. The area has a large number of private individual wells. Sampling of these wells showed contamination rates of 33–60 percent. The sewage lagoon had leaked on three previous occasions in the past 15 years, but this was the first time that related illness had been recognized.

The limestone terrain of the area combined with the numerous unchlorinated private wells appeared to be responsible for the magnitude of the outbreak. Intensive surveillance to ascertain the extent of gastroenteritis provided information on at least 759 cases that were associated with the ingestion of sewage-contaminated water. Diarrhea, nausea, vomiting, cramps, fever, chills, prostration, and headache were the reported symptoms; stool cultures were negative for bacterial enteric pathogens, suggesting a possible viral etiology.

7.1.2.3. Sewage Overflow and Seepage

On November 16, 1971, a trailer court tenant in Anchorage, Alaska, telephoned local health authorities to complain of water that was "dirty" and had a bad odor (30). A sanitary engineer investigating the complaint found raw sewage standing in the well house to the top of the well casings. Investigation revealed 89 of 114 persons exposed were ill; symptoms included nausea, vomiting, abdominal pain, fever, and diarrhea caused by *S. sonnei*. The water source was two approximately 242-ft-deep wells, steel cased and enclosed in a well house. The well house was 3 ft below ground level and the well casings extended only 1 ft above the well-house floor. Sometime prior to the morning of November 16, a "soft plug" had obstructed the borough sewer and sewage backed up through a drain in the well-house floor. As the sewage level rose above the casings, sewage spilled over the casings and was pumped into the trailer-park water system.

An outbreak of shigellosis occurred in November 1972 at a junior high school in Stockport, Iowa, where 208 cases of gastrointestinal illness were reported among the 289 students and 25 staff members (32). A similar illness affected 12 of 26 visiting basketball players. Rectal swab specimens were positive for *S. sonnei*. Epidemiologic investigation revealed the vehicle of infection to be the water supply, a shallow well in the schoolyard. Fluorescein dye introduced into a shower drain appeared in the well water within 30 minutes. A sample of tap water at the school showed high levels of coliforms and yielded *S. sonnei*.

7.1.3. Diseases Transmitted by Outbreaks in Groundwater Systems

7.1.3.1. Chemical Poisonings

In May 1972, a contractor built a warehouse and office structure just outside a small town in Minnesota, and a well was drilled to supply water. During the next $2\frac{1}{2}$ months, 11 of the 13 individuals employed by the contractor became ill; two were hospitalized. The hospitalization led to the discovery of elevated urine-arsenic levels in both patients. Samples obtained from the well on two occasions yielded arsenic concentrations of 21.0 and 11.8 mg/liter. Area residents reported that grasshoppers had been a serious problem in the late 1930s and that a grasshopper bait composed of arsenic, bran, and sawdust had been stored on the property now occupied by the warehouse and office. The bait was apparently stored on the ground and later buried in the area. Soil samples on the property revealed arsenic concentrations of 3000 and 12,600 mg/liter at varying soil depths (30).

Chemical spills have also affected groundwater quality and caused outbreaks. Accidental spillage of 10,000 gallons of 100 percent phenol in 1974 resulted in the chemical contamination of a number of wells in a rural area of southern Wisconsin (33). An illness characterized by diarrhea, mouth sores, burning of the mouth, and dark urine was reported by 17 persons exposed to the phenol-contaminated water. Gasoline spilled at a service station in Pennsylvania affected 10 individuals using private wells near the station; 10 mg/liter leaded gasoline was detected in the wells. Possible deliberate contamination of a well with arsenic caused an outbreak in Virginia in 1979; arsenic levels of 19–108 mg/liter were found in the groundwater after high arsenic levels were detected in the urine and blood of eight individuals using the well.

7.1.3.2. *Typhoid Fever*

All of the outbreaks of typhoid fever occurring in the United States during 1971–79 involved groundwater systems. The largest reported outbreak of typhoid fever in the United States since 1939 occurred at the South Dade Migrant Farm Labor Camp, Florida, in February and March 1973 (16, 34, 35). Epidemiologic investigation of the 210 cases implicated the camp's water supply as the vehicle of infection. Two shallow wells supplied water to the camp, and the water was disinfected prior to distribution. An engineering evaluation revealed that in early February, chlorination of the water was interrupted for a period of time. It was felt that contamination of the water supply occurred then. A young mentally retarded child who developed typhoid fever in January and who attended a day-care center located adjacent to the well was felt to be the index case and source of contamination. The aquifer was composed of solution channels and the wells had a history of intermittent contamination.

Between April 4, 1972 and May 22, 1972, five cases of typhoid occurred in a residential area near Yakima, Washington that was served by individual driven well points and septic tanks (30). Upon investigation, a typhoid carrier was identified in the area. Dye flushed through the sewage system in his home was traced within 36 hrs to numerous wells in the area including the ill family's well, which was 210 ft away. The water from this well yielded typhoid bacillus and coliforms. The soil in the area was extremely pervious gravel and at the time of the outbreak, the groundwater level was at or near its seasonal peak. Two additional typhoid fever outbreaks affecting a total of seven individuals occurred in Washington and Maryland; both were caused by use of contaminated, untreated groundwater from individual, private wells.

7.1.3.3. *Viral Gastroenteritis and Hepatitis*

Acute infectious nonbacterial gastroenteritis is caused in humans by at least two distinct groups of viruses (36, 37). The 70-nm rotavirus has been shown to be a major cause of acute diarrhea in infants and young children. The other viruses associated with acute nonbacterial gastroenteritis are less well characterized. Viruslike 27-nm particles have been seen in stool filtrates from specimens obtained during gastroenteritis outbreaks in Norwalk, Ohio, Honolulu, Hawaii, and Montgomery County, Maryland. The available epidemiologic data suggest that the 27-nm viruses have been responsible for some family and community outbreaks of acute gastroenteritis in older children and adults. At least three serologically distinct 27-nm particles associated with gastroenteritis have been identified. These and other viruses that are yet to be fully characterized may be the cause of some outbreaks of waterborne gastroenteritis of undetermined etiology. Rotaviruses have not been conclusively shown to be transmitted by drinking water, but an outbreak of 3172 cases of acute gastroenteritis affecting 30 percent of the population of a small Swedish town in January 1977 was suspected to have been caused by rotavirus (38).* Particles with a diameter of 75 nm, similar in appearance to rotavirus, were detected in fecal specimens of acutely ill and convalescent patients. No *Salmonella, Shigella, Yersinia*, cytopathogenic viruses, or viruses pathogenic for infant mice could be detected. All age groups were represented among the ill and there was no indication of transmission from children to adults. The disease was characterized by diarrhea, vomiting, headache, muscular pain, and fever. Symptoms disappeared after 1–3 days. Observations suggested that the water supply was contaminated with wastewater.

Six waterborne outbreaks of gastroenteritis caused by a Norwalk-like agent have been reported in the United States. All six outbreaks occurred in groundwater systems: three were caused by interrupted or inadequate chlorination, two were caused by use of contaminated, untreated groundwater, and one by backsiphonage of sewage through the distribution system. All outbreaks affected noncommunity water systems: five of the six outbreaks involved a transitory population. Coliform contamination of the water supply was evident in all of the outbreaks. These outbreaks illustrate that Norwalk-like agents may contribute to waterborne gastrointestinal disease and substantiate previous reports that outbreaks with a high attack rate, predominance of upper gastrointestinal symptoms, and relatively short duration of illness are compatible with viral gastroenteritis. Morens et al. (39) investigated such an outbreak at a resort camp in Colorado in 1976. Data obtained by questionnaire from 760 persons indicated that 418 (55 percent) had gastroenteritis at the camp or within a week of leaving it. Symptoms included vomiting (81 percent), diarrhea (65 percent) and fever (49 percent). The median duration of illness was 24 hrs. The attack rate increased with increasing consumption of water or iced beverages. The camp was supplied with water by a natural spring in a meadow at the base of a small hill. At the top of the hill was a private cabin with a septic tank approximately 50 ft above the spring. Interviews with maintenance personnel revealed that prior to the outbreak they discovered malfunctioning of the chlorinator and turned it off for several hours while making repairs. Fluorescein dye flushed into the cabin wastewater system rapidly appeared in the spring and in the camp tap water. The septic tank, set in fractured shale and decomposed granite, was unearthed and a hole in the leaching pipe several feet from the tank was found. Although routine laboratory tests did not reveal bacterial, viral, or parasitic pathogens, immune

*For the first time in the United States, rotavirus was identified as the cause of a waterborne outbreak of 1761 illnesses in Colorado in March 1981

electron microscopy detected 27-nm viruslike particles in two of five diarrheal stool filtrates. Oral administration of one of these bacteria-free filtrates to two volunteers induced a gastrointestinal illness similar to that observed in the camp visitors.

A survey of 57 groups visiting a camp in northeastern Pennsylvania between May 1, 1978 and June 16, 1978 revealed that 13 groups had experienced illness in over 15 percent of their members (40, 41). A total of 350 persons were reported to have been ill. All but one of the 13 groups with illness stayed at the camp between either May 12 and May 21 or June 5 and June 14. The attack rate among the eight groups that had illness and completed questionnaires ranged from 17–73 percent. Serum and stool specimens collected from ill members of the last group that visited the camp were negative for bacterial pathogens. However, three of five ill persons had a fourfold titer rise in antibody to a Norwalk-like agent; two controls were negative. The illness in this group was characterized by vomiting (81 percent), abdominal pain (74 percent), nausea (67 percent), and diarrhea (56 percent). An association was found between the quantity of camp water consumed and illness. A similar questionnaire, administered to two other ill groups, also showed an association between the quantity of camp water consumed and illness. Study of the water system demonstrated the presence of coliforms (38/100 ml), inadequate chlorination (no residual), and several areas of possible contamination.

On July 27, 1978, an outbreak of gastroenteritis was reported from another summer camp in northeastern Pennsylvania (40, 41). The cases were characterized by abdominal pain (80 percent), nausea (73 percent), and vomiting (53 percent). Headache (47 percent), diarrhea (38 percent), and chills (38 percent) were also prominent findings. The median duration of illness was two days. Review of the infirmary records revealed 73 cases of gastroenteritis during the first session of summer camp, which lasted from June 6 to July 23. This is approximately 10 times the rate reported from the previous year. A sharp increase in cases began 48 hrs after the arrival of the second-session campers; the attack rate was 61 percent. Food was not incriminated; however, consumption of five or more glasses a day of water or water-containing beverages was statistically associated with illness. Bacterial samples from the camp water supply revealed fecal coliforms from well water. Although the camp water supply was chlorinated, tests for residual chlorine level revealed no residual until July 28. No new cases were reported after July 29. Laboratory studies of stools from 10 patients and 10 controls revealed no bacterial pathogens. Three of three paired serum specimens, however, showed fourfold or greater rises to Norwalk-like agent by radioimmunoassay.

The three remaining waterborne outbreaks of viral gastroenteritis caused by a Norwalk-like agent occurred in (1) May 1978 in a school in Washington State affecting 467 individuals, (2) July 1979 in a camp at Island Lake, Pennsylvania affecting 151 individuals, (3) October 1979 in a camp at Brevard, North Carolina affecting 145 individuals.

There have been 68 waterborne outbreaks of viral hepatitis since 1946 in the United States (Table 7.16). Contaminated, untreated groundwater was respon-

Table 7.16. Waterborne Disease Outbreaks of Hepatitis A in the United States (1971–1979)

Cause	Community Systems	Noncommunity, Individual Systems
Contaminated surface water, untreated	1	11
Contaminated ground water, untreated	4	26
Inadequate or interruption of disinfection	3	5
Contamination through distribution system	11	2
Insufficient data to classify	3	2
	22	46

sible for most outbreaks of hepatitis A in noncommunity and individual water systems. Half of the outbreaks in community water systems occurred as the result of contamination of the distribution system, primarily through cross connections and backsiphonage. Outbreaks caused by hepatitis A virus or Norwalk-like agents have not been documented in correctly operated, conventional water treatment systems, except where distribution defects have been found as the source of the contamination.

Viral standards (42, 43) for drinking water have been proposed (less than one infectious viral unit per 378.5–3785 liter), but there has been no evidence of the spread of virus infection by drinking water that has been adequately treated to conventional bacteriological standards (44). The detection of viruses in water requires several cell culture systems and other methods; no single methodology can be used to determine whether a proposed viral standard is exceeded. The search for viruses in water supplies has focused on the recovery of enteroviruses. This group of viruses was once thought to cause epidemic gastroenteritis, but more recent studies have shown that these viruses are commonly excreted from the gastrointestinal tract in the absence of gastrointestinal symptoms (45). Enteroviruses are one of the most common causes of upper-respiratory-tract disease and are predominately recovered from specimens from that source (45). Disease caused by these viruses is most likely transmitted from person to person by respiratory aerosols (45). There has been evidence of waterborne outbreaks of viral gastroenteritis caused by 27-nm viruslike particles and rotavirus, but these viruses are not detectable by tissue culture systems used for enteroviruses. Gamble (44) critically examined the need for a viral drinking water standard and concluded that there was no evidence that drinking water contributes to the low-level transmission of virus infection; a viral standard was not justified; and in the absence of information relating the degree of viral contamination to disease, there was no logical basis on which to set a practical viral standard.

7.1.3.4 Yersiniosis

An outbreak (46) of 750 cases of waterborne gastrointestinal illness, in which *Yersinia enterocolitica* was isolated from well water, occurred among 1550 guests and 350 employees at a Montana ski resort from December 6, 1974 through January 17, 1975. An epidemiologic association was found between drinking water and illness, and chlorination of the wells stopped the outbreak. Two 18-m deep wells, developed in sand and gravel, supplied water to the resort. A sewer line was found to pass near the wells and samples collected from the wells after the outbreak yielded *Y. Enterocolitica* and coliform organisms from 1/100 ml to >16/100 ml. Routine bacteriologic surveillance during the previous three years had not detected coliform contamination of the wells. Although *Y. enterocolitica* was isolated from well water samples, the significance of this finding is unclear because rectal swab cultures from acutely ill persons were not examined for this organism. Christensen (47) reported the isolation of *Yersinia enterocolitica*, biotype 4, serotype 0:3 from an unchlorinated well water supply in a rural area of Denmark. The well water was suspected as the source of infection for an 8-month-old child hospitalized with acute gastroenteritis diagnosed as yersiniosis. A blood titer against serotype 0:3 was also demonstrated. The well water was used in the preparation of baby food, but it could not be established whether enrichment of *Yersinia* in the baby food or direct ingestion of water was responsible for the infection. No coliforms were found in the well water. However, the well was located in a low area easily contaminated by surface runoff.

Yersinia enterocolitica was recognized as a distinct species in 1964 and has been isolated with increasing frequency from man, animals, and foods such as market meat, vacuum-packed beef, mussels, oysters, and ice cream (48). It has also been found in surface waters, unchlorinated well waters in the United States (Washington, New York, Colorado), Canada, Norway, and Belgium, and chlorinated well water in Washington State (17, 49). Only 23 human infections were reported in 1966, but the number increased to over 4000 in 1974. The most common symptoms of *Y. enterocolitica* infection in man are fever, abdominal pain, and diarrhea. Infections in man probably have an alimentary origin, but the pathogenic mechanism of the organism is not known. In one experiment involving a human volunteer, a dose of 3.5×10^9 organisms was required to produce an infection (48).

The results of a survey in France (50–52) showed that *Y. enterocolitica* was frequently found in wild mammals, water, and fish. Strains of *Y. enterocolitica* similar to those found in small mammals were isolated from humans in the same area. Biochemical, serological, and phage typing suggests that there may be two distinct groups of *Y. enterocolitica*: the first represented by strains adapted to a specific host—human or animal—causing typical pathological disorders; the second represented by strains isolated from environmental sources such as food or water which do not have a clearly defined host. It is uncertain at this time whether this infection is an important cause of waterborne gastroenteritis of undetermined etiology.

7.2. INORGANIC CONTAMINANTS

The concentration of most bulk and trace inorganic constituents found in groundwater varies by area and, in many instances, over time. It is not possible in this chapter to provide a comprehensive summary of the health effects of all such constituents, but selected constituents based on their historical significance and occurrence in groundwaters of the United States are reviewed. Other references offer more comprehensive data on the human health effects of inorganic solutes in drinking water (5, 10), adverse effects associated with high exposures to metals and their compounds (53), and the relationship of the geochemical environment to health and disease (54–56).

7.2.1. Nitrate

A maximum contaminant level (10 mg/liter NO_3–N) for nitrate in drinking water has been established by the EPA for all public water systems serving at least 15 connections or 25 people for at least 60 days per year; the sole reported basis for the limit is to prevent infantile methemoglobinemia caused by acute exposure (7). Healthy adults have been reported to consume large quantities of nitrate in drinking water with relatively little, if any, effect (15). Possible effects of chronic ingestion and the potential formation of carcinogenic nitrosamines after ingestion of nitrate are not addressed by the EPA limit.

The development of infantile methemoglobinemia from nitrate in drinking water depends on the bacterial conversion of nitrate to nitrite before or after ingestion and occurs almost exclusively in infants less than three months old. Most reported cases have been associated with drinking water containing more than 22 mg/liter NO_3–N, but cases have occurred at concentrations below 10 mg/liter NO_3–N (5, 57). A number of variables are involved; not all infants develop methemoglobinemia when high nitrate water is ingested. Complicating the problem are nitrate-rich vegetables that may be given to infants at an early age and common drugs capable of producing methemoglobinemia. Methemoglobinemia can also occur in children and adults if sufficient nitrite is ingested directly. The EPA has not established a drinking water limit for nitrite.

It is estimated that about 1 percent of the U.S. population using public water supplies are consuming nitrate in excess of the EPA limit, primarily from groundwater systems (57–60). Some public water supplies in California have exceeded 10 mg/liter NO_3–N since 1935; in 1970, 18 California community water systems, serving populations ranging from 2000–32,000, exceeded the EPA limit for nitrate. A number of individual wells in Illinois, Kansas, Missouri, Nebraska, Oklahoma, Texas, and Wisconsin have also been reported to exceed the limit. A survey of 8844 wells in Illinois showed 27 percent to exceed 10 mg/liter NO_3–N. In one county in Texas some wells were found to have a nitrate concentration of 110–690 mg/liter NO_3–N.

The principal sources of human exposure to nitrate are vegetables, cured meats, and water, and in each instance the nitrate concentration can vary greatly (5, 61, 62). The average daily intake of nitrate from all sources in the United States is estimated to be 100 mg with drinking water contributing less than one percent of the total nitrate intake. The primary source of nitrate in the diet (86 percent) is vegetables and the second largest contributor of nitrate (9 percent) is cured meats.

Nitrate is reportedly toxic only at levels higher than those found in foods (5,

57, 63). Unless conditions exist for reducing nitrate to nitrite during digestion, nitrate is excreted without apparent adverse effects. Nitrite, however, whether ingested directly or produced by the bacterial conversion during digestion, is able to oxidate ferrous iron in hemoglobin to ferric iron and convert hemoglobin, the blood pigment that carries oxygen from the lungs to the tissues, to methemoglobin. Because the methemoglobin is incapable of binding molecular oxygen, the physiologic effect is oxygen deprivation or suffocation.

A methemoglobin reductase system normally present in the erythrocytes keeps methemoglobin at about 1–2 percent of the total hemoglobin in healthy adults (63). At methemoglobin levels of approximately 10 percent, the methemoglobin produces a slate-grey appearance of cyanosis and the disorder becomes clinically detectable (64).

Infants are generally more susceptible to the development of methemoglobinemia (63, 64). Exposure is greater because the infant's total fluid intake per body weight is three times that of an adult. The predominant form of hemoglobin at birth, fetal hemoglobin, is more susceptible to the development of methemoglobinemia. Other variables are also important. For example, one nonsusceptible infant 10 days of age was fed as much as 100 mg of nitrate per kilogram of body weight per day for an 8-day period without developing any signs of cyanosis; the methemoglobin level increased to 7.4 percent. Individual variation in effect is generally related to factors influencing the bacterial conversion of nitrate to nitrite, variation in the dose of nitrate, and certain protective factors. Variation in the dose of nitrate is influenced by the amount of water used for infant feeding and excessive boiling, which increases the nitrate concentration. Dietary intake of vitamin-C-rich juices or foods has been found to help maintain lower methemoglobin levels.

Many species of enterobacteria are capable of reducing nitrate to nitrite (57). In infants whose diet is normally carbohydrate, coliform organisms are felt to be primarily responsible for the conversion of nitrate to nitrite in the digestive tract. Coliform organisms can grow in the gastric pH range 5–7 normally found in infants less than three months old (65). Gastrointestinal illness and diarrhea may permit the bacteria responsible for the conversion of nitrate to nitrite to migrate from the lower intestine to the upper intestine and stomach, and thus, increase the likelihood of nitrite formation prior to absorption in the small intestine. Many cases of methemoglobinemia have been associated with individual, bacterially contaminated wells, which may be a source of nitrate-reducing bacteria and a cause of gastrointestinal illness (57). Contaminated infant formula, especially in powder form, may also be a source of nitrate-reducing bacteria (66).

There is a small, susceptible adult population that may develop methemoglobinemia after ingestion of nitrates (67). Adults who have had portions of the stomach surgically removed are susceptible if a low gastric pH cannot be maintained. Navajo Indians and Eskimos also have a higher than normal incidence of methemoglobinemia.

Nitrate in drinking water was first associated with methemoglobinemia in infants in 1945 (68). Since then, approximately 2000 cases have been reported in North America and Europe; approximately 8 percent of these cases resulted in death (69).

Although methemoglobinemia is not a reportable disease in the United States and its true incidence is not known, it appears that the incidence of methemoglobinemia associated with the use of high nitrate water from public water supplies in the United States is extremely small (60). Almost all of the 335 cases of methemoglobinemia associated with high-nitrate drinking water in the United States occurred in the 1940s and 1950s. Only one case of methemoglobinemia associated with a public water supply in the United States has been reported in the literature since 1960 (70). This case occurred in the early 1960s in a Colorado water supply where the nitrate concentration was 14 mg/liter NO_3–N. The family had just moved to the area and was unaware of the nitrate concentration of the water. A contributing factor was the preparation of the formula which included boiling the water for approximately 20 minutes. The infant was taken to a medical facility and improved after treatment by administration of methylene blue.

No cases of clinical methemoglobinemia were found in two epidemiologic studies (64, 71, 72) in areas of California and Illinois where water supplies contained as much as 20 mg/liter NO_3–N. There has never been a reported case of methemoglobinemia related to the ingestion of high nitrate water from public water supplies in California, even though several supplies have exceeded the EPA limit for many years (57, 72).

A recent study of hospital records from 1973 in Nebraska, where numerous individual wells exceed the nitrate limit, revealed only one case of methemoglobinemia possibly associated with high nitrate in an individual water supply (73). In the past 10 years, few investigators in the United States have been motivated to describe cases associated with individual water supplies (60). An infant in Texas developed methemoglobinemia associated with a farm well containing 77 mg/liter NO_3–N; another case in Oklahoma occurred in an infant using a private well with 110 mg/liter NO_3–N. Jones et al. (74) searched records at Children's Memorial Hospital in Oklahoma and found 10 cases of infantile methemoglobinemia from 1960–1971 associated with high nitrates in individual well water supplies. A survey of 37 states in 1979 revealed only two additional reports of cases of methemoglobinemia associated with individual wells over the past 10 years (60). This sporadic incidence of methemoglobinemia over the past 10–20 years suggests that factors other than nitrate concentration of water may be more important in the development of this disease in the United States today.

It is probable that changes in infant feeding practices and health education in high nitrate areas during this time have reduced exposure to high-nitrate water and may be responsible for the few clinical cases of methemoglobinemia associated with nitrate in drinking water in this country. Cases of infantile methemoglobinemia associated with nitrates in well water supplies, however, continue to be reported on a more routine basis in Europe (60).

7.2.2. Fluoride

Fluoride is found widely in water supplies and high concentrations of fluoride have been reported in areas of every continent (5, 10, 75). The concentration of fluoride in natural waters depends principally on the solubility of the fluoride-containing rocks with which the water is in contact. An extensive survey of community water supplies in the United States in 1969 showed that 8.1 million people in 2630 communities in 44 states were consuming water with more than 0.7 mg/liter of naturally occurring fluoride, with 1 million people in 524 communities receiving water with more than 2 mg/liter fluoride (76). Most of the communities with more than 0.7 mg/liter natural fluoride used groundwater sources in Arizona, Colorado, Illinois, Iowa, New Mexico, Ohio, Oklahoma, South Dakota, and Texas. Fleischer et al. (54) reported concentrations of up to 35 mg/liter fluoride in groundwater.

Small concentrations of fluoride in drinking water from 0.7–1.2 mg/liter, depending on the environmental temperature, are generally felt to have a beneficial effect on prevention of dental caries among children (5, 8, 10). The fluoride concentration in drinking water is not critical for caries protection. Rather, it is the amount of fluoride consumed during the tooth-forming years. As the uptake and deposition of fluoride is greatest before eruption and calcification of the teeth, its anticariogenic effect is greatest with children, especially those less than 8-years-old. Recent recommendations by the NRC (10) for adequate and safe intakes to protect against dental caries and osteoporosis are:

Age	Fluoride (mg)
<6 months	0.1–0.5
6–12 months	0.2–1.0
1–3 years	0.5–1.0
4–6 years	1.0–2.5
7–adulthood	1.5–2.5
adult	1.5–4.0

Two chronic effects are recognized as being caused by excess intake of fluoride in drinking water—mottling of tooth enamel or dental fluorosis and skeletal fluorosis. It is necessary to consider the severity of the effects since mild forms are sometimes considered beneficial (5). Mottling of tooth enamel may occur with fluoride concentrations of 0.8–1.6 mg/liter in drinking water. Skeletal fluorosis has been seen with consumption of water containing more than 3 mg/liter. Long-term consumption of water containing greater than 1 mg/liter fluoride may result in increased bone density and objectionable dental mottling in patients with long-standing renal disease or polydipsia, but increased bone density has been regarded as a beneficial rather than an adverse effect. Intake of fluoride for long periods in amounts greater than 20–40 mg/day may result in crippling skeletal fluorosis. Epidemiologic studies in areas where drinking water is naturally high in fluoride have found no adverse effects except in rare cases where the concentration is many times more than that recommended. Controlled studies with fluoridation at the 1 mg/liter level have reported no adverse effects. Studies reporting effects such as mongolism, cancer mortality, mutagenicity, and birth defects are felt to have been inadequately designed or found lacking in substance (5, 10).

7.2.3. Water Hardness

Water hardness is a general property of water that has been described and determined in several ways over the years. Its primary importance to consumers is in terms of washing. It is also a source of lime scale in boilers, hot-water tanks, and other water-using appliances. It is defined as the sum of polyvalent cations dissolved in water or as a measure of the amount of soap required to form lather. The two principal cations responsible for hardness are calcium and magnesium. Carbonate hardness occurs as a result of dissolved carbon dioxide in water which assists in dissolving of calcium and magnesium carbonate rock, limestone, and dolomite. Noncarbonate hardness results from the solution of compounds of calcium, magnesium, sulfate, chloride, nitrate, and other minor anions. Groundwaters generally have higher hardness than surface waters and are often softened by lime-soda or ion-exchange treatment.

More than 50 publications discussing the relationship between drinking water quality and cardiovascular disease have been published since Kobayashi (77) first reported a statistical correlation between cerebrovascular disease mortality and the acidity of water supplies in Japan some 24 years ago. Most of the associations of drinking water and cardiovascular disease have been obtained from correlational studies of mortality rates in areas having different water characteristics. In general, the studies have shown a negative association between cardiovascular disease mortality and water hardness, that is, lower cardiovascular mortality is found in areas where the hardness of drinking water is high (5, 10, 78). This association is presently considered statistical rather than causal. Although biologically plausible, the epidemiologic evidence is not specific; individual exposure and possible confounding by individual risk indicators have not been considered, the data are inconsistent, and the associations are relatively weak. In addition, hardness and total dissolved solids have also been found to be negatively associated with numerous other causes of death, including total mortality, cancer mortality, sudden death from other than arteriosclerotic heart disease, cirrhosis, and peptic ulcer and positively associated with fatal accidents, congenital malformations, and chronic obstructive pulmonary disease.

The negative association of water hardness with cardiovascular mortality was reported in most, but not all, studies involving large areas. However, there have been numerous failures to find similar negative associations within small geographic areas and in comparisons of communities. It is possible that the correlation of water quality with cardiovascular mortality is actually measuring the correlation of some other geographically related factor such as the demographic, socio-economic, and cultural characteristics. Some studies have adjusted for factors such as altitude, climate, latitude, humidity, air temperature, and socioeconomic status and found that they have influenced the correlation of water quality with cardiovascular mortality.

Very few studies have estimated risk or provided data for estimation of risk. In the United Kingdom, cardiovascular mortality for white males is about 40 percent higher in extreme soft water areas compared to hard water areas, whereas in the United States, the rates are only 15 percent higher. Neri et al. (79,

80) reported the maximum risk of the effect of water was 20 percent in Canada. Comstock (78, 81) calculated relative risks of 1.25 for total cardiovascular mortality for white males aged 45–64 in the United States and 1.19 in England and Wales.

If drinking water quality does indeed have an effect on cardiovascular disease, there must be something beneficial in hard water or detrimental in soft water. Since water hardness *per se* can hardly be related to disease, the most likely effect of a water factor would be through a deficiency of an essential element or an excess of a toxic element.

There are considerable data on the associations of calcium and magnesium with cardiovascular mortality. These constituents are the likeliest possibilities for the beneficial constituents in hard water. High calcium concentrations in hard water may be responsible for reducing or preventing the absorption of low levels of toxic metals, such as lead or cadmium, in either food or water. Another possibility is that hard water is a source of an essential element that is deficient in the diet, such as magnesium, which is important for myocardium function. Magnesium levels in tissue have been shown to be affected by levels in drinking water and low levels of magnesium in tissues appear to occur more often in soft water areas. It is possible that magnesium is not well-absorbed from some foods, and the magnesium content of hard water might be sufficient to prevent a deficiency where dietary magnesium is marginal.

It has also been suggested that trace metals present in soft water may be related to cardiovascular mortality. While corrosion of plumbing is known to be a source of metals in tap water, data indicate that factors other than hardness may be more important in determining corrosion, and concentrations of cadmium, chromium, copper, lead, or zinc in water are not necessarily associated with soft water. More research is needed to better define actual exposures to various trace constituents at the individual tap and their relationship to cardiovascular disease.

7.2.4. Sodium

Sodium, an ubiquitous constituent of natural water, is derived geologically from the leaching of surface and underground salt deposits, the intrusion of salt water into freshwater aquifers, and the incorporation of evaporated ocean spray near the seacoast into rainfall. The use of sodium chloride as a deicing agent on highways has increased progressively from 0.5 million tons in 1947 to 9 million tons in 1970. Runoff from highways and storage deposits has contributed to increased sodium concentrations in community well water supplies. Types of water treatment, especially softening and corrosion control, also contribute to increased sodium levels in water. In areas where the groundwater is considered hard, ion-exchange softening in which sodium ion is exchanged for calcium and magnesium is an important contributor of sodium to drinking water. Ion-exchange softening is employed by individuals who wish to soften water from municipal supplies or private wells and by smaller community water utilities using wells high in hardness. Sodium concentrations are increased, as two atoms of sodium are added to the water for every atom of calcium or magnesium removed. In terms of actual weights, 100 mg of calcium is replaced by about 115 mg of sodium, but the increase in the sodium concentration of the water is a direct function of the operating efficiency of the softening unit (5, 10).

The concentration of sodium found in drinking water from a survey of 2100 water systems that supplied about 50 percent of the U.S. population ranged from 0.4–1900 mg/liter (5). About 42 percent of the water systems had sodium concentrations greater than 20 mg/liter and 5 percent had sodium greater than 250 mg/liter. The higher concentrations were in groundwater supplies.

Sodium is the most abundant cation found in extracellular fluid and is essential to the regulation of the acid–base balance. It functions in the electrophysiology of cells, is important for extracellular osmolarity, and is required for propagation of impulses in excitable tissues and for active nutrient transport (10). The suggested adequate, safe intake of sodium ranges from 1100–3300 mg/day for normal adults and 115–750 mg/day for infants (82). Estimates of the sodium intake of U.S. adults range from 1600–9600 mg/day (83). Sodium is added to many foods during processing and sodium chloride is added during cooking and prior to consumption. Intake from food is generally the major source of sodium, but in instances where high sodium occurs in drinking water, a significant portion of the total sodium intake may be contributed by water. In Britain, the use of water high in sodium for reconstituting powdered infant formula has been attributed to unexpected infant deaths and infant brain damage from hypernatremia (5).

Evidence suggests that chronic excessive intake of sodium may be associated with adult hypertension in developed countries (5, 10). The size of the U.S. population predisposed to hypertension when exposed to elevated sodium intake is not known, but about 20 percent of the U.S. population has hypertension. The American Heart Association advocates a sodium-restricted diet for the long-term management of hypertension (5, 10). Sodium-restricted diets are also required in the treatment of congestive cardiac failure, renal disease, cirrhosis of the liver, toxemia of pregnancy, and Meniere's disease and may be required for patients on prolonged corticosteroid therapy. Approximately 3 percent of the U.S. population are currently on prescribed low-sodium diets of either 2000, 1000, or 500 mg/day (5, 10). The fraction of sodium that can be contributed by drinking water varies according to diet. Since most foods contain some sodium, it is difficult to provide a nutritionally adequate diet without a daily intake of 440 mg of naturally occurring sodium in food. Where water supplies contain more than 20 mg/liter sodium, total dietary sodium restriction of less than 500 mg/day is difficult to achieve and maintain. Since some 40 percent of the U.S. community water supplies exceed 20 mg/liter sodium (84), many state health departments require water utilities to inform physicians and customers of the sodium concentration in drinking water to assist in the maintenance of low-sodium diets (7, 9).

It has been estimated that 40 percent of the U.S. population would benefit from a total daily intake of less than 2000 mg of sodium (5). The contribution of water to the desired total intake of sodium would be 10 percent or less for concentrations of 100 mg/liter or less in drinking water. The NRC feels that sodium should be maintained at the lowest practicable levels in water and that practices such as deicing of highways and water softening, which increase the concentrations of sodium in water supplies, should be discouraged (5, 10).

7.2.5. Arsenic

Arsenic is a geochemical contaminant and natural sources account for much of the arsenic found in surface and groundwaters. Natural sources include erosion of surface rock, natural deposits, volcanoes, and hot springs (5, 10). The presence of arsenic in soils is also related to its use as a pesticide (7). Arsenic concentrations in soils where arsenical sprays have been used for long periods range from one to several hundred mg/liter (7). Other environmental sources include coal-fired power plants and nonferrous smelting operations. Arsenic is found in many foods; crustaceans and other shellfish may contain up to 170 ppm (7).

In the United States, the arsenic concentration of drinking water ranges from a trace to approximately 0.1 mg/liter; higher levels (up to several mg/liter) have been found in wells in Indiana, Utah, Alaska, Oregon, and California (5, 7, 10, 85, 87, 88). Wells in Nova Scotia and the southwest coast of Taiwan have also been found to contain high levels of arsenic (89, 90). European bottled mineral waters have been found to contain up to 0.19 mg/liter of arsenic (91).

The chemical forms of arsenic consist of trivalent and pentavalent inorganic compounds and organic forms. In general, soluble trivalent arsenic compounds are more toxic than pentavalent, and inorganic compounds are more toxic than organic arsenicals (5, 10). Inorganic arsenic is readily absorbed through the gastrointestinal tract, lungs, and to a lesser extent, the skin; some organic forms are also absorbed through the intestine and skin. Arsenic trioxide is only slightly soluble in water and is not well absorbed; pentavalent arsenic, whether inorganic or organic, is better absorbed than the trivalent form.

Arsenic affects tissues that are rich in oxidative systems, primarily the alimentary tract, kidneys, liver, lungs, and epidermis. It is damaging to capillaries resulting in hemorrhage into the gastrointestinal tract, sloughing of mucosal epithelium, renal tubular degeneration, hepatic fatty changes, and necrosis (5, 10). In chronic exposures, arsenic accumulates primarily in bone, muscle, skin, and, to a lesser extent, in liver and kidneys (5, 10).

Epidemiologic studies have suggested an association between chronic exposure to arsenic and cutaneous lesions, cardiovascular disease, and cancer (5, 10). In Taiwan, an increased prevalance of skin cancer, hyperpigmentation, and keratosis and a higher rate of a peripheral vascular disorder resulting in gangrene of the extremities were seen in a population consuming well water with 0.6 mg/liter or greater concentration of arsenic (90, 92). Residents of Antofagasta, Chile, where the drinking water contained 0.8 mg/liter arsenic, showed an increased incidence of bronchial and pulmonary disease, cardiovascular pathology (Raynaud's syndrome), hyperpigmentation of the skin, and cutaneous lesions such as leukoderma, melanoderma, hyperkeratosis, and squamous-cell carcinoma (93–95). After water treatment that reduced arsenic levels to 0.08 mg/liter, the incidence of cutaneous lesions in Antofagasta decreased from 313 to 19 per 100,000 people per year. High incidences of human skin cancer have also been reported in Reichenstein, Silesia (96) and Cordoba, Argentina (97), where populations have been exposed to high levels of arsenic in drinking water.

7.2.6. Selenium

Selenium is found in water primarily as a result of leaching from rocks and soils that are high in selenium (5, 10). Water-soluble selenium has been identified in soils, some salt deposits, and in other geologic materials. There is a wide variation in the concentration of selenium depending on geological location. Groundwaters may contain significant amounts of selenium in areas where there is an excess of selenium in rocks and soils. Selenium can reach toxic concentrations in wells drilled through seleniferous shales. In seleniferous areas of Wyoming, wells contain enough selenium to be poisonous to man and livestock (5). A high concentration of selenate in well water on an Indian reservation in Wyoming was associated with loss of hair and nails in children (98). Members of an Indian family in Utah developed listlessness, weakened nails, and loss of hair after three months exposure to well water with 9 mg/liter selenium (7). Water from some springs and shallow wells were reported to contain more than 0.100 mg/liter selenium, but only a few micrograms per liter were found in deep wells (5). Some farm wells near Denver, Colorado had high selenium concentrations and 0.210 mg/liter selenium was reported in well water on a South Dakota Indian reservation (5). Water consumed by most populations, however, is relatively low in selenium and rarely constitutes a significant source of selenium (5, 10).

Except for limited reports of acute industrial or accidental exposures, effects on humans must be extrapolated from animal data (5, 10). Inorganic and organic forms of selenium are readily absorbed from the gastrointestinal tract of animals. Selenite and selenate are distributed largely to the liver, kidneys, muscle mass, gastrointestinal tract, and blood. Chronic administration of selenium results in an increased concentration in the testes.

Inhalation causes acute respiratory distress. Skin exposure causes severe local irritation and dermatitis, the severity of response depending on the chemical form of selenium. Chronic exposure of humans to high levels of selenium by inhalation or by ingestion results in central nervous system and gastrointestinal disturbances and dermatitis. In animals, acute exposure to selenium causes respiratory distress, diarrhea, pulmonary edema, hemorrhage, liver, and renal necrosis; chronic exposure results in death from gastroenteritis, myocardial damage, hydrothorax, pulmonary edema, renal and liver damage. Sodium selenite is toxic to rats at concentrations of 6–9 mg/liter in drinking water; concentrations of 1 mg/liter are without observed toxic effect. The toxicity of selenium can be altered extensively by interactions with sulfate, methionine, cystine, mercury, lead, zinc, cadmium, copper, arsenic, and vitamin E, but little is known about these interactions (5, 10).

Selenium is essential for domestic animals, but the margin of safety is relatively narrow (5, 10). A low level of selenium is essential to prevent myopathies, liver injury, and congenital abnormalities in domestic and laboratory animals and poultry. There is evidence that selenium may be an essential trace element for humans. An estimated adequate and safe intake of selenium for adults is felt to be between 0.500 and 0.200 mg with lower intakes for children and infants (10). The NRC (10) feels that the maximum no-observed-adverse effect level for selenium in water is in the range of 0.100–0.500 mg/liter. The current EPA interim standard of 0.01 mg/liter barely provides a minimum nutritional amount of selenium assuming consumption of two-liters/day of water as the standard. There has been concern that selenium is a carcinogen. The NRC (10) does not feel that the current scientific data justify this concern and suggests reevaluation of the current EPA standard based on minimum requirements of selenium.

7.3. ORGANIC CHEMICALS

7.3.1. Occurrence

Community and individual drinking water wells throughout the United States have been contaminated by synthetic organic compounds, often in concentrations (Table 7.17) much higher than are found in water supplies using the most contaminated surface water sources (2–4, 10, 99–102). There are no established

Table 7.17. Organic Compounds Detected in Drinking Water Wells (4)

Compound	Highest Reported Concentration (μg/liter)	Compound	Highest Reported Concentration (μg/liter)
Trichloroethylene	27,300	1,1-Dichloroethylene	280
Toluene	6,400	1,2-Dichloroethane	250
1,1,1-Trichloroethane	5,440	Bis (2-ethylhexyl) phthalate	170
Acetone	3,000	Dibromochloropropane	137
Methylene chloride	3,000	Trifluorotrichloroethane	135
Dioxane	2,100	Dibromochloromethane	55
Ethyl benzene	2,000	Vinyl chloride	50
Tetrachloroethylene	1,500	Chloromethane	44
Cyclohexane	540	Butyl-benzyl-phthalate	38
Chloroform	490	Lindane	22
Di-*n*-butyl-phthalate	470	1,1,2-Trichloroethane	20
Carbon tetrachloride	400	Bromoform	20
Benzene	330	1,1-Dichloroethane	7
1,2-Dichloroethylene	323	α-BHC	6
Ethylene dibromide	300	Parathion	4.6
Xylene	300	δ-BHC	3.8
Isopropyl benzene	290		

drinking water standards for most of these compounds (7, 8) and consumers generally cannot rely on unusual tastes and odors to identify contaminated wells.

Twenty-four states have reported finding one or more synthetic organic compounds in community groundwater supplies (2–4). Large-scale contamination of groundwater has been reported in New Jersey and California (2) and hundreds of wells have reportedly been closed because of organic contamination in high concentrations (2–4). In January 1980, 39 wells providing drinking water to some 400,000 people in the San Gabriel Valley, California were closed because of contamination with up to 600 μg/liter trichloroethylene (4). Approximately 100 drinking water wells surrounding a municipal landfill in Jackson Township, New Jersey were closed (3, 4) because of contamination with toluene (6400 μg/liter), methylene chloride (3000 μg/liter), acetone (3000 μg/liter), ethyl benzene (2000 μg/liter), trichloroethylene (1000 μg/liter), benzene (330 μg/liter), and chloroform (33 μg/liter). Local officials in Bedford, Massachusetts closed four wells that supplied 80 percent of the community's drinking water after discovering up to 2100 μg/liter dioxane and 500 μg/liter trichloroethylene in 1978 (4). Sixteen individual wells near an industrial waste handling facility built in 1972 to process waste oil from the Tamano spill in Casco Bay, Maine were closed in 1977 after it was discovered that they were contaminated with trichloroethane, trichloroethylene, freon, acetone, xylene, and dimethyl sulfide (3, 4). A number of domestic, stock, and irrigation wells near the Rocky Mountain Arsenal in Colorado have been abandoned because of contamination of 30-square-miles of a shallow water table aquifer with chemical byproducts from the manufacture of pesticides and herbicides (3, 4). In the 1930s, wells near a chemical company that produced and used creosote to treat wood products in St. Louis Park, Minneapolis, Minnesota were closed when the water tasted like tar; recently, some wells over a mile away were closed after suspect carcinogens and chemical byproducts of creosote production were detected in the groundwater (3, 4).

Thirty-nine communities using groundwater were included in EPA surveys for selected organics in drinking water (3, 4). Although the cities were not chosen because of potential groundwater contamination, 11 volatile chlorinated hydrocarbons were found in both treated and untreated groundwater (Table 7.18). Trichloroethylene was the chemical most frequently found. Eighteen states have begun monitoring (4) for some synthetic organic compounds in drinking water wells (Table 7.19). In general, the frequency of detection of the organic compounds found in these monitoring programs has not been greatly different from the EPA surveys, but concentrations of contaminants have been higher, as wells were sampled because of potential problems.

Several states have completed extensive sampling of groundwater. A survey that included all counties in Michigan identified 268 sites where groundwater was contaminated, 381 sites where contamination was suspected, and more than 50,000 sites where there is an activity with the potential to contaminate groundwater (103). A survey by the Massachusetts Commission of Water Supply found that at least one-third of the communities in the state have been affected to some degree by chemical contamination; individual and community wells were

Table 7.18. Selected Organic Compounds in the Drinking Water of Communities Using Groundwater as a Source of Water Supply (EPA Surveys) (3, 4)

Compound	Number of Communities Sampled	Percent with Detectable Levels	Concentration Range (μg/liter)	Concentration Mean (μg/liter)
Trichloroethylene	25	36	0.1–53.0	6.8
Carbon tetrachloride	39	28	0.2–13.0	3.8
Tetrachloroethylene	36	22	0.2–3.1	2.1
1,1,1-Trichloroethane	23	22	1.3–3.0	2.1
1,1-Dichloroethane	13	23	0.2–0.5	0.3
1,2-Dichloroethane	25	4	0.2	0.2
Trans-dichloroethylene	13	15	0.2–1.9	1.1
Cis-dichloroethylene	13	31	0.1–37.0	9.4
1,1-Dichloroethylene	13	8	0.2	0.2
Methylene chloride	38	3	7.0	7.0
Vinyl chloride	25	4	9.4	9.4

restricted or closed in 22 towns (104). All public water supply wells in Nassau County, New York were analyzed for six synthetic organic chemicals (4, 105). Tetrachloroethylene was found in 15 percent of the wells at concentrations up to 375 μg/liter; 1,1,2-trichloroethylene in 13 percent of the wells at concentrations up to 300 μg/liter; chloroform in 11 percent of the wells at concentrations up to 67 μg/liter; 1,1,1-trichloroethane, carbon tetrachloride, and trifluorotrichloroethane occurred in less than 10 percent of the wells at concentrations of 310, 21, 135 μg/liter, respectively. Twenty-three of 380 community wells were closed in Nassau County. A similar analysis in Suffolk County, New York

Table 7.19. Organic Compounds Reported in Eighteen State Surveys (4)

Compound	Number of Wells Sampled	Percent of Wells with Detectable Concentrations
Trichloroethylene	2894	14
Carbon tetrachloride	1659	18
Tetrachloroethylene	1586	13
1,1,1-Trichloroethane	1585	19
1,1-Dichloroethane	787	18
1,2-Dichloroethane	1218	7
Dichloroethylenes	781	23
Methylene chloride	1195	2
Vinyl chloride	1109	6

resulted in the closing of 13 of 120 community wells (4). A study of 30 public water systems elsewhere in New York for 112 organic contaminants (105) showed that the industrial chemicals bis(2-ethylhexyl)phthalate, toluene, and di-*n*-butyl phthalate to be present in 92, 85, and 54 percent of 39 wells in concentrations up to 170, 10, 470 μg/liter, respectively.

7.3.2. Health Effects

It is difficult at this time to know which of the organic groundwater contaminants are the most prevalent and represent important exposures. Existing data have been compiled primarily from surveys that have fortuitously included groundwater supplies and from the monitoring of wells with suspected contamination. In most cases, only selected chemicals have been analyzed. These data may not be representative of the actual overall problem and may be due to aggressive groundwater monitoring activities as well as the degree of industrial contamination in the affected areas. Surveys have recently been conducted to determine the geographical extent of the problem and the number of people affected in the United States, but these data are not yet available (4). It should, nevertheless, be remembered that any listing of organic compounds in groundwater is limited by the current state of the art of analytical technology and available funds to conduct surveys.

Some of the organic compounds detected have been identified as human or animal carcinogens, but not all have been tested for carcinogenicity (Table 7.20). Exposure to multiple contaminants has generally not been considered and adequate epidemiologic studies have not been conducted for populations consuming contaminated groundwater. Potential adverse health effects and risks must, therefore, be based on available data from acute and chronic toxicity studies in experimental animals, poisonings and accidental ingestion, clinical exposures, and occupational exposures. The available data from these sources have been summarized for nine organic compounds felt to be important groundwater contaminants at this time (3, 4, 10).

7.3.2.1. Trichloroethylene

Trichloroethylene (TCE) has been reported with varying frequency in groundwater from 14 states; TCE was detected in 79 percent of 19 wells tested in Delaware, but in only 2 percent of 1200 wells tested in Connecticut (4). A summary of 2984 samples of well water in 18 states showed that 14 percent of the well water supplies contained measurable levels of TCE.

TCE is slightly soluble in water and can be formed during the chlorination of well water. It is an industrial solvent used extensively for degreasing metal (3). In 1973, over 451 million pounds were produced in the United States (3). TCE is also used in dry-cleaning operations, organic synthesis, refrigerants, and fumigants (5). Most septic tank cleaning fluids contain TCE (4).

Table 7.20. Carcinogenicity of Selected Organic Compounds Detected in Groundwater (4)

1. Compounds carcinogenic in humans
   a. Benzene
   b. Vinyl chloride
2. Compounds carcinogenic in one or more species of animals
   a. Carbon tetrachloride
   b. Chloroform
   c. Dibromochloropropane
   d. 1,2-Dichloroethane
   e. Dioxane
   f. Ethylene dibromide
   g. Lindane
   h. Tetrachloroethylene
   i. 1,1,2-Trichloroethane
   j. Trichloroethylene
3. Compounds not tested for carcinogenicity in animals
   a. Bis (2-ethylhexyl) phthalate
   b. Butyl benzyl phthalate
   c. Chloromethane
   d. Cyclohexane
   e. Dibromochloromethane
   f. 1,1-Dichloroethylene
   g. 1,2-Dichloroethylene
   h. Di-*n*-butyl phthalate
   i. Methylene chloride
   j Toluene
   k. Trifluorotrichloroethane
   l. Xylene

The pharmacokinetics of TCE have been extensively studied in humans and experimental animals exposed by inhalation but not ingestion (5, 10). Stewart and Dodd (106) found that TCE penetrated intact human skin and it is anticipated that TCE is readily absorbed from the gastrointestinal tract, since it is a small, uncharged lipophilic molecule. Approximately 55 percent of inhaled TCE is retained in humans (10).

As evidenced from its use as an anesthetic, exposure to TCE results in depression of the central nervous system, incoordination, and unconsciousness (10). Acute exposure of experimental animals to TCE also results in depression of the central nervous system and inhibits cardiac function (10). There have been accounts of poisonings of individuals who ingested TCE; clinical signs and symptoms were primarily gastrointestinal upset, narcosis, and occasional cardiac abnormalities. Long-term studies of the toxicity of TCE in a variety of species of experimental animals and a limited number of human inhalation studies indicate that the toxicity of TCE is similar for acute and chronic exposures, but TCE does have the potential for accumulation in the body (10).

Epidemiologic studies of occupational exposures to TCE have reported primarily irritant and depressant effects; current studies are assessing whether there is an association between occupational exposure and cancer (10).

TCE was tested for carcinogenicity in a chronic feeding study of oral doses of 1097 and 549 mg/kg for male and female rats, 2339 and 1169 mg/kg for male mice, and 1739 and 839 mg/kg for female mice. Dose-related hepatocellular carcinomas were seen in male and female mice, but not in rats. However, questions have been raised about the study design and dose selection, feeding schedule, and housing of experimental animals in this study (10). Questions have also been raised about the true carcinogenicity of TCE, as the industrial grade TCE used in this study contained contaminants such as 0.2 percent epichlorohydrin which Donohue et al. (107) believe to be responsible for the carcinogenic response. A second investigation of the carcinogenic potential involved inhalation exposure of mice and rats for 6 hrs/day, 5 days/week to 100, 300, and 600 ppm TCE. A preliminary report of this study indicated a slight increase in hepatocellular carcinomas and adenomas in male mice at the highest vapor concentration; no liver tumors were observed in rats (10).

A review of these data by NRC resulted in a drinking water snarl of 105 mg/liter TCE for a 24-hr exposure and 15 mg/liter TCE for a 7-day exposure (10). These are based on a minimum effect level of 300 mg/kg, an uncertainty factor of 100, and the assumption that the only source of exposure is from a 2-liter/day intake of water. The snarls also ignore the possibility that TCE may be carcinogenic in the short term. A snarl for chronic, noncarcinogenic toxicity was not established, as the lowest dose of 549 mg/kg in the only chronic animal study produced renal injury and decreased life span (10).

7.3.3.2. Carbon Tetrachloride

Carbon tetrachloride has been found in chlorinated groundwaters serving as public drinking water in many states. It is felt that its presence is due primarily to impurities in chlorine used as a disinfectant (3). Twenty-eight percent of 39 treated groundwaters sampled in EPA surveys were found to be contaminated with carbon tetrachloride, but only 7 percent of untreated groundwater sources showed the presence of carbon tetrachloride. Wells in New Jersey and New York have been found to contain concentrations of 400 and 135 μg/liter, respectively (4).

Carbon tetrachloride is highly persistent. Over 1 billion pounds were produced in the United States in 1973 (3). It is used in the manufacture of chlorofluoromethanes, grain fumigants, cleaning agents, fire extinguishers, and solvents (3, 5). It is relatively insoluble in water.

Carbon tetrachloride is readily absorbed through the gastrointestinal tract and the lungs (5). Intestinal absorption is enhanced by fats, oils, and alcohol. Some individuals exposed to carbon tetrachloride will develop liver damage with little or no renal involvement, but others will exhibit renal failure with no hepatic disease. Chronic exposure usually results in nausea, vomiting, and nervous system symptoms such as headache, drowsiness, and excessive fatigue. Acute effects observed in both experimental animals and humans include renal dysfunction and hepatic nodular hyperplasia and cirrhosis. Biochemical and biologic lesions from chronic exposure of experimental animals include hepatic cirrhosis and necrosis, renal damage, changes in blood enzymes, and increased serum bilirubin. Carcinogenic bioassays of carbon tetrachloride after administration by several routes of exposure, including oral, have produced hepatomas in mice, rats, and hamsters. The hepatomas in most cases were associated with regenerative nodular hyperplasia or post-necrotic cirrhosis. There was no evidence of tumors in other organs within the time limits of the experiment.

Based on animal data that indicate 0.25 mg/kg to be the lowest dose that produces a toxic effect, an uncertainty factor of 1000, and assuming all exposure to carbon tetrachloride comes from a 2-liter/day intake of drinking water, a 24-hr snarl of 14 mg/liter and a 7-day snarl of 2 mg/liter were calculated by NRC (10). These calculations do not consider carcinogenic effects and no value was calculated for chronic exposure because carbon tetrachloride is a carcinogen in animals.

7.3.3.3. Tetrachloroethylene

Tetrachloroethylene was found in 22 percent of 36 community groundwater supplies sampled in the EPA surveys (3). Levels of 1500, 740, and 717 μg/liter have been detected in wells in New Jersey, Connecticut, and New York, respectively (4). Tetrachloroethylene is a solvent widely used in dry cleaning and degreasing operations. Over 700 million pounds were produced in the United States in 1973 (3).

The pharmacokinetics of tetrachloroethylene have been extensively studied in humans exposed to vapors but not orally (10). Tetrachloroethylene is rapidly absorbed through the lungs and skin and it is expected that it would be rapidly and completely absorbed from the gastrointestinal tract, as it is a small, uncharged, lipophilic molecule. Tetrachloroethylene accumulates in the body, primarily in adipose tissue at 3–4 times the rate of trichloroethylene (10).

Tetrachloroethylene is a potent depressant of the central nervous system, but residual organ damage has not commonly been observed in humans who have been exposed to large quantities of this compound (10). Tetrachloroethylene was formerly used widely as an intestinal anthelminthic compound. Toxicologic investigations with experimental animals have produced similar results as observed in humans.

With the exception of a recently conducted study of carcinogenic effect, investigations of the chronic toxicity of tetrachloroethylene have all involved inhalation exposure. Oral exposure of male and female mice and rats with doses of tetrachloroethylene ranging from 386–1072 mg/kg/day for up to 78 weeks resulted in a high incidence of nephrotoxicity but little evidence of hepatotoxicity. A high incidence of hepatocellular carcinomas at both high and low dose levels were reported in both genders of mice but not in rats; the NRC urges that these results be interpreted with caution because of limitations of the experimental design (10).

Using a dose of 490 mg/kg as a "minimum toxic dose," an uncertainty factor of 100, and assuming that the sole source of exposure of tetrachloroethylene will be from 2 liters/day of drinking water, NRC calculated a 24-hr snarl of 172 mg/liter and a 7-day snarl of 24.5 mg/liter (10). This ignores the possibility that tetrachloroethylene may be carcinogenic.

7.3.3.4. 1,1,1-Trichloroethane

1,1,1-Trichloroethane or methyl chloroform at a mean concentration of 2.1 μg/liter was found in 22 percent of 23 community water supplies using groundwater in EPA surveys (3, 4). Concentrations of over 5000 μg/liter were reported from wells in New York and Maine; 1600 μg/liter was detected in a well in Connecticut (4). Methyl chloroform is widely used as an industrial cleaner and degreaser of metals, spot remover, adhesive, and vapor pressure depressant. In 1978, over 600 million pounds were produced in the United States (3). Dioxane is commonly added to promote its stability (10).

The pharmacokinetics of inhaled methyl chloroform have been extensively studied, but there is little information on ingested methyl chloroform (10). It is felt that it is readily absorbed from the gastrointestinal tract, since it penetrates intact human skin. Methyl chloroform is one of the least toxic of the alkyl halocarbons and the relative lack of toxicity of methyl chloroform in comparison to certain other alkyl halocarbons has been attributed to its rapid elimination and stability (10).

Depression of the central nervous system is the primary effect in humans after acute high level exposure (10). Clinical experience suggests that acute high-level inhalation of methyl chloroform can adversely affect the human cardiovascular system. Overall results of animal experimentation are similar to findings in humans, and the toxicity of methyl chloroform appears to be similar for acute or chronic exposure (10). Extremely large quantities, up to 3000 mg/kg, given orally to mice and rats for 78 weeks showed little apparent histopathological change of any organ in either species; decreased weight gain, obvious ill health, and decreased survival time did occur (10). The only study of carcinogenic potential of ingested methyl chloroform failed to reveal evidence of carcinogenicity (10).

The NRC (10) has calculated a 24-hr snarl of 490 mg/liter and a 7-day snarl of 70 mg/liter; these are based on a minimum oral effect level of 1.4 mg/kg, an uncertainty factor of 100, and the assumption that the sole daily exposure is from 2 liters of drinking water. A chronic snarl of 3.8 mg/liter was based on the lowest dosage level of the chronic toxicity studies (a no-observed-adverse effect level was not administered), an uncertainty factor of 1000, and the assumption that 20 percent of the total intake of methyl chloroform is provided by drinking water.

7.3.3.5. 1,2-Dichloroethane

1,2-Dichloroethane (ethylene dioxide) is used in the manufacture of vinyl chloride and tetraethyl lead, as an insecticidal fumigant, in tobacco flavoring, in soap and scouring compounds, in wetting and penetrating agents, as a metal degreaser, and as a constituent of paint, varnish, and finish removers (3, 5).

Data from EPA surveys show that 4 percent of the 25 groundwater supplies sampled contained ethylene dioxide at a mean concentration of 0.2 μg/liter (3,

4). Concentrations of up to 300 μg/liter have been reported from wells in Massachusetts (4).

Ethylene dioxide is rapidly absorbed after oral or pulmonary exposure. A number of human fatalities resulting from ingestion of ethylene dioxide have occurred and the symptoms and signs of poisoning (nausea, vomiting, hemorrhagic lesions of most organs) have been similar whether the compound has been ingested or inhaled (10). Toxic manifestations of chronic exposure to low levels via inhalation or skin contact in occupational settings resemble those in individuals poisoned by acute exposure (10). These effects include changes in mood and appetite and depression among other neurological abnormalities, nausea, vomiting, gastric pain, and liver and kidney dysfunction. No acute or chronic snarls have been developed because there is little data on the quantities of ethylene dioxide consumed in nonfatal cases of ingestion. It is difficult to relate findings from inhalation or dermal exposure to chronic oral intake.

There is recent information that high doses of ethylene dioxide may be carcinogenic in experimental animals (10). A study of oral exposures ranging from 47–299 mg/kg/day for up to 78 weeks showed a dose-related incidence of squamous-cell carcinoma of the forestomach in male but not female rats, cancer in female mice at several other sites including mammary glands and endometrium, and alveolar/bronchiolar adenomas in both male and female mice.

7.3.3.6. 1,1-Dichloroethane

1,1-Dichloroethane is not reported to be produced commercially in the United States, but is imported for use as a solvent and cleaning agent in specialized processes. Very little information is available on the toxicity of 1,1-dichloroethane (3). 1,1-Dichloroethane has been reported to be present in well water from Alabama, Colorado, Florida, Massachusetts, North Carolina, Pennsylvania, and Tennessee (3).

7.3.3.7. Dichloroethylenes

Dichloroethylenes are a group of three isomers. *Cis* 1,2-dichloroethylene and *trans* 1,2-dichloroethylenes have not had wide industrial usage, but 1,1-dichloroethylene has (3). In 1976, it was estimated that 120,000 metric tons of 1,1-dichloroethylene were produced in the United States (3). 1,1-dichloroethylene is used for the production of methyl chloroform.

Very little information is available on the toxicity of the *cis* and *trans* 1,2-dichloroethylene. 1,1-dichloroethylene has been studied and exposure to high levels of this isomer can have adverse effects on the kidney and liver (3).

7.3.3.8. Methylene Chloride

Methylene chloride is used in the manufacture of paint and varnish removers, insecticides, solvents, pressurized spray products, and Christmas tree bubble lights. In 1973, over one-half billion pounds of methylene chloride were produced in the United States (3). Exposure to methylene chloride produces an elevation of the blood carboxyhemoglobin in humans. High levels of methylene chloride result in the formation of high levels of carboxyhemoglobin and impairment of the central nervous system (3). Three percent of the wells surveyed by the EPA have revealed the presence of methylene chloride (3, 4). Concentrations as high as 3000 μg/liter have been recorded.

Based on animal data that indicate 1 mg/kg as the minimum effect level, an uncertainty factor of 1000, and assuming all exposure to methylene chloride comes from 2 liters/day of drinking water, a 24-hr snarl of 35 mg/liter, and a 7-day snarl of 5 mg/liter were calculated by NRC (10). Because of the lack of adequate chronic exposure data, a chronic snarl was not calculated.

7.3.3.9. *Vinyl Chloride*

Vinyl chloride has been used for over 40 years in producing polyvinyl chloride. In 1975, nearly 5 billion pounds of vinyl chloride were produced in the United States (3). Vinyl chloride has been detected in nearly 16 percent of the untreated groundwater, but in only 4 percent of treated groundwater in EPA surveys (3, 4). It appears that because of its volatility much of this contamination never reaches the consumer at the water tap. Vinyl chloride is considered a human carcinogen. Some 48 cases of hepatic angiosarcoma have been diagnosed in vinyl chloride workers around the world (5). In experimental animal studies, vinyl chloride has induced a variety of tumors in addition to angiosarcoma; the oral route of administration has been more effective in producing angiosarcoma than has exposure by inhalation (5).

Acute exposure of humans to vinyl chloride has been shown to produce central nervous system dysfunction, sympathetic-sensory polyneuritis, and organic disorders of the brain (5). Lesions of the skin, bones, liver, spleen, and lungs have also been reported after chronic exposure (5).

7.4. RADIONUCLIDES

Radioactivity in public water supplies may be naturally occurring or from man-made sources such as nuclear power plants, fallout from nuclear weapons testing, and the use of radioisotopes in scientific laboratories, medicine, and industry (7, 108). Radium-226 is the most important of the naturally occurring radionuclides. It may occasionally be found in surface water due to man's activities, but it is usually found in groundwater as a result of geological conditions. In contrast, man-made radioactivity is found primarily in surface water.

It has been estimated that 1.1 million people in the U.S. consume water that contains more than 3 pCi/liter of radium-226 (5, 7). About 1 million people in the midwest, primarily in Iowa, Illinois, Wisconsin, and Missouri, are exposed to significant levels of radium-226 and radium-228 in groundwater (5). In Illinois and Iowa, some 500,000 people have well water supplies with radium-226 levels of 3–6 pCi/liter, about 300,000 people with 6–9 pCi/liter, and about 120,000 with 9–80 pCi/liter (109). The major contributors of α-emissions in groundwater are radium-226 and radium-228 daughter products; they also produce the major α-particle dose to the body, particularly the skeleton (5, 7).

The abundance of the radioactive gas radon-222, which is formed by the decay of radium-226, is not highly correlated with the radium concentration of water (5). Radon concentration is generally 1 pCi/liter in surface water, but activity concentrations in groundwater can be a few thousand times greater (5). Some mineral water may contain 500,000 pCi/liter (5). Consumption of water containing 1 μCi of radon-222 will result in a stomach dose of about 20 mrads (5). Because radon is a noble gas that is quickly released from water, it is possible that in some areas of high concentration it might cause significant inhalation exposure when water is used for showers, vaporizers, or humidifiers.

The limits established by the EPA for radioactivity in drinking water are 5 pCi/liter of combined radium-226 and radium-228, 15 pCi/liter of gross α-particle activity, and a total dose equivalent of 4 mrem/year for man-made radioactivity (7, 108). Uranium and radon are excluded from current regulations but may be included in the future (108). The 5 pCi/liter of radium produces a bone dose of 150 mrem/year and the dose for other α-particle emitters is estimated at no more than $^1/_5$ the value for radium (7). Risk estimates indicate that continuous consumption of water containing radium-226 or radium-228 at 5 pCi/liter may cause between 0.7 and 3 cancers per year per million exposed persons (7).

REFERENCES

1. A. Zaporozec, *Ground Water* **17,** 199 (1979).
2. *Proposed Ground Water Protection Strategy*, U.S. Environmental Protection Agency, Washington, D.C., 1980, 51 pp.
3. *Planning Workshops to Develop Recommendations for a Ground Water Protection Strategy*, U.S. Environmental Protection Agency, Washington, D.C., 1980, 170 pp.
4. *Contamination of Ground Water by Toxic Organic Chemicals*, Council on Environmental Quality, Washington, D.C., 1980, 84 pp.
5. Safe Drinking Water Committee, National Research Council, *Drinking Water and Health*, National Academy of Sciences, Washington, D.C., 1977, 939 pp.
6. G. F. Craun and L. J. McCabe, *J. Am. Water Works Assoc.* **67,** 593 (1975).
7. *National Interim Primary Drinking Water Regulations*, U.S. Environmental Protection Agency, EPA-570/9-76-003, Washington, D.C., 1977, 159 pp.
8. Interim Primary Drinking Water Regulations; Amendments, 40 CFR Part 141, *Federal Register*, **45,** 168, (Aug. 27, 1980), p. 57332.
9. *State Program Plans for Drinking Water*, U.S. Environmental Protection Agency, Drinking Water Branch, Region I, Boston, Mass., 1979.
10. Safe Drinking Water Committee, National Research Council, *Drinking Water and Health, Vol. 3*, National Academy Press, Washington, D.C., 1980, 415 pp.
11. *National Secondary Drinking Water Regulations*, U.S. Environmental Protection Agency, EPA-570/9-76-000, Washington, D.C., 1979, 37 pp.
12. A. S. Benenson, Ed., *Control of Communicable Disease in Man*, American Public Health Assn., Washington, D.C., 1975, 413 pp.
13. J. Mydlo, F. Fedak, S. Kramer, O. Niznikova, M. Manchacek, and E. Belicova, *Cesk. Epidemiol. Mikrobial. Immuniol.* **19,** 191 (1970).

14. T. Mittermayer, A. Pauerova, E. Belicova, A. Nickyova, and H. Smilnicka, *Cas. Lek. Cesk.* **115,** 354 (1976).
15. G. F. Craun and L. J. McCabe, *J. Am. Water Works Assoc.* **65,** 74 (1973).
16. G. F. Craun, *J. Water Pollut. Control Fed.* **46,** 1384 (1974).
17. G. F. Craun, *J. Water Pollut. Control Fed.* **50,** 1362 (1978).
18. P. A. Blake, M. L. Rosenberg, J. Florencia, J. B. Costa, L. P. Quintino, and E. J. Gangarosa, *Am. J. Epidemiol.* **105,** 344 (1977).
19. J. T. A. George, J. G. Wallace, H. R. Morrison, and J. F. Hasbourne, *Br. Med. J.* **3,** 208 (1972).
20. G. F. Craun, *Am. J. Public Health* **69,** 817 (1979).
21. Center for Disease Control, *Morb. and Mort. Wkly. Rpt.* **21,** 59 (1972).
22. G. F. Craun, *J. Am. Water Works Assoc.* **73,** 360 (1981).
23. G. F. Craun, *J. Environ. Health* **41,** 259 (1979).
24. G. F. Craun and R. A. Gunn, *J. Am. Water Works Assoc.* **71,** 422 (1979).
25. G. F. Craun, L. J. McCabe, and J. M. Hughes, *J. Am. Water Works Assoc.* **68,** 420 (1976).
26. *Water-Related Disease Outbreaks Annual Summary, 1979*, Center For Disease Control, Atlanta, GA., Sept. 1981, 23 pp.
27. J. B. Weisman, G. F. Craun, D. N. Lawerence, R. A. Pollard, M. S. Saslow and E. J. Gangarosa, *Am. J. Epidemiol.* **103,** 391 (1976).
28. M. L. Rosenburg, J. P. Koplan, I. K. Wachsmuth, J. G. Wells, E. J. Gangarosa, R. L. Guerrant, and D. A. Sack, *Ann. Intern. Med.* **86,** 714 (1977).
29. H. Inoue, *J. Jpn. Assoc. Infect. Dis.* **50,** 336 (1976).
30. L. J. McCabe and G. F. Craun, *J. Am. Water Works Assoc.* **67,** 95 (1975).
31. Center for Disease Control, *Morb. and Mort. Wkly. Rep.* **27,** 183 (1978).
32. W. B. Baine, C. A. Herron, K. Bridson, W. H. Barker, Jr., S. Lindell, G. F. Mallison, J. C. Wells, W. T. Martin, M. R. Kosuri, F. Carr, and E. Voelker, Sr., *J. Infect. Dis.* **133,** 588 (1976).
33. M. A. Horwitz, J. M. Hughes, and G. F. Craun, *J. Infect. Dis.* **133,** 588 (1976).
34. M. S. Saslaw, J. L. Nitzkin, R. Feldman, W. Baine, K. Pfeiffer, and M. Pearson, *Am. J. Public Health* **65,** 1184 (1975).
35. Pfeiffer, K. R., *J. Am. Water Works Assoc.* **65,** 803 (1973).
36. H. B. Greenburg, J. Valdesusa, R. H. Yolken, E. Gangarosa, W. Gary, R. G. Wyatt, T. Konno, H. Suzuki, R. M. Channock, and A. Z. Kapikian, *J. Infect. Dis.* **139,** 564, (1979).
37. G. F. Craun, *J. Water Pollut. Control Fed.* **52,** 1833 (1980).
38. E. Lycke, J. Blomberg, G. Berg, A. Ericksson, and L. Madsen, *Lancet* **2,** 1056 (1978).
39. D. M. Morens, R. M. Zweighaft, T. M. Vernon, G. W. Gary, J. J. Eslien, B. T. Wood, R. C. Holman, and R. Dolin, *Lancet* **1,** 964 (1979).
40. Center for Disease Control, *Morb. and Mord. Wkly. Rep.* **27,** 403 (1978).
41. G. F. Craun, *J. Water Pollut. Control Fed.* **51,** 1751 (1979).
42. J. L. Melnick, C. P. Gerba, and C. Wallis, *Bull. WHO* **56,** 499 (1978).
43. J. L. Melick and C. P. Gerba, *J. Infect. Dis.* **139,** 736 (1979).
44. D. R. Bamble, *Lancet*, **1,** 425 (1979).

45. T. F. Smith, *Ann. Intern. Med.* **91,** 119 (1979).

46. K. V. Eden, M. L. Rosenburg, M. Stoopler, B. T. Wood, A. K. Highsmith, P. Skaliy, J. G. Wells, and J. C. Feeley, *Public Health Rep.* **92,** 245 (1977).

47. S. G. Christiansen, *Acta.* Vet. *Scand.* **20,** 154 (1979).

48. G. K. Morris and J. C. Feeley, *Bull. WHO* **54,** 79 (1976).

49. T. F. Wetzler, "*Yersinia Enterocolitica* in Water Supplies," In *Proceedings 6th Annual Water Quality Technology Conference, Am. Water Works Assoc.*, Denver, Col., 1978, 2B-2, p. 1.

50. H. Bercovier, *Med. Mal. Infect.* **6,** 425 (1976); *Microbiol. Abs.* **12,** B8158 (1977).

51. J. M. Alonso, H. Bercovier, M. Bourdien, and H. H. Mollaret, *Med. Mal. Infect.* **6,** 425 (1976); *Microbiol. Abs.* **12,** B8158 (1977).

52. H. H. Mollaret, *Med. Mal. Infect.* **6,** 442 (1976); *Microbiol. Abs.* **12,** B8160 (1977).

53. L. Friberg, G. F. Nordberg, and V. B. Vouk, Ed., *Handbook on the Toxicology of Metals*, Elsevier/North-Holland Biomedical Press, Amsterdam, 1979.

54. Subcommittee on the Geochemical Environment in Relation to Health and Disease, National Research Council, *Geochemistry and the Environment, Vol. I*, National Academy of Sciences, Washington, D.C., 1974, 113 pp.

55. Subcommittee on the Geochemical Environment in Relation to Health and Disease, National Research Council, *Geochemistry and the Environment, Vol. II*, National Academy of Sciences, Washington, D.C., 1977, 163 pp.

56. Subcommittee on the Geochemical Environment in Relation to Health and Disease, National Research Council, *Geochemistry and the Environment, Vol. III*, National Academy of Sciences, Washington, D.C., 1978, 200 pp.

57. International Standing Committee on Water Quality and Treatment, International Water Supply Assoc., *Aqua* **1,** 5 (1974).

58. G. F. Craun, D. G. Greathouse, and D. H. Gunderson, *Int. J. Epidemiol.* **10,** 309 (1981).

59. L. J. McCabe, J. M. Symons, R. D. Lee, and G. C. Robeck, *J. Am. Water Works Assoc.* **62,** 670 (1970).

60. G. F. Craun, *J. Environ. Health* **44,** 20 (1981).

61. *Nitrites, Nitrates, and Methemoglobinemia*, National Institute of Environ. Health Sciences, Research Triangle Park, N.C., 1970, 43 pp.

62. J. W. White, Jr., *J. Agri. and Food Chem.* **23,** 886 (1975).

63. Am. Acad. Pediatrics Committee on Nutrition, *Pediatrics*, **46,** 475 (1970).

64. E. F. Winton, R. G. Tardiff, and L. J. McCabe, *J. Am. Works Assoc.* **63,** 95 (1971).

65. M. Cornblath and A. F. Hartman, *Pediatrics* **33,** 421 (1948).

66. Z. Knoteck and P. Schmidt, *Pediatrics* **34,** 78 (1966).

67. O. Batik, "A Study of the Occurrence of Methemoglobinemia in Areas Where Drinking Water is High in Nitrate," U.S. Environmental Protection Agency, Cincinnati, Ohio, 1979.

68. H. H. Comly, *J. Am. Med. Assoc.* **129,** 112 (1945).

69. G. Walton, *Am. J. Public Health* **41,** 986 (1951).

70. J. Vigil, S. Warburton, W. S. Haynes, and L. R. Kaiser, *Public Health Rep.* **80,** 1119 (1965).

71. J. R. Goldsmith, S. N. Rokaw, and L. A. Shearer, *Int. J. Epidemiol.* **4,** 207 (1975)

72. L. A. Shearer, J. R. Goldsmith, C. Young, O. A. Kearns, and B. R. Tamplin, *Am. J. Public Health* **62,** 1174 (1972).
73. R. S. Grant, "Well Water Nitrate Poisoning Review: A Survey in Nebraska 1973–1978," Nebraska Dept. of Health, Omaha, Nebraska, 1979.
74. J. H. Jones, S. H. Thomas, G. W. Schoenhals, N. R. Grantham, and H. D. Riley, Jr., *J. Okla. State. Med. Assoc.* **66,** 60 (1973).
75. World Health Organization, *Fluorides and Human Health*, WHO Monograph Series No. 59, 1970, 364 pp.
76. *Natural Fluoride Content of Community Water Supplies*, U.S. Dept. of Health, Education, and Welfare, Washington, D.C., 1969.
77. J. Kobayashi, *Ber. Ohara. Inst. Landwirt. Biol.* **11,** 12 (1957).
78. National Research Council, *Geochemistry of Water in Relation to Cardiovascular Disease*, National Academy of Sciences, Washington, D.C., 1979.
79. L. C. Neri, D. Hewitt, and G. B. Schreiber, *Am. J. Epidemiol.* **99,** 75 (1974).
80. L. C. Neri, D. Hewitt, and G. B. Schreiber, *J. Am. Water Works Assoc.* **67,** 403 (1975).
81. G. W. Comstock, *Am. J. Epidemiol.* **94,** 1 (1971).
82. Food and Nutrition Board, National Academy of Sciences, *Recommended Dietary Allowances*, 9th Rev. Ed., National Academy of Sciences, Washington, D.C., 1980.
83. L. K. Dahl, *N. Eng. J. Med.* **258,** 1152 (1958).
84. J. M. White, J. C. Wingo, L. M. Alligood, G. R. Cooper, J. Gutridge, W. Hydaker, R. T. Benack, J. W. Dening, and F. B. Taylor, *J. Am. Diet Assoc.* **50,** 32 (1967).
85. P. D. Whanger, P. H. Weswig, and J. C. Stoner, *Environ. Health Perspect.* **19,** 139 (1977).
86. J. R. Goldsmith, M. Deane, J. Thom, and G. Gentry, *Water Res.* **6,** 1133 (1972).
87. J. M. Harrington, J. P. Middaugh, D. L. Morse, and J. Housworth, *Am. J. Epidemiol.* **108,** 377 (1978).
88. J. W. Southwick, A. E. Western, M. M. Beck, T. Whitley, and R. Issacs, "Community Health Associated With Arsenic in Drinking Water in Millard County, Utah," U.S. Environmental Protection Agency, EPA-600/1-81-064, Washington, D.C., 1981
89. D. A. Grantham and J. F. Jones, "Arsenic Contamination of Water Wells in Nova Scotia, " In Conference of the Atlantic Canada Section, Am. Water Works Assoc. and Federation of Assoc. on Canadian Environment, Halifax, Nova Scotia, Sept. 12–15, 1976.
90. W. P. Tseng, *Environ. Health Perspect.* **19,** 109 (1977).
91. B. C. J. Zoeteman and F. J. J. Brinkman, "Human Intake of Minerals from Drinking Water in the European Communities," in R. Amavis, W. J. Hunter, and J. G. P. M. Sheets, Eds., *Hardness of Drinking Water and Public Health*, Pergamon Press, New York, 1977.
92. W. P. Tseng, H. M. Chu, S. W. How, J. M. Fong, C. S. Lin, and Shu Yeh, *J. Nat. Cancer Inst.* **40,** 453 (1968).
93. J. M. Borgono and R. Greiber, "Epidemiological Study of Arsenicism in the City of Antofagasta," In D. D. Hemphill, Ed., *Trace Substances in Environ. Health – V*, University of Missouri, Columbia, MO., 1972.

94. R. Zaldivar, *Beitz Pathol. Bd.* **151,** 384 (1974).

95. J. M. Borgono, P. Vicent, H. Venturino, and A. Infante, *Environ. Health Perspect.* **19,** 103, 1977.

96. L. Geyer, *Arch. Dermatol Syphilol.* **43,** 221 (1898).

97. R. M. Bergoglio, *Prensa. Med. Argent.* **51,** 994 (1964).

98. O. A. Beath, *Sci. News Letter* **81,** 254 (1962).

99. R. S. Woodhull, *J. Am. Water Works Assoc.* **73,** 188 (1981).

100. A. A. Guerrera, *J. Am. Water Works Assoc.* **73,** 190 (1981).

101. J. C. Petura, *J. Am. Water Works Assoc.* **73,** 200 (1981).

102. F. B. DeWhalle and E. S. K. Chian, *J. Am. Water Works Assoc.* **73,** 206 (1981).

103. Michigan Department of Natural Resources, *Assessment of Groundwater Contamination in Michigan*, Lansing, Michigan, 1979.

104. Special Legislative Commission on Water Supply, Commonwealth of Massachusetts, *Chemical Contamination*, Boston, Mass., 1979. 72 pp.

105. N. K. Kim and D. W. Stone, *Organic Chemicals and Drinking Water*, State Department of Health, Albany, New York, 1980.

106. R. D. Stewart and H. C. Dodd, *Am. Ind. Hyg. Assoc.* **25,** 439 (1964).

107. E. V. Donahue, J. McCann, and B. N. Ames, *Cancer Res.* **38,** 431 (1978).

108. *Radioactivity in Drinking Water*, U.S. Environmental Protection Agency, EPA-570/9-81-002, Washington, D.C., 1981, 76 pp.

109. R. E. Rowland, H. F. Lucas, Jr., and A. F. Stehney, "High Radium Levels in the Water Supplies of Illinois and Iowa," In T. L. Cullen and L. P. Franca, Eds., *Proc. Int. Symp. Areas of High Natural Radioactivity*, Academia Brasileira de Ciencias, Rio de Janeiro, 1977.

8

MICROBIOLOGICAL ASPECTS OF GROUNDWATER POLLUTION DUE TO SEPTIC TANKS

C. Hagedorn

Department of Agronomy
Mississippi State University, Mississippi State, Mississippi

8.1. SEPTIC-TANK SYSTEMS IN WASTE DISPOSAL

Domestic wastewaters contain bacteria, viruses, protozoa, and helminths pathogenic to humans (1). In addition, these infectious agents are widely distributed in many waste effluents and are commonly present in high numbers. Therefore, untreated domestic wastes embody a potential health hazard, and proper wastewater purification and disposal is an important concern (2). Septic-tank soil-absorption treatment systems are the principal disposers of waste effluent to the soil environment (3). The Office of Water Supply (4), in a U.S. Environmental Protection Agency (EPA) report to Congress, estimated that three billion cubic meters of domestic wastewater are annually placed in soil subsurface horizons. Soil percolation of septic wastes is required for the purification of drainfield effluent before it replenishes groundwater utilized for individual and public water supply wells. However, many shallow groundwater supplies have been polluted by contaminated recharge waters. Several investigators have attributed the decline in water quality to indiscriminate use of septic tank systems in soils unsuited for adequate domestic waste purification (5–10). Septic-tank soil-absorption systems are not confined to rural areas of low population density. There are four counties (Nassau County, New York, Suffolk County, New York, Dade County, Florida, and Los Angeles County, California) in the United States with more than 100,000 housing units served by septic tanks, as well as 23 additional counties with more than 50,000 household installations (3).

8.2. BACTERIAL MODIFICATION OF SEPTIC-SYSTEM EFFLUENTS

On-site sewage treatment systems are composed of two individual units, the septic tank which functions under anaerobic conditions and the percolation field which operates aerobically under unsaturated flow. The septic tank is designed to slow down the movement of raw sewage and to promote the removal of solids either by settling or liquefaction. The organic load and fecal bacterial populations are reduced only to a limited extent. As an example of the minimum purification afforded by passage through the septic tank, Ziebell et al. (11) enumerated selected fecal bacterial populations in effluents from five systems and observed mean population densities of $3.4 \times 10^6/100$ ml for total coliforms, $4.2 \times 10^5/100$ ml for fecal coliforms, $3.8 \times 10^3/100$ ml for fecal streptococci, and $1.0 \times 10^4/100$ for *Pseudomonas aeruginosa.* Also, BOD_5 reductions of only 30–50 percent were noted in all cases. Therefore, distribution of the septic-tank effluents into unsaturated soil appears to be necessary to complete the treatment process. The soil must furnish the bulk of the wastewater treatment by the processes of physical filtration, chemical reaction, and biological transformations (12).

The most comprehensive study on soil absorption of septic-system effluents has been conducted by the Small Scale Waste Management Project Group at the University of Wisconsin. In the report of investigations on the purification efficiency of 19 subsurface soil disposal systems, Bouma et al. (13) concluded that septic systems which exhibited proper hydraulic functioning also served to purify septic tank effluents. Bacterial filtration was determined by dissecting a drainfield and enumerating indicator organisms present in the soil at various distances below the drainfield trench. The large populations of total coliforms, fecal coliforms, and enterococci present in the septic-tank effluent were reduced to levels associated with control soil samples within two feet below the percolation trench. In addition, the most abrupt population declines occurred in the "biological mat or clogged zone" located at the interface of the drainfield trench and the soil. The importance of this clogged zone in the proper functioning of septic-tank drainfield systems was emphasized. Since this clogged layer, which is only a few centimeters in thickness, was highly efficient in trapping and holding bacterial species present in the wastewater, it served as the primary barrier to subsurface escape of fecal organisms. However, if the mat became too thick or dense and restricted the hydraulic functioning of the system, then the effluent could not enter the soil, became ponded in the trench, and subsequently spilled out onto the soil surface. Shallow wells were installed strategically in and near the absorption fields and groundwater samples were assayed for indicator organism density and total bacterial counts. Although total bacterial counts were higher in samples taken from wells adjacent to the absorption field and downfield in the direction of groundwater flow, in most cases, no indicator organisms were detected in these samples. Thus, these results indicated that groundwater collected adjacent to these treatment systems rendered a minimal health hazard.

A reduction in coliform and streptococcal indicator bacteria numbers with depth below the distribution trench was also observed by Ziebell et al. (11). Upon dissection of a "typical" absorption field located in a sandy soil, indicator bacteria enumerations showed that the high levels of organisms present in the septic-tank effluent were increased approximately fivefold in the soil of the clogged zone at the base of the trench. However, soil samples collected 30-cm lateral to and 8-cm below the trench produced coliform levels 100-fold less than the septic-tank effluent, and samples collected 30-cm lateral to and 38-cm below the trench yielded values 3000-fold less than in the tank effluent. Therefore, in a properly functioning absorption field, the fecal indicator and potentially pathogenic bacteria were almost completely removed after a relatively short travel through unsaturated soil.

Viraraghavan and Warnock (14) constructed an experimental tile line into which household septic-tank effluent was diverted to further investigate the efficacy of the soil-absorption system. Porous cup sampling probes were placed in a row approximately 0.76-m adjacent and parallel to the test tile line and groundwater sampling wells were installed at increasing distances downgradient. The results indicate that unsaturated soil removes a high percentage of the indicator organisms present in the sewage effluent. Also, groundwater samples show that the fecal populations that were recovered from the row of wells decreased with increased distance from the septic tile. More recently, Brown et al. (15) reported the results of a two-year study on the migration of effluent bacterial populations into undisturbed soils with 7.6, 41, and 80 percent sand. These soils were enclosed in lysimeters of 153 $\times$ 204 $\times$ 180-cm deep and septic effluent was metered equally into a 15-cm diameter drain tile enclosed in a 30-cm-square layer of gravel located 30 cm below the soil surface of each lysimeter. Weekly soil leachate samples were collected through porous cups located 120 cm below the septic tile line and the fecal coliform density was determined in the leachate samples. Of the 545 composites analyzed, only 26 were positive for fecal coliforms and the majority of these occurred near the beginning of effluent application to the tile lines. In some cases, however, fecal coliforms exhibited higher counts in isolated locations distant from the tile line and thus appeared to move through the soil or survive in root channels and soil cracks. However, Dazzo and Rothwell (16) reported that porcelain porous cups retained coliforms since as much as 65 percent of the cups in their experiments yielded coliform-free samples. Therefore, the results of the two previous reports should be viewed with caution.

In the Wisconsin Project (13), absorption field dissection results produced a similar distribution of fecal indicator populations in the unsaturated soil beneath the drainfield trench. This delineated a soil transition zone from high bacterial populations in the clogged region that were equal to or exceeded septic-tank effluents to reduced population levels from deeper zones that were equivalent to unpolluted soil. In addition, the size of the transition zone was an indicator of the soil depth required for retention of the wastewater microorganisms and, hence, the depth of soil necessary for purification of the effluent. Based on this, approximately 30–90-cm of soil beneath the base of the drainfield trench was adequate for complete bacterial removal of septic-tank effluents provided the soil had both a layer permeable to effluent flow and another region adequately restrictive to form a clogged zone. Therefore, published guidelines for the installation of conventional septic systems (17) that recommend from 120–150-cm of "suitable" soil as an adequate zone for the protection of groundwater fall well in line with experimental data.

Similar results have been reported for a wide diversity of microorganisms associated with septic-filter systems. Kristianson (18) demonstrated that rods and cocci dominated the bacterial flora of sand filter trenches while cocci and small rods penetrated the entire profile in a slightly clogged system but were largely retained in a clogged system. This was supported by earlier studies (19) that showed that, as clogging increased, anaerobic and facultative bacteria predominated while aerobic types were found in large numbers only in highly aerated systems.

8.3. BACTERIAL TRANSPORT THROUGH SOILS

The soil depths required for removal of enteric bacteria from septic-tank effluents have been derived from a somewhat limited body of information. Indeed, research investigations on the purification efficiency of a much wider variety of soils subjected to many different waste application techniques result in less close agreement on the degree of bacterial retention by soil. The land application of human wastes includes topics that range from the disposal of raw excreta in pit latrines to application of oxidation-pond effluents into rapid-infiltration basins. In addition, investigations of these various disposal methods serve as the body of knowledge necessary to further define the conditions in which wastewaters are purified by land application, and, as such, have been the subject of several reviews (20–23).

Although a wide diversity of disposal methods have been studied (Table 8.1) and large variations were observed in the results, some pertinent conclusions can be ascertained from this information: (1) coliforms and other microorganisms move only a few dozen cm with the percolating waters in unsaturated soil layers, although much greater distances are possible under saturated flow conditions; (2) with all soil water conditions, the degree of bacterial retention by the soil is inversely proportional to the size of the component particles in the unstructured matrix; (3) the physical straining or filtration of organisms by soil particles are the main limitation to travel through soils, and sedimentation of bacterial clusters occurs throughout the zone of saturated flow; (4) adsorption is a factor in the retention of bacteria by soil and becomes more effective in soils with increasing clay content; (5) death of the microorganism plays an important role during the longer retention periods or unsaturated flow conditions. Thus, both physical removal and inactivation influence the change in bacteriological quality of wastewaters percolating through the soil.

8.4. BACTERIAL MIGRATION FROM SEPTIC-TANK ABSORPTION FIELDS

Precipitation has been shown to contain less than one coliform/100 ml sample, possibly derived from dust, pollen grains, or insects. Yet once this water reaches the earth, contamination occurs from numerous sources in the urban and rural environments (41). Septic systems submerged in stormflow saturated soils represent one potentially large source of fecal contamination; however, the magnitude of such contributions have only been assessed to a limited extent. Reneau et al. (42) investigated the interactions between soils of various suitability for septic-tank effluent disposal and seasonal perched water tables on the bacterial contamination of a watershed from septic sources. The watershed covered an area of 80 ha and contained within its boundaries approximately 30 single-family dwellings and an elementary school. The watershed soils were classified according to their suitability for septic-system effluent disposal as suitable, marginal, and unsuitable; of the total septic systems placed in this area, 17 percent were in suitable, 41 percent in marginal, and 42 percent in unsuitable soils. Nine surface and seven subsurface sampling sites were established throughout the watershed. Samples were collected periodically during the first halves of the years 1972 and 1973, when streamflow was continuous and seasonal water tables were present. The bacterial contamination of the watershed from septic systems was evaluated from the total fecal coliform density in the water samples. Bacterial enumeration of a surface water sample that was subjected to minimal contamination yielded fecal coliform most probable numbers (MPN) ranging from 3–930/100 ml whereas a surface sample in direct contact with septic effluent contained from 5.3×10^2–2.4×10^7 fecal coliform MPN/100 ml. Also, surface samples not in contact with any effluent source had an average fecal coliform MPN of 2.3×10^2/100 ml, while the

Table 8.1. A Summation of the Results of Selected Studies on the Transport of Bacteria Through Soil in Relation to Land Application of Domestic Wastewaters

Nature of Pollution	Organisms	Medium	Maximum Distance Traveled	Time of Travel	Reference
Sewage trenches intersecting ground water	*Bacillus coli*	Fine sand	19.8 m (65 ft)	27 wks	Stiles and Crohurst (24)
Sewage trenches intersecting ground water	*Coliforms*	—	70.7 m (232 ft)	—	Warrick and Muegge (25)
Sewage in pit latrine intersecting ground water	*Bacillus coli*	Fine and coarse sand	24.4 m (80 ft)	—	Caldwell (26)
Sewage in bored latrine intersecting ground water	*Bacillus coli*	Sand and sandy clay	10.7 m (35 ft)	8 wks	Caldwell and Parr (27)
Sewage in pit latrine intersecting ground water	*Bacillus coli*	Fine and medium sand	3.1 m (10 ft)	—	Caldwell (28)
Primary and treated sewage in infiltration basins	*Coliforms*	Fine sandy loam	0.6–4 m (2–13 ft)	—	Butler et al. (29)
Diluted primary sewage injected subsurface	*Coliforms*	Aquifer	30 m (98 ft)	33 hrs	McGauhey and Krone (20)
Canal water in infiltration basins	*Escherichia coli*	Sand dunes	3.1 m (10 ft)	—	Baars (30)

Subsurface injection	*Enterococci*	—	15 m (44 ft)	—	Fournelle (31)
Secondary sewage in infiltration basins	*Coliforms*	Sandy gravels	0.9 m (3 ft)	—	McMichael and McKee (32)
Tertiary treated wastewater in percolation beds	*Fecal coliforms and fecal streptococci*	Coarse gravels	457.2 m (1500 ft)	15 da	Merrell (33)
Primary sewage injected subsurface	*Coliforms*	Sand and pea gravel aquifer	30.5 m (100 ft)	35 hr	Krone et al. (34)
Secondary sewage injected subsurface	*Fecal coliforms*	Fine to coarse sand aquifer	30.5 m (100 ft)	—	Wesner and Baier (35)
Tertiary treated wastewater in percolation beds	*Coliforms*	Sand and gravel	830 m (2723 ft)	—	Anan'ev and Demin (36)
Inoculated water and diluted sewage injected subsurface	*Bacillus stearothermophilis*	Crystalline bedrock	28.7 m (94 ft)	24–30 hrs	Allen and Morrison (37)
Tertiary treated wastewater in infiltration basins	*Coliforms*	Fine to medium sand	6.1 m (20 ft)	—	Young (38)
Secondary sewage in infiltration basins	*Fecal coliforms*	Fine loamy sand to gravel	9.1 m (30 ft)	—	Bouwer et al. (39)
Primary sewage in infiltration basins	*Fecal streptococci*	Silty sand and gravel	183 m (600 ft)	—	Schaub and Sorber (40)

watershed outflow produced a mean fecal coliform MPN of $5.6 \times 10^3/100$ ml. Subsurface waters generally displayed a decrease in total and fecal coliform MPN with increased distance from individual septic-tank soil-absorption installations. Contamination of the watershed outflow was particularly noticeable during or following a rainstorm when septic-tank effluent was flushed by movement of runoff water into surface flow. Thus, this watershed, which contained septic tank systems installed in unsuitable soil, yielded runoff waters of marginal bacterial quality.

Reneau and Pettry (43) continued this research based on the prior observation that coliform and fecal coliform numbers decreased with increasing subsurface distance from a septic drainfield. Three septic tank drainfields located on different soils were each instrumented with a series of piezometers at selected distances from the drainfield in the direction of groundwater flow. The soils, located on the western edge of the coastal plain in Virginia, were considered marginally suitable for the disposal of domestic wastes. In addition, these soils, underlain by layers restrictive to water movement, developed perched water tables during periods of seasonally high rainfall. Total and fecal coliforms present in the groundwater were enumerated and the results indicated that although there was a large reduction in bacterial numbers with horizontal distance from the drainfield, high bacterial concentrations were present at all sampling locations. Thus, at 28 m from the drainfield in the Beltsville soil, up to 4.3×10^2 fecal coliform MPN/100 ml were recovered; in the Goldsboro soil, 2.9×10^3 fecal coliform MPN/100 ml were recovered at 13.5-m distance, while in the Varian soil 1.1×10^5 fecal coliform MPN/100 ml were recovered at 6.1 m from the drainfield. In all cases, the maximum coliform density was observed in the saturated zone directly above the restrictive layers present in these soils and abrupt decreases in bacterial numbers were observed below these layers. Since few coliforms were observed at the deeper depths in these field sites and then only on a limited number of sampling dates, the authors concluded that the groundwater supply was not being contaminated by vertical movement of fecal bacteria. Yet very large surface areas were required to remove the fecal indicator organisms from effluent as it penetrated horizontally through the soil and all possible routes of groundwater recharge were not investigated. This demonstrated the extended distances that indicator organisms could be transported under saturated flow conditions as compared to unsaturated flow situations.

Artificial drainage has been shown to improve the hydraulic functioning of septic systems located in soils underlain by slowly permeable soil horizons and subjected to seasonally high water tables. Yet health guidelines that specify the minimum distance between the disposal area and intercepting tile lines or drainage ditches were established intuitively with little data support. To address this situation, Reneau (44) examined the influence of artificial drainage on the penetration of total and fecal coliforms into wet, tile drained soils and drew conclusions to assess the potential health hazards. Sampling wells were installed (at three household systems) to depths of approximately 1.5 m in the area between the disposal field and the drainage tile. The wells were arranged in two rows positioned perpendicular to the septic system drainfield and were composed of four wells per row, placed at increasing distance from the disposal trench, with the final well located on the opposite side of the drainage tile to collect uncontaminated samples. The drainage tiles were located 11.9 m from the disposal trench in site 1, 11.6 m at site 2, and 19.2 m at site 3. Water samples collected during high water table periods displayed a rapid decrease in total and fecal coliforms with increased distance from the disposal area. However, waters in the drainage tiles contained considerably greater total and fecal coliform populations than the control well samples, indicating that bacteria in saturated soil could travel long distances through horizontal penetration. In addition, an adjacent stream receiving the tile flow contained moderately high concentrations of total and fecal coliforms and reflected the influx of septic-effluent bacterial populations. Based on the data from this study, the author constructed several hypothetical situations to demonstrate the discharge of fecal organisms into receiving streams from various drainage systems and emphasized the necessity for maintaining an adequate distance between the disposal area and intercepting tile lines.

Viraraghavan (45) positioned a row of 2-m-deep groundwater sampling wells at distances of 0, 3, 6, 9.2, 12.2, and 15.2-m downgrade from the end of an experimental septic tile line. Viraraghavan and Warnock (14) had used this same experimental set-up in their earlier report and noted efficient removal of fecal microorganisms from septic effluents during periods of low water tables. However, the experimentation in the 1978 study was conducted during periods of elevated watertables when only 15 cm of unsaturated soil occurred below the drainfield trench. The results of enumeration of total coliforms, fecal coliforms, and fecal streptococci from groundwater samples indicated that these populations also experienced a decline with distance from the septic tile line. Yet fecal coliform numbers at a distance of 15.2 m from the tile line were between 10^2–10^4/100 ml sample.

These reports have established that fecal microorganisms present in septic effluents have the potential to penetrate saturated soil, and under certain conditions, to be transported various distances through the soils presumably via saturated flow. However, the hydrologic processes responsible for transporting fecal organisms through the soil were not investigated in these static system studies.

A preliminary experiment to examine the dynamics of the subsurface migration of fecal organisms from failing septic systems was conducted by Hagedorn et al. (46). Antibiotic-resistant strains of *Escherichia coli* and *Streptococcus faecalis* were placed (as biological tracers) into gravel pits positioned in a Veneta soil series located on 2 percent slope. Sampling wells were located radially around the pit at 0.5 and 1 m, and additional wells were located at up to 15 m from the pit in the direction of groundwater flow. The characteristics of tracer penetration through the saturated soil profile was determined by enumerating the resistant organisms present in the groundwater samples at successive intervals after amendment. The results demonstrated that tracer organisms were intercepted and carried from the gravel pit by saturated flow since successive peaks of bacterial numbers at increasing distance from the pits appeared only after periods of precipitation creating elevated water tables. Although the bacterial density peaks were lessened at increasing sampling distances from the pit, 10^3–10^4 cells/ml were detected at even the furthest sampling distance. Also, a second rainfall event produced a secondary set of peaks in the enumeration of tracer vs. time; the bacterial numbers were not as great, however. In addition, the organisms survived in the soil water at relative high concentrations for up to 32 days after introduction into the soil.

In a more detailed study, Rahe et al. (47) established simulated drainfield installations on two hillslope soils and monitored downslope bacterial penetration through saturated zones. Antibiotic-resistant *E. coli* were introduced into saturated soil through a horizontal injection line representing a drainfield tile, and groundwater was collected at successive periods of time from an array of sampling piezometers located at various depths and distances downslope. In both sites, the flow of tracer bacteria through the soil was characterized by a rapid penetration of the saturated soil with high numbers of organisms contributing to the peak flow. The bulk of bacterial transport through the soil occurred in specific depth-zones, with the primary transmitting zone located directly above the water-restrictive layer at one site and near the surface layers in the other. These results demonstrated that pathogenic species potentially present in septic effluents could be disseminated through soil in high numbers, over large distances, and after relatively short periods of time. Since bacterial translocation was largely confined to specific zones, Rahe et al. (47) proposed that these highly permeable areas actually represented regions in the soil that contained a large volume of macropores and that partial displacement through these pores accounted for the rapid flow rates.

Using a longitudinal dispersion equation to interpret the dispersal of *E. coli* cells through a porous medium, Hagedorn and McCoy (48) found that the distinctive trends in bacterial numbers recovered over time supported the hypothesis that nearly complete displacement of the amended material had occurred with incoming water. Rapid subsurface movement of the amended bacteria was observed over short time periods for several different soils. The elevated water tables flowed downslope and in the majority of situations entered into a stream or intermittent waterway, while at a few sites, the water table surfaced downslope and resulted in surface flow into streams and reservoirs. The public health concerns in such situations where drainfield systems may become immersed by elevated water tables are obvious.

McCoy and Hagedorn (49) demonstrated that introduced *E. coli* populations were transported downslope in specific zones in a saturated soil profile and at an apparent maximum velocity of 17.0 cm/min. A mathematical relationship was derived that described the overall decrease in numbers of organisms with increased distance through the soil. Their results emphasized the need for a large zone of unsaturated soil around waste treatment systems to prevent the release of large numbers of fecal microorganisms.

8.5. VIRUS REMOVAL FROM WASTE EFFLUENTS

Numerous studies have shown that many soils are effective at removing viruses although most of this data has been gathered from laboratory experiments. Much confirmatory field work is still needed. Because viruses are electrically charged colloidal particles that can adsorb to charged surfaces, considerable experimentation concerning this aspect of virus retention has been described (50; see also Chapter 4). Current knowledge about the soil filtration or removal of viruses as well as adsorption mechanisms has largely been derived from column studies (51, 52) or laboratory batch experiments (53, 54). These batch experiments have proven useful in demonstrating how some factors (i.e., ionic strength) affect virus adsorption although many of these studies have been inadequate because the laboratory conditions were often too artificial. Virus saturation in small systems has often made certain soil characteristics appear limiting when, in fact, the results could not be repeated with soil columns because virus saturation is difficult to attain in soil systems (52).

There is some evidence from both field and column studies which indicates that the most important characteristic in virus retention is effluent flow velocity (55). For adsorption to be effective, close contact between soil and virus particles is essential and waste disposal in soil must be done in a manner so that this contact can occur. Situations must be avoided where subsurface transport over long distances might result, such as the macropore movement of fecal coliforms described by Rahe et al. (47). Soil disposal systems must either be designed to accommodate close contact between virus and adsorbent or the flow rate must be controlled to allow sufficient time for virus adsorption to occur.

There is a deficit of information regarding virus movement from on-site waste disposal systems. Brown et al. (15) described a two-year study where septic effluent was monitored for coliphages in three undisturbed soils. Regardless of application rates, the coliphages were rarely detected in soil leachates, and 120-cm of any of the soils tested appeared to be sufficient to minimize any possibility of groundwater contamination by coliphages. This study, however, was conducted in central Texas, a region where seasonally perched water tables or extended freezes do not regularly occur. Green and Cliver (56) described mound systems with sandy fill material which were efficient in the removal of poliovirus. They detailed the essential design criteria, such as dose rate, fill spacing, operation temperature, and system conditioning that must be considered for effective virus removal. With proper design and operation, mounds incorporating a 60-cm sandy fill yielded effluents that presented no health hazards from human enteric viruses.

It is possible that application rates and system modifications designed to contain fecal bacteria will be sufficient for fecal viruses as well. It would appear, however, inappropriate to use this as a guideline without considerably more information on enteric virus adsorption kinetics, survival times, elution conditions, and transport mechanisms (see Chapter 4).

8.6. RESEARCH NEEDS

Lawes et al. (57) reported nearly 100 years ago that a large portion of the water added to moist soils moves immediately through open channels and interacts only slightly with the water in the soil itself. Thomas and Phillips (58) conclusively demonstrated that the flow of water through macropores is important in soil and groundwater recharge and in solute movement through soils. Recent interest in water movement through soil has resulted in reports that show that partial displacement occurs in structured soils when water from irrigation or rainfall flows through macropores (59, 60).

In view of this information, it would appear that the primary research need for on-site waste disposal is the development of reliable soil suitability indices for various disposal systems. Only the Wisconsin Project (13) approaches this end, and its instigation was greatly facilitated by a waterborne disease outbreak at a resort that affected several members of the Wisconsin legislature (J. Bouma, personal communication). Soil physical properties such as the formation, distribution, and proportion of pore space represented by macropores, the assessment of true saturated hydraulic conductivity rates, and hydrologic relationships of water entering and moving through soil must be quantified (see Chapter 2). Once basic interactions are identified, they can be used to develop a program wherein a predictive capacity could be included to rapidly evaluate different soils under a variety of moisture conditions.

Septic-tank systems rank highest in the total volume of wastewater discharged directly into groundwater systems and are the most frequently reported source of groundwater contamination (3, 61). One-third of all waterborne disease outbreaks in the United States from 1971–1976 were traced to the consumption of water from untreated groundwater sources (62; see Chapter 7) and approximately 60 million consumers rely on the absence of pathogens in untreated groundwater (61). With the suburban expansion around metropolitan areas into once-rural areas, the trend is set for the exposure of larger numbers of individuals to untreated drinking water. Suburban growth frequently exceeds the expansion of municipal facilities (including water) into newly subdivided localities, and this makes it essential that on-site waste disposal systems are developed that do not contribute to the deterioration of water quality.

There are several additional concerns that should mandate some research effort. One important consideration is the development of alternative systems that will function in soils where a conventional septic tank and drainfield are not suitable. Elevated mound systems are one promising possibility (13) and investigations should be expanded on evaluating the longevity, efficiency, and cost of these modified on-site units. Research is needed on techniques to upgrade the performance or efficiency of conventional systems. Examples of such are the use of interceptor tile lines around drainfields to draw off water and pressure distribution to equally disperse effluent over the treatment zone. Deep-plowing to break up the restrictive layers beneath a drainfield is an option that may increase waste retention times and reduce flow through macropores and spaces between the soil peds. Using a fill material as a drainfield bed after removing the original soil is another possibility.

Virtually all of the health-related research in water quality has been centered on fecal indicator bacteria. Efforts are needed to assess the movement potential in soil of other organisms, especially viruses, protozoan cysts, and helminth ova. Many waterborne disease outbreaks are suspected to have a viral etiology when no other causative agent can be identified, and the dynamics of viral translocation through soil may or may not resemble that of fecal bacteria. Critical examination of alternative systems or modifications should include a determination of the retention efficiency of both enteric bacteria and viruses.

Numerous reports in this review have shown the contamination that can occur in surface waters and on a watershed basis through the inefficient operation of conventional systems under unsuitable conditions (42, 59). The National Clean Water Act (Sect. 208 of PL92-500) will require, when standards are enforced, a level of water quality that will not allow a continued deterioration of groundwater sources. This is especially true in areas with higher populations (suburban), and even without legislation, alleviation of threats to potable water should be addressed for sanitary and health-related reasons.

REFERENCES

1. W. D. Burge and P. B. Marsh, *J. Environ. Qual.* **7,** 1 (1978).
2. A. W. Hoadley and S. M. Goyal, "Public Health Implications of the Application of Wastewaters to Land," in R. L. Sanks and T. Asano, Eds., *Land Treatment and Disposal of Municipal and Industrial Wastewaters.* Ann Arbor Sci. Pub., Ann Arbor, MI., 1976, p. 101.
3. J. J. Geraghty and D. W. Miller, *J. Am. Water Works Ass.* **70,** 162 (1978).
4. Office of Water Supply, U.S.EPA Rpt. No. 570/9-77-002, (1977).
5. D. Brooks and I. Cech, *Water Res.* **13,** 33 (1979).
6. J. E. Hackett, *Groundwater* **3,** 27 (1965).
7. J. W. Maynard, *Can. J. Pub. Health* **60,** (1969).
8. G. G. Robeck, *Groundwater* **7,** 33, (1969).
9. S. S. Sandhu, W. J. Warren, and P. Nelson, *Appl. Environ. Microbiol.* **37,** 744 (1979).
10. G. J. Wall and L. R. Webber, *Can. J. Public Health* **61,** 47 (1970).
11. W. A. Ziebell, D. H. Nero, J. F. Deiningor, and E. McCoy, In *Home Sewage Disposal, Proc. Natl. Home Sewage Disposal Symp.*, Am. Soc. Ag. Engr., St. Joseph, MI, 1974, p. 58.

12. S. N. Goldstein, V. D. Wenk, M. C. Fowler, and S. S. Poh, Nat. Tech. Info. Serv., U.S. Dept. Comm., Springfield, VA, 1972, p. 10.
13. J. Bouma, W. A. Ziebell, W. G. Walther, P. G. Olcott, E. McCoy, and F. D. Hole, University of Wisconsin Ext. Cir. No. 20, Madison, WI, (1972).
14. T. Viraraghavan and R. G. Warnock, *J. Water Poll. Contr. Fed.* **48,** 934 (1976).
15. K. W. Brown, H. W. Wolf, K. C. Donnelly, and J. F. Slowley, *J. Environ. Qual.* **8,** 121 (1979).
16. F. B. Dazzo and D. F. Rothwell, *Appl. Microbiol.* **27,** 1172 (1974).
17. U.S. Public Health Service, *Manual of Septic Tank Practice*, U.S.H.E.W. Publ. No. 526 (1967).
18. R. Kristiansen, *J. Environ. Qual.* **10,** 361 (1981).
19. A. James, *J. Appl. Bact.* **27,** 197 (1964).
20. P. H. McGauhey and R. B. Krone, Sanitary Engr. Res. Lab. Rep. No. 67-11, Part V., Univ. Calif., Berkeley, CA (1967).
21. J. C. Romero, *Groundwater* **8,** 37 (1970).
22. C. P. Gerba, C. Wallis, and J. L. Melnick, *J. Irri. Drain., Div. ASCE* **101,** 157 (1975).
23. M. Hutchinson, In J. H. Cole, Ed., *Groundwater Pollution in Europe*, Water Info. Center Inc. Port Washington, NY., 1974, p. 167.
24. C. W. Stiles and H. R. Crohurst, *Pub. Health Rpt.* **38,** 1350 (1923).
25. L. F. Warrick and O. J. Muegge, *J. Am. Water Works Assoc.* **22,** 215 (1930).
26. E. L. Caldwell, *J. Infect. Dis.* **61,** 269 (1937).
27. E. L. Caldwell and L. W. Parr, *J. Infect. Dis.* **61,** 149 (1937).
28. E. L. Caldwell, *J. Infect. Dis.* **62,** 273 (1938).
29. R. G. Butler, G. T. Orlob, and P. H. McGauhey, *J. Am. Water Works Assoc.* **46,** 97 (1954).
30. J. K. Baars, *Bull. World Health Org.* **16,** 727 (1957).
31. M. J. Fournelle, *U.S. Pub. Health Dept.* **72,** 203 (1957).
32. F. C. McMichael and J. E. McKee, Pub. No. 33, State of California Qual. Control Board (1965).
33. J. C. Merrell, Jr., Water Pol. Res. Series Pub. No. WP-20-7. Fed. Water Poll. Control Admin., Cincinnati, OH, (1967).
34. R. B. Krone, G. T. Orlob, and C. Hodgkinson, *Sewage Indus. Wastes* **30,** 1 (1958).
35. G. M. Wesner and D. C. Baier, *J. Am. Water Works Assoc.* **62,** 203 (1970).
36. N. I. Anan'ev and N. D. Demin, *Hyg. Sanit.* **36,** 292 (1971).
37. M. J. Allen and S. M. Morrison, *Groundwater* **11,** 6 (1973).
38. R. H. F. Young, *J. Water Poll. Contr. Fed.* **46,** 1296 (1973).
39. H. Bouwer, J. C. Lance, and M. S. Riggs, *J. Water Poll. Cont. Fed.* **46,** 844 (1974).
40. S. A. Schaub and C. A. Sorber, *Appl. Environ. Microbiol.* **33,** 609 (1977).
41. E. E. Geldreich, L. C. Best, B. A. Kenner, and D. J. VanDonsel, *J. Water Poll. Contr. Fed.* **40,** 1861 (1968).
42. R. B. Reneau, Jr., J. H. Elder, Jr., D. E. Pettry, and C. W. Weston, *J. Environ. Qual.* **4,** 249 (1975).

43. R. B. Reneau, Jr. and D. E. Pettry, *J. Environ. Qual.* **4,** 41 (1975).

44. R. B. Reneau, Jr., *J. Environ. Qual.* **7,** 23 (1978).

45. T. Viraraghavan, *Water, Air Soil Poll.* **9,** 355 (1978).

46. C. Hagedorn, D. T. Hansen, and G. H. Simonson, *J. Environ. Qual.* **7,** 55 (1978).

47. T. M. Rahe, C. Hagedorn, E. L. McCoy, and G. F. Kling, *J. Environ. Qual.* **7,** 487 (1978).

48. C. Hagedorn and E. L. McCoy, Water Resources Res. Inst., No. 65, Oregon State Univ., Corvallis, OR (1979).

49. E. L. McCoy and C. Hagedorn, *Water Air Soil Poll.* **11,** 467 (1979).

50. G. Bitton, *Water Res.* **9,** 473 (1975).

51. G. G. Robeck, N. A. Clarke, and K. A. Dostal, *J. Am. Water Works Assoc.* **54,** 1275 (1962).

52. J. C. Lance, C. P. Gerba, and J. L. Melnick, *Appl. Environ. Microbiol.* **32,** 520 (1976).

53. G. F. Carlson, Jr., F. E. Woodward, D. F. Wentworth, and O. J. Sproul, *J. Water Poll. Control Fed.* **40,** R89 (1968).

54. S. M. Goyal and C. P. Gerba, *Appl. Environ. Microbiol.* **38,** 241 (1979).

55. J. C. Lance, *Trans. ASAE.* **21,** 1114 (1978).

56. K. M. Green and D. O. Cliver, in *Proc. of the Nat'l. Home Sewage Disposal Symp.*, Madison, WI, 1974, p. 137.

57. J. B. Lawes, J. H. Gilbert, and R. Warrington, *On the Amount and Composition of the Rain and Drainage Waters Collected at Rothamsted.* William Clowes and Sons, Ltd. London, 1882.

58. G. W. Thomas and R. E. Phillips, *J. Environ. Qual.* **8,** 149 (1979).

59. G. W. Thomas, R. E. Phillips, and V. L. Quisenberry, *J. Soil Sci.* **29,** 32 (1978).

60. V. L. Quinsenberry and R. E. Phillips, *Soil Sci. Soc. Am. J.* **42,** 675 (1978).

61. M. J. Allen and E. E. Geldreich, *Groundwater* **13,** 45 (1975).

62. G. F. Craun, L. J. McCabe, and J. M. Hughes, *J. Am. Water Works Assoc.* **68,** 420 (1976).

9

LAND DISPOSAL OF SEWAGE EFFLUENTS AND RESIDUES

J. C. Lance

Water Quality and Watershed Research Laboratory
United States Department of Agriculture
Agricultural Research Service, Durant, Oklahoma

Land application of municipal wastes is an effective way to conserve water, nutrients, and energy resources and to avoid pollution of streams and lakes. However, land treatment systems must be carefully designed to minimize health

Table 9.1. Potential Groundwater Pollution Problems Associated with Land Treatment[a]

Type of Land Treatment System	Bacteria	Virus	Nutrients	Heavy Metals	Trace Organics
Irrigation	+	+	+	−	−
Rapid infiltration	++	++	++	+	+
Overland flow	−	−	−	−	−
Digested sludge on cropland	+	+	++	−	−
Sludge compost on cropland	−	−	+	−	−

[a]Little or no potential problem = −. Moderate potential = +. Considerable potential = ++.

hazards and groundwater pollution associated with the pathogens and pollutants in municipal wastewater and sludge. A wide array of physical, chemical, and biological processes must be managed effectively so that municipal wastes can be assimilated by soils. Sewage water is generally applied to land by three application systems: (1) irrigation systems for cropland, (2) high-rate land filtration or groundwater recharge systems, and (3) by overland flow or spray runoff. Solids from sewage treatment plants are usually applied to cropland as digested sludge or composted sludge.

The potential for groundwater pollution depends upon the kind of land treatment system and the degree of pretreatment (Table 9.1). This chapter will focus on these potential groundwater pollution hazards and design and management practices that can be used to minimize them.

9.1. BACTERIAL MOVEMENT IN LAND TREATMENT SYSTEMS

As wastewater moves through the soil, bacteria are removed by filtering, adsorption, and die-off. The same mechanisms impede the leaching of bacteria from sludge because of rain or irrigation. Straining by soil pores is probably the most important barrier to bacterial penetration. Butler et al. (1) found that removal of bacteria from liquid percolating through a given depth of soil was inversely proportional to the particle size of the soil and was highest during the first few centimeters of travel. He concluded that removal was by filtering. Development of an organic mat on the surface enhanced the removal of coliforms.

Aulenbach et al. (2) reported that 99 percent of the coliforms were removed by movement of secondary effluent through 3 m of sand. Bitton et al. (3) concluded that the filtering effect of the soil increases as the soil water content decreases. This indicates that bacterial movement through the soil to groundwater in slow-rate infiltration systems should not be a problem, since the water is applied in fairly small amounts. Similarly, movement of bacteria through sludge-treated soils would not be expected. However, Ziebell et al. (4) reported that short-circuiting of septic-tank effluent through large air-filled pores in a highly aggregated clay soil allowed movement of fecal bacteria through a 50-cm-long soil column. This kind of movement could be a problem in areas where highly aggregated soils which develop extensive cracks occur over shallow water tables.

Bouwer et al. (5) found that fecal coliform concentrations were usually less than 100/100 ml in water from a 9.1-m-deep well below loamy sand recharge basins intermittently flooded with secondary sewage effluent. This was about a 10^4 reduction in fecal coliform concentrations. Bacterial concentrations peaked when water infiltrated during the first day of the flooding cycle reached the well. Fecal coliforms were never detected in renovated water samples from a well 91 m from the recharge area.

Lance and Whisler (6) found that reduction of fecal coliforms by filtration through 250-cm-long columns filled with loamy sand was proportional to the concentration of fecal coliforms applied at the surface. The columns reduced fecal coliform concentrations by about 10^3 during nine-day flooding periods at infiltration rates of 40–50 cm/day. Therefore, some bacteria can penetrate to the groundwater in groundwater recharge systems. However, rapid infiltration of secondary effluent in the sands and gravels of the Salt River bed near Phoenix, Arizona produced renovated water that was suitable for unrestricted irrigation, lakes with primary contact recreation, and other purposes (7).

9.2. VIRUS MOVEMENT IN LAND TREATMENT SYSTEMS

9.2.1. Virus Adsorption Studies with Small Soil Samples and Short Soil Columns

A number of laboratory studies with many soils have shown that most soils have a high-adsorption capacity toward viruses. Gerba et al. (8) summarized data from a number of laboratory experiments showing varying degrees of adsorption of viruses by soils, but, in general, some movement of viruses through most soils.

In many of the earlier studies, viruses suspended in tap water or distilled water were applied to soil columns. Robeck et al. (9) allowed poliovirus suspended in dechlorinated tap water to percolate through 60 cm of California dune sand. With virus doses of from 10,000–60,000 PFU/ml (plaque-forming units/milliliter), the sand was capable of removing 99 percent of the viruses in 98 days of operation. Virus removal was found to depend on the flow rate through coarse sand; removal greater than 90 percent occurred when flow rates were below 1.2 m/day. Higher flow rates resulted in greater virus breakthrough. Drewry and Eliassen (10) found that when virus suspended in distilled water was applied to 40–50-cm-long columns of sterile soil, over 99 percent of the virus was removed. Radioactive tagging experiments indicated that most of the virus was retained in the top 2 cm of the column. In Hawaii, 10-cm columns of two acid soils from Oahu and Lahaina removed 99.3 percent of the poliovirus applied in distilled water, whereas Tantalus cinder removed 78 percent of the applied virus (11).

Laboratory studies with soils and batch experiments with suspensions of soils or clays have shown that the ionic composition of the water containing the added viruses is a very important factor affecting virus adsorption (12,13). A likely mechanism of adsorption is that an increase in the ionic strength of the solution reduces the thickness of the double layer of charges around the viruses and soil particles, allowing them to move close enough together to be bound by London-van der Waals forces (14). Similarly, viruses can desorb when the ionic strength of the suspending medium is decreased (15, 16).

These laboratory studies using small soil samples or relatively short soil columns give much qualitative information on factors affecting soil adsorption of viruses. However, such laboratory systems cannot measure the effect of some factors such as flow velocity of the water and wetting and drying cycles. For example, short columns (60-cm long) do not give an accurate assessment of soil-virus interactions during wetting and drying cycles because capillary retention inhibits drainage. Therefore, field studies and long soil columns that more nearly represent field systems are needed to predict the potential of a land treatment system for groundwater pollution or to make quantitative predictions of virus movement through various soils.

9.2.2. Virus–Soil Interactions in Long Soil Columns and Field Sites

Lance et al. (17) constructed 250-cm-long soil columns by packing polyvinyl-choride pipes with sand taken from groundwater recharge basins of the Flushing Meadows project, an experimental groundwater recharge system located in the Salt River bed near Phoenix, Arizona. A number of experiments on the removal of nitrogen, phosphorus, and coliforms from secondary sewage effluents by the soil columns established that the columns were good models of the field system (18). When the soil columns were flooded with effluent in the same manner as the basins of the Flushing Meadows project, most of the laboratory-grown polio-viruses suspended in the secondary effluent were adsorbed in the top 5 cm of the soil columns. Ceramic soil water samplers werre used to monitor virus concentrations at various depths in the soil column. Wang et al. (19) showed that the ceramic samplers with pore sizes of 20 μm gave high virus recoveries from sewage water or tap water. The virus concentration in the sewage water was reduced by about 99 percent (from a starting concentration of 3×10^4 PFU/ml) during the first 2 cm of travel, but an additional 158 cm of travel was required for another 99 percent reduction (18). The virus adsorption profile at the end of the 27-day period was almost identical to the pattern observed on the first day. Thus, a soil that had been intermittently flooded with sewage for 10 years was not saturated by applying heavy loads of virus (thousands of times more viruses than are normally present in sewage effluent). When viruses were extracted from soil samples taken from Flushing Meadows groundwater recharge basins, most were found in the upper 2.5 cm of soil (20). This confirmed that virus adsorption patterns in the soil columns and field soils were similar.

Flooding the columns with distilled water immediately after the application of the sewage–water mixture caused considerable virus desorption and movement (17). Most of the viruses were adsorbed again lower in the column, but a few were detected in the column outflow after 2 days of flooding with distilled water. However, this happened after about 60 cm of distilled water had been applied under continuous flooding. Only a few viruses were detected in profile samples when 1 or 3 mM $CaCl_2$ solution was applied instead of distilled water and none was detected below the 80-cm depth. Desorption of virus by distilled water was greatly reduced when the soil was dried for one day between the application of virus-enriched sewage and distilled water; no viruses were desorbed when the drying time was increased to 5 days. No viruses were detected below the 80-cm depth when at least one day of drying was allowed between the application of virus-enriched sewage and of distilled water. Thus, virus movement due to rainfall would be slight unless heavy rains occurred during the first day of a drying period following sewage application. Heavy rains during the first few hours of the drying period could cause some movement but not enough to pass viruses through 250 cm of the Flushing Meadows sand.

When a soil column was flooded with virus-enriched secondary effluent and later with virus-enriched primary effluent, the adsorption patterns were quite similar (21). Thus, the additional organic carbon in the primary effluent (69.5 mg/liter for the primary vs. 10 mg/liter for the secondary) had no added effect on virus adsorption. Viruses also would be adsorbed near the soil surface in fields treated with primary sewage effluents.

Increasing the concentration of poliovirus added to the sewage water from 0.9×10^2 to 2.6×10^4 PFU/ml increased the number of viruses found at each depth in the soil. However, the percentage of the applied virus found at each depth was the same for all concentrations (22). Thus, the maximum depth of penetration by the virus was independent of the concentration of applied virus. This means that reducing virus concentrations in the sewage effluent by some pretreatment like chlorination would not prevent penetration of some viruses to a particular depth. Pretreatment could reduce the number of viruses penetrating to that depth. However, chlorination would not be needed to reduce virus movement in most agricultural soils because little virus movement would be expected except in very sandy soils.

Viruses were not detected below 160 cm when a column packed with coarse sand from the Salt River bed was flooded with virus-enriched sewage at an infiltration rate of 0.6 m/day (22). However, some viruses broke through the column when the infiltration rate was increased to 1.2 m/day. Virus adsorption was not greatly affected by further increases in the infiltration rate. Apparently virus adsorption is not affected by increases in the flow rate up to some breakthrough point. After that breakthrough point was reached, some viruses moved through, but 99 percent were still removed at infiltration rates of 12 m/day. These experiments on flow velocity suggested that permeability may be the most important soil characteristic affecting virus movement. Also, the method of application of the sewage water greatly affects virus movement. Lance and Gerba (unpublished data) found that virus penetration in soil columns filled with sand was greatly reduced when water was applied sparingly to allow only unsaturated flow instead of saturated flow.

Since most soils have considerable capacity to adsorb viruses and virus penetration is limited by unsaturated flow, movement of viruses through soils to groundwater under normal irrigation practices usually would not be a problem. The limited data available from field studies support this concept. Viruses were not isolated (with one questionable exception) from leachate samples collected at 150-cm depth below grass sod and sugarcane fields in Hawaii that were regularly irrigated with secondary effluent (23). Studies at Roswell, New Mexico failed to yield viruses in 151–333-liter samples taken from shallow wells (3–37 m) beneath farmland that was intermittently flooded with secondary sewage effluent containing 960–5070 PFU/liter (24).

On the other hand, Wellings et al. (25) isolated viruses in water from tile lines 150 cm beneath a field that was sprinkler irrigated with secondary sewage effluent and from 3- and 6-m-deep wells after a three-month period when the cumulative rainfall was 71 cm. The soil was an Immokalee sand with little or no silt or clay. The combination of heavy rains, a very permeable soil, and a shallow water table thus resulted in some viruses moving to the groundwater.

The soil column studies indicated that viruses would not move through to the groundwater below rapid infiltration systems if the water moved through at least 2.5 m of sand at a flow rate less than 1 m/day. In field tests, data support the concept that the potential for virus penetration to groundwater below rapid-infiltration systems depends upon soil permeability. Viruses were not detected in water sampled at 6- to 9-m depths below the Flushing Meadows project where soils are sandy and sewage infiltration rates were 30–70 cm/day (26). Vaughn and Landry (27) found little virus detention at a rapid infiltration site with flow rates of 18–24 m/day, but only limited virus penetration through the same sand and gravel test basin when the flow rate was reduced to 1.4 m/day. Viruses have been detected in the groundwater below field sites at Fort Devens, Massachusetts (28), Vineland, New Jersey (29) and several sites in Long Island, New York (30). All of these sites are located on sandy gravel soils with high infiltration rates. However, even when viruses have been detected in wells below rapid-infiltration systems, the overall virus removal by the land treatment systems approached the virus removal obtained by extensive tertiary treatment of sewage (31). Thus, virus removal by the least effective land treatment system approached that obtained by tertiary treatment. The lab and field studies suggest that rapid infiltration systems can be designed and managed to prevent virus penetration to the groundwater.

Schaub et al. (32) measured the removal of enteric viruses and tracer bacteriophages from sewage water applied to overland flow systems. The tracer bacteriophage f_2 concentration was reduced by 30–60 percent while enteric virus concentrations were reduced by 68–85 percent during travel down a 36-m slope. Viruses should not penetrate to the groundwater below overland flow systems because very impermeable soils are required for these systems.

Laborabory studies described above indicate that viruses from sludge would not be expected to penetrate to the groundwater because viruses applied in sludge are probably less mobile than those applied in sewage water due to adsorption of viruses on sludge solids. Limited field data support this concept. Viruses added to sludge placed on lysimeters in Denmark did not move through the 135-cm soil depth even though they were viable for 23 weeks (33). The total rainfall for the test period was 30 cm. Viruses were not detected in 189- to 378-liter groundwater samples at a site in Florida where sludge was spread over agricultural land (34).

Since viruses are inactivated during composting, the application of sludge compost to land should not contaminate the groundwater with viruses. Hurst (20) found that viruses survived for only a few days after soils flooded with sewage water were allowed to dry. However, laboratory and field studies showed long survival times (several months) under anaerobic conditions and at temperatures near freezing (20, 35). Since aerobic conditions are needed for high virus inactivation rates, viruses could survive for many weeks if they penetrate to the groundwater where anerobic conditions prevail (see Chapter 4). Under certain conditions, poliovirus was shown to travel at least 90 m laterally through an aquifer (36). Viruses can be adsorbed in aquifers if the groundwater does not move through gravel layers or fractured rock.

9.3. NUTRIENT MOVEMENT TO GROUNDWATER

9.3.1. Nitrogen Movement

A wide variety of chemical and biological reactions can be used to remove nitrogen from sewage water or sludge (37). These reactions can be divided into two main groups—reactions that result in: (1) removal of nitrogen from the soil and (2) storage of nitrogen in the soil. Nitrogen can be transfered from the soil to the atmosphere in gaseous form by biological denitrification, volatilization of ammonia, and chemodenitrification. Nitrogen incorporated into vegetation is removed when the crop is harvested. Nitrogen may be immobilized in the soil by adsorption of NH_4–N on the soil cation-exchange sites, fixation by clay minerals, adsorption by organic matter, and incorporation into microbial tissue. Some of the reactions immobilize nitrogen temporarily while others fix it in a stable form. Reactions responsible for nitrogen removal from sewage water are illustrated in Fig. 9.1. Similar reactions occur when sludge is applied, but much of the nitrogen in sludge is organic nitrogen that must be mineralized before it can be taken up by plants or transformed by other reactions.

9.3.1.1. Plant Uptake of Nitrogen

Nitrogen uptake by plants grown on land treated with municipal wastes is an effective way to reduce the nitrogen content of the wastewater. The volume of water that can be applied in excess of that used by the crop depends on the amount of nutrients removed from the field in the harvested plant materials. A

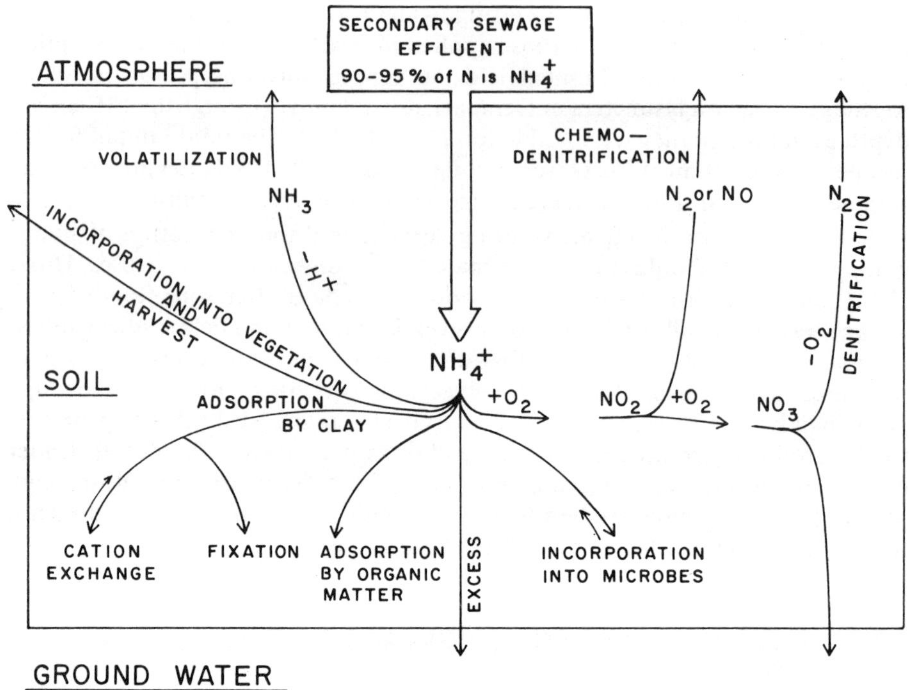

Figure 9.1. Reactions responsible for nitrogen removal during land application of secondary sewage effluent (48).

crop should be selected that has a long growing season and a high nitrogen requirement. The best crop for such a system from the standpoint of nitrogen removal is a flood-tolerant perennial grass because it has both characteristics. Sopper (38) reported that reed canarygrass irrigated with 51 mm of secondary effluent per week removed 453 kg/ha of nitrogen during one year, whereas corn silage irrigated at the same rate removed only 178 kg/ha. This illustrates the advantage of using a crop with a long growing season. Lance (37) noted that in the southern half of the United States, bermudagrass can be overseeded with small grains during the winter to provide an actively growing crop during most of the year.

9.3.1.2. Biological Denitrification

Biological denitrification is a reaction capable of removing large quantities of nitrogen from soils irrigated with sewage water or fertilized with sludge. Soils contain an abundance of denitrifying bacteria that can use free oxygen or can use nitrate as a substitute hydrogen acceptor if free oxygen is absent (39). Most of the nitrogen from nitrate is released as N_2 by denitrification if the soil pH is above 6.0 (40). However, oxides of nitrogen such as N_2O and NO may be produced in more acid soils. Denitrification proceeds slowly below pH 5.5 and below 50° F 10° C) and practically ceases at 36° F (2° C) (41).

Three conditions that must be met in order to remove nitrogen from soil by biological denitrification are: (1) the NH_4–N that makes up most of the nitrogen content of secondary effluent must be oxidized to nitrate, for example, nitrification must occur before denitrification (organic nitrogen present in sludge or sewage water must be minearalized to NH_4–N before nitrification); (2) the nitrate must be in a zone where oxygen is absent; (3) organic carbon must be provided in that zone as an energy source for the denitrifying bacteria.

Oxygen is the factor that most often limits nitrification in soils. Nitrification requires 4.57 mg of oxygen for every milligram of nitrogen oxidized. Because oxygen transport is limited in saturated soils, the soil moisture content often regulates nitrification. The optimum temperature of nitrification is between 30° C and 35° C, but nitrification proceeds slowly at 2° C and stops somewhere near the freezing point of water (57). Nitrification is also severely limited below pH 5. Nitrifiers exist in soils for long periods of time under adverse conditions such as prolonged flooding, even when nitrification is at a virtual standstill due to the shortage of oxygen (42). However, freezing can greatly reduce the population and 2-3 weeks are required to build up substantial numbers because nitrifiers multiply more slowly than most other bacteria.

The formation of reduced zones can happen on a massive scale when the soil is flooded and all of the oxygen is driven out or consumed by oxidative reactions. However, some denitrification can occur in a reduced microenvironment in an otherwise well-drained aerobic soil. For example, NH_4-N can be oxidized to NO_3–N which then diffuses into a reduced microzone inside an aggregate near a plant root or near pieces of decomposing plant or animal residue. The extent of denitrification by this mechanism is limited to less than 30 percent of the total nitrate, whereas all of the nitrate could be denitrified in a water-logged soil (39).

The effect of moisture content on denitrification can be divided into three general environments. At some moisture content well below field capacity, nitrogen loss is insignificant. Bremner and Shaw (43) observed that little loss occurred below 60 percent of water-holding capacity even if other conditions for denitrification were favorable. Therefore, denitrification was directly inhibited by the low-moisture content.

The second general environment is the well-aerated soil with reduced microzones where denitrification occurs. These zones increase in number as moisture content increases and are more numerous in fine-textured soils than in sandy soils.

When the moisture content is increased above field capacity, a third environment is established where denitrification is quite rapid. Meek and MacKenzie (44) found that production of N_2 in incubated soil samples did not change when the soil water varied from 34–41 percent, but increased sharply when the moisture content reached 44.5 percent. The soil contained 48 percent moisture at saturation and 32 percent at field capacity. Stefanson (45) reported that changes of 2 percent to 6 percent in water content above field capacity sometimes doubled the amount of denitrification. Increases in soil moisture content above field capacity result in saturation of large pores and a predominantly reduced environment. Under such conditions, the denitrification rate would usually be governed by the amount of organic carbon available if the nitrogen were in the nitrate form.

Woldendorp (46) and Stefanson (45) reported that living plants stimulate denitrification. The effects of plants could be due to the lowering of oxygen content in the rhizosphere and to an increased carbon supply from root excretions and dead root tissue.

If the first two conditions are met (i.e., nitrification and reduced conditions), the extent of denitrification is limited by the food supply (usually organic carbon) available to the denitrifying bacteria. The supply needed varies with the carbon source and the proportion of carbon used by microorganisms other than denitrifiers. If glucose is supplied for the denitrifiers, the equation can be written as:

$$C_6H_6O_6 + 4NO_3 \longrightarrow 2N_2 + 6H_2O + 6CO_2 \qquad [9.1]$$

This equation indicates that a minimum of 1.3 mg C would be needed for every milligram of NO_3–N. If the equation was written with methanol as the carbon source, at least 0.7 mg C per 1 mg of NO_3–N would be required. In practice, St. Amant and McCarty (47) found that 1.3 mg C per 1 mg of NO_3–N was needed for 90 percent dentirification using methanol. Lance et al. (48) found that 75 percent removal could be obtained when water containing 18 ppm NO_3–N from soil columns intermittently flooded with sewage water was mixed with more sewage water and recycled. The recycled water contained 0.75-mg organic carbon per mg NO_3–N. As a rule of thumb, a 1:1 ratio of organic carbon to NO_3–N is needed for 80–90 percent denitrification. The organic carbon available for denitrification in soils is often limited because most of the organic material is oxidized by the time NO_3–N is formed and moves into a reduced zone.

9.3.1.3. *Ammonia Volatilization*

Nitrogen from soils treated with wastes can be transferred to the air as ammonia, NH_3. Little nitrogen is present as NH_3 at pH 7 and below, but as the pH increases, more nitrogen is transformed into NH_3. However, volatilization of ammonia requires considerable air–water contact and this is not provided below the soil surface. Some nitrogen can be lost this way when sewage is applied with sprinklers. These losses are small because the pH of secondary sewage effluent usually is between 7 and 8 and less than 10 percent of the nitrogen exists as NH_3. Ammonia losses are extensive when sludge is applied to the soil surface, but sludge usually is incorporated into the soil.

9.3.1.4. *Chemodenitrification*

Nitrites may break down in the soil by nonbiological reactions to yield gaseous nitrogen. This process is called chemodenitrification. It is unlikely that chemodenitrification removes much nitrogen from sewage water or sludge applied to land because most of the reactions require a pH below 5.0 and high nitrite concentrations (39).

9.3.1.5. *Nitrogen Immobilization Reactions*

Other reactions do not remove nitrogen from the soil but immobilize it in the soil for varying lengths of time. Nitrogen is stored primarily as a result of immobilization of NH_4–N since NO_3–N is quite mobile and generally moves with soil water. Ammonium nitrogen may be incorporated into microbial cell tissue. This nitrogen can be released by decay of microbial cells, but some of the nitrogen immobilized in this way is stable (49). However, the carbon content of secondary sewage effluent is not sufficiently high to build enough microbial cells to tie up a large fraction of the nitrogen in the sewage. Microbes contain 5–15 g carbon per gram of nitrogen, but secondary effluent usually contains less than 1 g carbon per 1 g nitrogen. Most of the nitrogen in sludge is already immobile because the digestion process incorporates most of the nitrogen into microbial cells, but nitrogen release begins when sludge is mixed with soil.

The ammonium ion can be fixed an a stable form by expanding clay minerals if it is trapped between the layers of minerals such as montmorillonite or vermiculite. However, soils with a high NH_4–N fixing capacity are relatively uncommon in land treatment sites because well-drained soils that have a fairly low clay content are usually used for land disposal of sewage.

Ammonia reacts with soil organic matter to form complexes that are resistant to leaching and decomposition. The exact chemical nature of this organic matter–ammonia complex is not well understood. Broadbent et al. (50) reported that fixation of ammonia was slight in acid conditions, but increased rapidly as the pH increased above seven. Burge and Broadbent (51) reported that 33 mg of nitrogen was fixed for each gram of carbon, but Young (52) found that 36 mg of nitrogen was fixed for each gram of carbon. NH_4–N fixation by organic matter probably would be of little importance at the low NH_4–N concentrations and near neutral pH of wastewaters normally applied to soils with a low organic matter content, but it could be significant where sludge applications provide organic matter and higher NH_4–N concentrations (53).

Adsorption of NH_4–N by the soil cation-exchange complex provides a mechanism for temporary nitrogen storage. The ammonium ions in wastewater are adsorbed by negatively charged clay and organic colloids in the soil. The cation-exchange capacity (CEC) of the soil depends upon the amount and type of clay and the amount of organic matter. The CEC may range from 1 meq to 2 meq/100 g of soil for very sandy soils with little clay or organic matter to more than 100 meq/100 g for soils high in clay or organic matter, or both. The fraction of the CEC that may be used to adsorb NH_4–N depends upon the concentration of other cations in the water applied because these cations (particularly, divalent cations such as Ca^{2+} and Mg^{2+} compete with NH_4–N for exchange sites. This fraction, called the exchangeable ammonium percentage, can be estimated if the CEC and the concentration of the principal competing divalent cations are known. The ammonium adsorption ratio (AAR) must be calculated first by

$$\mathrm{AAR} = \frac{\mathrm{NH_4\text{–}N}}{(\mathrm{Ca}^{2+} + \mathrm{Mg}^{2+})^{1/2}} \qquad [9.2]$$

All concentrations must be expressed as meq/g. The exchangable ammonium percentage increases as the AAR increases, but the relationship must be determined experimentally. An experimentally derived equation relating the potassium adsorption ratio and exchangable potassium percentage can be used for estimating the exchangable ammonium percentage since potassium and ammonium ions behave similarly in cation-exchange reactions (54). The equation for potassium may be applied by substituting ammonium for potassium to give

$$\text{Exchangable ammonium percentage} = \frac{100(0.0360) + 0.1051\ \text{AAR})}{1 + (0.0360 + 0.1051\ \text{AAR})} \qquad [9.3]$$

The NH_4–N adsorbed by the soil CEC is only temporarily immobilized because it can be readily oxidized to NO_3–N when oxygen is available. However, this adsorption is extremely important because it retains nitrogen within the root zone for a time. In high-rate flooding systems, NH_4–N is held until oxygen becomes available for nitrification during dry periods.

9.3.1.6. Nitrogen Removal in Irrigation Systems

Nitrogen removal reactions can be managed in various ways to prevent nitrogen from moving through the soil to the groundwater. In irrigation systems, most of the nitrogen removal is obtained when plant material is harvested. Some nitrogen loss could occur from denitrification, but Brar et al. (55) concluded from a study of a land treatment site at Pennsylvania State University that such losses would be minimal under ordinary agricultural practices due to limited organic carbon supplies.

The most reliable way to prevent nitrate movement to groundwater below irrigation systems is to balance the applied nitrogen against the nitrogen incorporated into the vegetation. A range of nitrogen uptake values that might be expected for various agronomic crops compiled from considerable data in the literature for above-average yield is shown in Table 9.2 (56). Since a 1-m depth of secondary sewage effluent (20 mg/1N) applied on 1 ha of land contains about 200 kg nitrogen, application of sewage at rates equivalent to the water requirements of crops would not result in leaching of nitrate to groundwater. Sometimes, additional nitrogen fertilization is needed because the wastewater does not contain enough nitrogen.

Sewage must be applied in excess of crop water requirements if a limited area of cropland is available for taking the flow from a sewage treatment plant. It may also be desirable to apply sewage year-round to avoid storage. In these situations, denitrification probably could be encouraged by controlling the soil moisture content so that nitrogen could be applied in excess of crop needs. At certain times, particularly when plants are not actively growing, water should be applied for several days to produce anaerobic conditions conducive to denitrification rather than applying frequent small amounts that result in essentially aerobic systems. The periods of saturation must be alternated with dry periods so that oxygen enters the soil for nitrification but is then removed by denitrification

Table 9.2. Maximum Yield and Nitrogen Uptake of Various Crops (Based on 2.5 Percent N Content, Except 3 Percent for Alfalfa and 1.2 Percent for Corn) for Three Regions of the United States (56)

	North		Central		South	
Crop	Yield	Uptake	Yield	Uptake	Yield	Uptake
	mt/ha					
Alfalfa	10	.300	18	0.540	—	—
Coastal bermudagrass	—	—	—	—	24	0.600
Kentucky bluegrass	9	0.225	12	0.300	—	—
Bromegrass	10	0.250	18	0.450	—	—
Reed canarygrass	10	0.250	18	0.450	—	—
Tall fescue	10	0.250	18	0.450	24	0.600
Quackgrass	10	0.250	—	—	—	—
Corn (fodder)	15	0.180	—	—	—	—

during flooding. Plowing crop residues into the soil before long flooding periods would also stimulate denitrification because of the added organic carbon and oxygen demand.

9.3.1.7. Nitrogen Removal in High Rate Land Filtration Systems

The tremendous quantity of nitrogen applied in high-rate land systems distinguishes this system from the other land application methods. As much as 20,000 kg N/ha may be applied with high-rate systems in one year. Denitrification is the only reaction capable of removing most of the nitrogen applied at these rates.

The first requirement for denitrification presented was that the NH_4–N of the sewage must be oxidized to nitrate. This means that nitrogen must pass through both an oxidized environment for nitrification and a reduced zone for denitrification within the same soil profile. The amount of nitrogen involved makes the use of reduced microzones impractical. However, nitrification and denitrification can be accomplished in the same profile by alternating flooding and drying cycles (5, 18). The NH_4–N is adsorbed by the soil cation-exchange complex during one flooding period and is held until the dry period when oxygen enters the soil and nitrification occurs. Some nitrate is denitrified in reduced microzones during the dry period. However, most of the nitrate is leached from the soil during the next flooding period and the incoming sewage water provides organic carbon for denitrification while the NH_4–N in the sewage water is adsorbed. Nitrification is 98 percent complete when the lengths of the flooding and drying periods are manipulated to supply enough oxygen for the NH_4–N stored during the flooding period and the exchange capacity of the soil is not exceeded. If more nitrogen is applied than can be nitrified during the dry period, it will gradually accumulate on the soil exchange complex and the NH_4–N concentration of the reclaimed water will eventually build up (5, 57). However, the flooding period must be long enough to establish a reduced soil profile for denitrification to occur. Both field and laboratory results indicated that short, frequent flooding periods of 2 days flooding alternated with 5-days drying resulted in complete nitrification but no denitrification (5, 18). For the loamy sand used in those experiments, a schedule of 9 days flooded and 5 days drying provided enough oxygen during the dry period for complete nitrification and the reducing conditions needed for denitrification during the flooding period. The optimum schedule of flooding and drying depends upon the soil and wastewater used.

The difficulty with this approach has been that nitrate formed during the dry period is leached through the soil in a concentrated peak, and the concentration of organic carbon in the sewage water is too low for complete denitrification. The nitrate concentrations reached 75 ppm to 100 ppm when laboratory soil columns were intermittently flooded with secondary effluent (18). When laboratory columns and field recharge basins were managed for maximum infiltration, the net nitrogen removal was only 30 percent (5, 18). Infiltration rates were 0.3 m/day to 0.6 m/day. Nitrogen removal by the soil columns was increased to 90 percent by increasing the average dissolved carbon content of the sewage water to 150 ppm by the addition of dextrose (58). This showed that the C to NO_3–N ratio of the nitrate peak determined the extent of denitrification. Since the addition of dextrose or some other organic material is expensive, it is desirable to utilize the organic carbon of the sewage water to support denitrification. The nitrate must be mixed with a large volume of the incoming sewage water to provide sufficient carbon for complete denitrification.

Two methods were developed with soil columns for mixing nitrate with incoming sewage. Seventy-five to eighty percent of the nitrogen was removed (48). One method was to adjust the infiltration rate so that nitrate was mixed with sewage inside the soil columns. Infiltration rates were varied between 15 and 60 cm/day by packing the columns to different bulk densities. The percentage of nitrogen removal increased exponentially as the infiltration rate decreased (Fig. 9.2). The nitrate concentration of water from the columns with infiltration rates of less than 15 cm/day was consistently around 5 ppm, with little evidence of a nitrate peak. The change in infiltration rate affected nitrogen removal by changing the flow characteristics of the system. Nitrate formed in the soil during the dry period was leached in a narrow concentrated band when the infiltration rate was high. A decrease in infiltration rate caused the nitrate to be leached in a more diffuse wave, allowing considerably more mixing of the nitrate with the infiltrating sewage water. Mixing the nitrate with the incoming sewage water provided a more favorable C to NO_3–N ratio for the denitrification than that obtained when nitrate was leached in a concentrated band. Salt tracer studies confirmed that the amount of mixing depended upon the infiltration rate.

A definite relationship between infiltration rate and nitrogen removal probably exists for all soils, but it may be different for each soil. Some methods for changing infiltration rates are adjusting the depth of ponding, compacting the soil surface, and adjusting the suspended solids content of the sewage water.

This method for increasing nitrogen removal by a rapid-infiltration system was field-tested at the Flushing Meadows project in the Salt River bed near Phoenix, Arizona (7). The nitrogen removal was about 30 percent of the applied

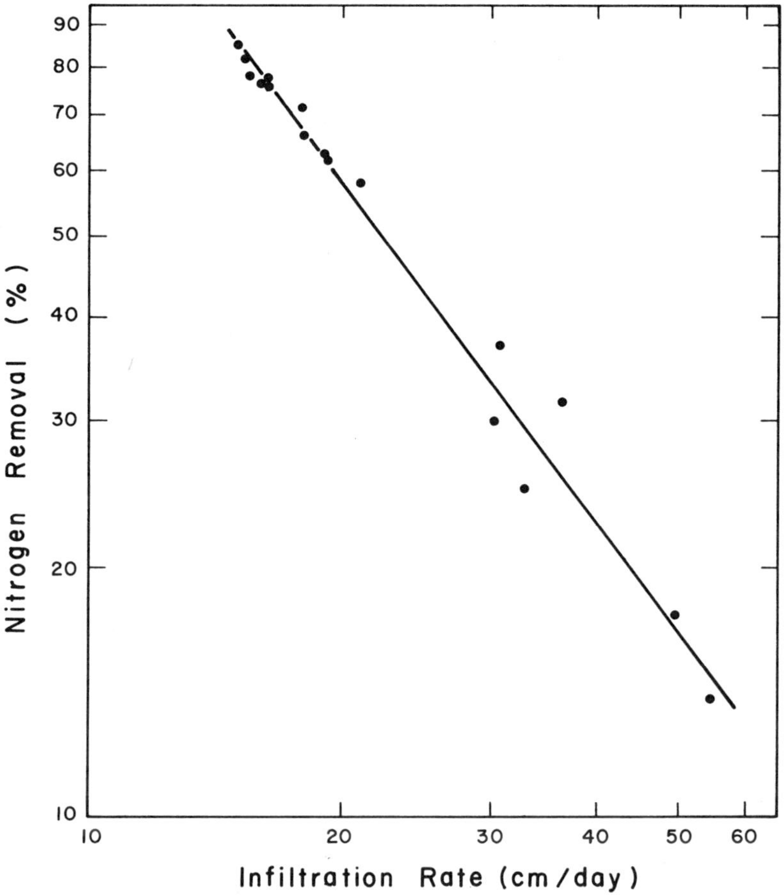

Figure 9.2. The effect of infiltration rate on nitrogen removal by a soil column flooded with secondary sewage effluent.

nitrogen at a hydraulic loading rate of about 120 m/year. The hydraulic loading rate was reduced to an average of about 65 m/year (during a three-year period) by reducing the ponding depth in the infiltration basins from 30 to 15 cm/day and adjusting the flooding and drying cycles to insure complete nitrification of the applied NH_4–N. These management changes increased the nitrogen removal to about 65 percent and the total N content of the renovated water was reduced to 9.6 mg/liter.

The N removal at any infiltration rate can be increased by applying primary sewage effluent to the rapid-infiltration basins instead of the secondary sewage effluent. Lance et al. (59) showed that applying primary effluent with a C to N ratio of 3 to 1 to soil columns resulted in an N removal of 46 percent compared with 29 percent removal of N from secondary effluent with a C to N ratio of 1:1. The higher organic matter content of the primary effluent allowed more denitrification during rapid infiltration. N removals of about 90 percent were reported for a land filtration system in Hollister, California, where primary effluent was applied at low infiltration rates (60).

The second method developed to increase nitrogen removal to 75–80 percent by soil columns was to collect the high-nitrate water, mix it with two parts of sewage water, and recycle it through the columns. The high-nitrate water, represented by the nitrate peaks, was mixed with sewage and recycled throughout the remainder of the flooding period by this method. Nitrogen removal ranged from 75–80 percent when 20–25 percent of the total volume of water draining from columns with infiltration rates of 25 to 30 cm/day was recycled (48). It may be possible to use this method with a land filtration system equipped with tile drains for collecting high nitrate water and a holding pond for mixing it with sewage water, but the method has not been field-tested.

9.3.1.8. Nitrogen Removal in Overland Flow Systems

As wastewater trickles over the grassed slopes in overland flow systems, most of the nitrogen is removed. Sedimented wastewater at Melbourne, Australia, (61) screened cannery waste at Paris, Texas (62), and comminuted raw sewage at Ada, Oklahoma (63) have been effectively treated in field systems by this method. Nitrogen removal was reported as 60 percent at Melbourne, 85 percent at Paris, and 80 percent at Ada. Nitrogen removal was due to plant uptake, denitrification, and volatilization of ammonia. Nitrate movement to the groundwater is not a problem in overland flow systems because only a small fraction of the applied wastewater moves through the soil.

9.3.1.9. Nitrogen Removal During Land Application of Sludge

When sludge is applied to cropland, the amount of nitrogen available in any one year must be balanced against the amount used by the crop. Sludges contain from 1 to 6 percent nitrogen that is partly inorganic and partly organic (64). Since much of the NH_4–N in liquid sludge is volatilized during drying, the total N content of liquid sludge is usually higher than that of dried sludge. The inorganic N that is primarily NH_4–N is available immediately while the organic N must be mineralized before it can be taken up by the crop or leached. Generally, 30–60 percent of the nitrogen in fluid sludges is NH_4–N (65). The organic N (40–70 percent of the total) is mineralized slowly and the rate of this microbial conversion depends upon local conditions such as temperature, water content, and the amount of inorganic N added. For Wisconsin soils, Keeney et al. (66) suggested a decay series of 15 percent availability in the first year, 6 percent of the remaining nitrogen in the second year, 4 percent in the third year, and 2 percent in the fourth year. Pratt et al. (67) suggested a decay series of 35, 10, 6, and 5 percent for years 1 to 4 for sludge applied under warmer conditions to California soils.

If liquid sludge is applied to the soil surface, nitrogen loss by volitalization can range from 30 percent to essentially all of the NH_4–N (64). However, liquid sludge ordinarily would be incorporated into the soil to avoid losses from runoff following rainfall. Some nitrogen could still be lost by volatilization and denitrification, but crop removal would account for most of the removal of available nitrogen.

Nearly all of the nitrogen in sludge compost is in the organic form that must be mineralized before it is available for crops or it is mobile. The nitrogen content of sludge compost is less than that of the original digested sludge due to volatilization losses during composting. Research indicated that compost from the Beltsville Aerated Pile Method (about 1 percent total N) mineralizes at the rate 10, 5, and 2 percent for the first, second, and subsequent years, respectively (68). If a user depended upon sludge compost as the sole N source for a crop requiring 100 kg N/ha, the application of compost needed would be 100 mt/ha for the first year, about 60 mt/ha for the second, third, and fourth years, and about 30 mt/ha thereafter until a mineralization equilibrium was established to supply adequate available nitrogen to the crop. Since large compost applications are required to supply enough available nitrogen, the application of compost probably would be limited by regulations governing the amount of heavy metals applied to soils before enough excess nitrogen for leaching could be applied.

9.3.2. Phosphorus Movement

Phosphorus (P) movement through soils is quite limited due to various adsorption and precipitation reactions. For example, Kao and Blanchar (69) found that a soil fertilized with P for 82 years had nearly double the total P content of unfertilized soil, but the ability of the fertilized soil to adsorb additional P was the same as the unfertilized soil. Similarly, phosphorus in sewage water applied in irrigation systems is fixed in the soil. For example, Dowdy et al. (70) found that P concentrations in soil water below corn plots irrigated with secondary sewage effluent were always 0.1 mg/liter at both the 60- and 125-cm depths in spite of an accumulated addition of 710 kg/ha of P over a four-year period. Phosphorus removal was also essentially complete at the Muskegon County site and projections indicated that phosphorus removal should continue for at least 50 years (53). Kardos and Hook (71) studied four field sites that had been irrigated year-round with sewage water for 9–11 years and found that no site had leaching losses of more than 3 percent of the total applied P.

A given soil does not have a fixed capacity to remove P because adsorption and precipitation of P is dependent upon many factors such as the phosphorus concentration in the soil solution, soil pH, temperature, time, and the concentrations of other ions in the wastewater or sludge (53). The complex reactions governing phosphorus removal can be considered in two general categories: inital adsorption reactions and much slower precipitation reactions that regenerate additional adsorptive surfaces. As mentioned above, the soil can retain P applied at the application rates used for wastewater irrigation systems and sludge applications to cropland. However, the very high application rates used on the sandy soils in rapid-infiltration systems can overwhelm the soil's P removal capacity and cause substantial quantities of P to be leached to the groundwater.

Bouwer et al. (7) showed that reclaimed water from wells below groundwater recharge basins of the Flushing Meadows project still contained about 60 percent of the original P concentration in the wastewater. During five years of flooding, the amount of P applied was about 10,000 kg/ha/year. Using soil columns as models of the Flushing Meadows project, Lance (72) found that the P removed by the sand columns was proportional to the flow rate through the soil. P removal could be increased from about 10 percent to about 80 percent by decreasing the flow rate from 45 cm/day to 10 cm/day. When flow rates were lowered at the Flushing Meadows project by reducing the water depth in the recharge basins, P removal was increased from 50 percent to about 80 percent (7). Therefore, high rate-infiltration systems can be managed so that most of the P is removed by the soil. However, the removal capacity of some soils can be reduced by many years of treatment with large volumes of wastewater. Levin et al. (60) found that after 30 years of wastewater application at an annual rate of 15.6 m, the P removal by 7 m of vertical travel was about 30 percent.

9.4. HEAVY METALS IN LAND TREATMENT SYSTEMS

Since the concentration of heavy metals in municipal wastewater is quite low, heavy metals should not be a limiting factor in long-term wastewater irrigation practices. Brown (73) reviewed the fate of heavy metals applied to cropland during irrigation with wastewater and found no evidence of leachate losses of metals from those soils. The metals applied to the land were generally concentrated near the soil surface by adsorption and precipitation mechanisms. Heavy metals could move through the sandy gravel soils used at some rapid-infiltration sites. However, concentrations in wastewaters are usually low because most of the heavy metals are removed in the sludge unless some particularly heavy concentrations are present from industrial releases. Source control would be the logical solution in those cases.

Application of high metal sludges on land can result in soil enrichment in heavy metals. Generally, buildup does not result in groundwater contamination but heavy metals may be taken up by plants and enter the food chain through direct ingestion by humans or indirectly through animals. Cadmium (Cd) is the element of greatest concern to human health because it is taken up readily by most crops and plant uptake is not limited by phytoxicity at the concentrations normally found in sludge. In order to limit any potential health risks, the amount of sludge applied to land is restricted by regulatory agencies (68).

Compared to other trace metals, boron (B) is rather weakly adsorbed by soils (74). It is weakly adsorbed by clay minerals and iron and aluminum oxides (75) and can be desorbed if as much as about three pore volumes of water is applied (76). Therefore, high-rate land filtration systems could be particularly susceptible to the leaching of B. For example, Bouwer, et al. (7) found that B concentrations in the wells beneath groundwater recharge basins were about the same as B concentrations in the sewage effluent.

9.5. LEACHING OF DISSOLVED SALTS IN LAND TREATMENT SYSTEMS

Sewage effluents usually contain about 100–300 mg/liter more salt than the water entering the municipal supply system (77). In arid areas where evapotranspiration rates are high, salts concentrate in the root zone and must be removed by leaching. The total dissolved solids (TDS) content of most sewage effluents tends to be in the range of 500–1000 mg/liter. The amount of extra water that needs to be applied to a crop to leach these salts out of the root zone is calculated from the salt-balance equation, which in its simplest form is written as

$$C_i D_i = C_d D_d \qquad [9.4]$$

where C_i = salt concentration of irrigation water; D_i = amount of irrigation water infiltrated into soil; C_d = salt concentration of leachate; D_d = amount of leachate.

Values of C_d can be taken from relations between soil-water salinity and crop yield (78), so that the leaching requirement D_d/D_i can be calculated. For most field irrigation systems, the actual value of D_d/D_i tends to be well in excess of the leaching requirement to maintain salt balance in the root zone. The salt content of the leachate will then be correspondingly less. From a groundwater protection standpoint, D_d/D_i should be as close to the leaching requirement as possible. This will then produce a leachate with maximum salt concentration which would minimize dissolution of soil minerals and enhance precipitation of salt (mostly calcium carbonate and calcium sulphate) in the vadose zone. A minimum value of D_d would also minimize the rate at which the leachate moves down to the groundwater, thus maximizing the amount of salt stored and "in transit" in the vadose zone. This will delay and perhaps even reduce the salt load on the underlying groundwater. In high-rate land infiltration systems, D_i and D_d are about the same because the amount of water lost by evaporation is small in comparison to the amount applied. This results in little salt concentration in the leachate. For example, Bouwer et al. (5) found that the water moving below the Flushing Meadows project contained only about 2 percent more salt than the sewage effluent applied at the surface at the rate of 100 m/year.

9.6. ORGANIC COMPOUNDS IN LAND TREATMENT SYSTEMS

Movement of high concentrations of organic compounds through the soil to the groundwater is very unlikely because of the large organic degradation capacity of the soil microorganisms. Laboratory and field studies showed that about 80 percent of the total organic carbon was removed from sewage water in rapid-infiltration systems even when the annual loading rate was as high as 20,000 kg/ha (5, 59). At the slower leaching rates used for irrigation systems and sludge applications to cropland, the degradation of organic compounds should be more extensive. Thus the readily degradable organic carbon (proteins, carbohydrates, etc.) in sewage effluent and sludge is almost completely decomposed in the soil environment. However, chlorinated organics, detergents, pesticides, and other "trace organics" found in the sewage effluent or added to the soil as a result of agricultural practices are not all degraded in the soil and can move down to the groundwater and persist in the aquifer. These compounds tend to be present in low concentrations (nanograms to micrograms per liter), but some of them can be carcinogenic or otherwise toxic. Three phenomena of importance that limit the transport of trace organics in a soil percolation system are volatilization, sorption, and biodecomposition (79). Volatile trace organics can be removed by mass transfer from the liquid to the gaseous phase and then to the atmosphere. Volatilization losses would be expected in surface application systems and systems that allow considerable unsaturated flow through the soil. Therefore, volatilization losses during land treatment of organic wastes would be extensive except during long flooding periods used in high-rate land filtration systems. However, even in those systems, some volatilization can occur during application and drying periods.

Sorption processes retard the movement of trace organics through the soil and the amount of removal depends upon the soil sorption capacity for the various organic compounds that range from none to very high.

Trace organics may be biodegradable but are often below the minimum concentration to which a compound can be degraded by microorganisms because of energy limitations. Biodegradation of such materials can occur generally only if they are used as secondary substrates. This requires the presence of a primary organic substrate in a concentration higher than its minimum concentration and the presence of a microorganism that is capable of decomposing both the primary and secondary substrate. The chances of this decomposition occurring are high in soils where considerable organic substrate has been added because complex microbial populations are found in soils. The opportunities for degradation are more limited in high-rate land filtration systems where the detention time for the organic compounds may be limited to a few days.

Bouwer et al. (80) studied the movement of trace organics through a soil column flooded with sewage water at rates used in a high-rate land filtration system. Many trace organic contaminants were removed during soil percolation whereas others such as chloroform were not attenuated significantly by passage through the 250-cm long soil columns. Similar results were noted when samples from wells at the 23rd Avenue Rapid-Infiltration Project in Phoenix, Arizona were analyzed for trace organics (81, 82).

It appears that movement of trace organics below land treatment systems probably would not be a serious problem except possibly for high-rate land filtration systems. Trace organics move through these systems in very low concentrations and it is not known if these concentrations are harmful. More research is needed on the fate and movement of trace organics in the underground environment.

9.7. PRETREATMENT BEFORE LAND APPLICATION

Land application should be practiced as part of a total system that may include other treatment methods before or after land application or both. Treatments usually precede land application

Pretreatment may be helpful by removing pathogens, promoting infiltration by removal of suspended solids, preventing overloading of the system by removing nutrients, or making the system more aesthetic by reducing and stabilizing organic loads. The degree of pretreatment is determined by the capacity of the land treatment system to remove these various components. Therefore, pretreatment must be site-specific.

9.7.1. Pathogen Removal by Pretreatment

The large variety of pathogens present in domestic wastewater includes pathogenic bacteria, viruses, protozoa, and parasitic worms. These pathogens may reach humans by vegetables contaminated by wastewater used for irrigation, by farm workers working in contaminated crops or soils, by farm workers or residents in the area exposed to aerosols containing pathogens, or by pathogens moving through the soil into groundwater supplies. Some of these hazards can be alleviated by pretreatment, but management practices and exclusion of certain uses are often more practical than pretreatment.

Foster and Engelbrecht (83) assessed the pathogen removal efficiency of conventional sewage treatment systems (Table 9.3). In general, primary treatment reduces the populations of helminth ova but has little impact on the numbers of bacteria or amoebic cysts. Virus removal is probably significant because most viruses are associated with suspended particles, although some laboratory studies have failed to demonstrate significant virus removal. Thus,

Table. 9.3. Removal of Pathogens from Sewage Water by Conventional Treatment Systems[a]

Pathogen	Primary (%)	Activated sludge (%)	Trickling filter (%)
Salmonella	15	96–99	84–99.9
Mycobacterium	48–57	Slight–87	66–99
Amoebic cysts	No removal in 3 hours	No apparent removal	11–99.9
Helminth ova	72–98	No apparent removal	62–76
Virus	3 to extensive removal	76–99	0–84

[a]Adapted from Foster and Engelbrecht (83).

primary treatment removes only those pathogens heavy enough to settle out of the water or associated with suspended solids. However, primary treatment may have a significant impact on pathogen survival on crops and soils due to the removal of suspended solids that protect pathogens from adverse conditions.

Sewage treatment is not principally designed to affect pathogen removal except by disinfection. The sole purpose of disinfection is to reduce the level of any remaining pathogens to negligible levels from the public health standpoint. While most disinfectants are highly effective in removing bacteria from wastewater, far greater difficulty is encountered in the inactivation of viruses. The most commonly used disinfectant for wastewater is chlorine, although the use of other halogens, UV light, and photodynamic inactivation is currently being investigated (84, 85).

It is difficult to substantially reduce virus numbers in primary and secondary sewage by chlorine application, but in advanced wastewater treatment systems involving physical–chemical treatment, where the amount of organic matter, ammonia, and turbidity are greatly reduced, large reductions in virus concentrations are possible (86). For example, Nupen et al. (87) reported a 10^7 reduction in concentrations after breakpoint chlorination of tertiary treated effluents.

Some viruses will be present in wastewater plant effluents. Sproul (88) and Gerba et al. (89) previously estimated the percentage removal of viruses by various wastewater treatment systems. This information has been used to calculate the viruses remaining after various treatment sequences (Table 9.4). It should be emphasized that this information is largely based on work done with enteroviruses, and other viral groups may be removed more or less efficiently by these treatment processes. Based on the estimates of Clarke et al. (90) that raw sewage in the U.S. contains on the average 7,000 PFU/liter, Sproul (88) estimated that 350 viruses/liter would remain after activated sludge treatment, including chlorination and physical–chemical treatment involving phosphate precipitation and 6 PFU/liter would remain after activated carbon adsorption and chlorination. In other parts of the world, virus concentrations would be much higher. Calculations using the data of Buras (91) show that as many as 25,000 PFU of virus/liter may be present after secondary treatment and chlorination (Table 9.4).

Since pathogens are also concentrated in sludges, pretreatment of sludge may be desirable before land application. Anaerobic digestion destroys many pathogens, but some have been shown to survive lengthy digestion and air drying. Decomposition of organic matter during composting produces temperatures in the 60–70° C range that destroys such pathogens (92). However, the action of ammonia and antibiotics during composting may also be important. Forced-aeration composting was found to be more effective in destroying pathogens in cool, wet climates—that is, more than 100 cm of rainfall annually and winters with many hours of temperatures well below freezing—than windrow composting. Composting also stabilizes the organic content of sludge, thereby greatly improving the physical and aesthetic qualities of sludge.

Table 9.4. Estimated Enterovirus Reduction by Various Sewage Treatment Methods

Treatment	Percent of Virus Removal Expected	Viruses Remaining in Effluent (PFU/liter)	
		Clarke et al. (90)[a]	Buras (91)[b]
Primary treatment			
Sedimentation	0	7000	500,000
Chlorination	50	3500	250,000
Secondary treatment			
Stabilization ponds	90	700	50,000
Trickling filters	50	3500	250,000
Chlorination (after trickling filters)	50	1750	125,000
Activated sludge	90	700	50,000
Chlorination (after activated sludge)	50	350	25,000
Tertiary treatment (after sedimentation)			
Excess lime precipitation	90–99.99	700–0.7	50,000–50
Alum precipitation (after lime)	90	7–0.07	5,000–5
Chlorination (after lime and alum)	99–99.99	0.07–0.000007	50–0.0005
Activated carbon adsorption	0–50		

[a]Based on an estimate of 7000 PFU/liter.
[b]Based on a finding of 500,000 PFU/liter in wastewater in Israel.

9.7.2 Nutrients Removed by Pretreatment

Conventional treatment systems are relatively ineffective in removing nitrogen and phosphorus from wastewater. Primary treatment usually results in only slight removal while secondary treatment results in 10–20 percent removal. Therefore, conventional treatment has little effect on the nutrient load applied to land treatment systems. Various tertiary treatment systems can be used to remove most of the nutrients at considerable expense. Lagoon or oxidation ponds vary in their effectiveness, depending upon the detention time and management system. For example, the storage lagoons of the Muskegon project reduced the total nitrogen (N) content from 8.2 to 5.6 mg/liter and total phosphorus (P) from 2.4 to 1.4 mg/liter. The Deer Creek Lake lagoons reduced the total N from 50.6 to 5.4 mg/liter and P from 6 to 0.4 mg/liter (93). Some of the reduction was due to dilution. Lagoons can significantly reduce nutrient loads when used as pretreatment before land application. If storage is required, nutrient removal will depend largely upon the detention time in the lagoon.

9.7.3. Heavy Metal Removal by Pretreatment

Removal of heavy metals from wastewater may be necessary to avoid injury to plants and to prevent harmful accumulations of metals in food chain crops and soils or movement to groundwater. Most heavy metals are removed by conventional treatment and even primary treatment is effective in producing wastewater acceptable for land treatment (53). This is due to the association of most of the metals with the suspended solids. However, this may not be true for some effluents with large amounts of industrial wastes containing high metal concentrations. The association of heavy metals with suspended solids means that most of the metals are concentrated in the sewage sludge. The heavy metal content is one of the primary factors limiting the load of sludge that can be applied for land treatment. Thus, various pathogens and pollutants are still present in municipal wastes that are available for land treatment, even if the waste is extensively pretreated prior to land application.

9.8. SUMMARY

Land application can be an economical and effective way to utilize municipal wastes if land treatment systems and management practices are carefully designed to avoid contamination of water resources and food supplies. Relatively thin layers of soils remove bacteria primarily by straining and viruses by adsorption. Exceptions are very coarse sands and gravels that have very high-infiltration rates and may allow direct channeling to the groundwater. The most important soil characteristic affecting pathogen removal by soils seems to be soil permeability. The pathogen removal capacity of a soil can be changed by changing soil permeability and perhaps other alterations. More research is needed to understand basic pathogen–soil interactions and to develop methods to increase pathogen removal capacity of soils.

Contamination of groundwater by nutrients can be prevented if the capacity of land treatment systems to assimilate nitrogen is not exceeded. For wastewater irrigation and sludge utilization systems, crop removal of nitrogen is the process that must be managed to prevent movement of nitrogen through the soil to the groundwater. In high-rate groundwater recharge systems, denitrification must be promoted to eliminate much of the nitrogen. Most soils have a tremendous capacity for phosphate removal, but some very sandy gravel soils could allow phosphorus movement to the groundwater.

Movement of heavy metals in soils is quite limited, but quantities of metals applied to soils must be restricted to limit soil enrichment and subsequent crop contamination. In arid areas, irrigation of land treatment systems must be managed to minimize salt movement to groundwater by enhancing the precipitation of salts in the vadose zone. The movement of trace organics in land treatment systems is minimal except in high-rate land filtration systems. More research is needed on the movement and degradation of trace organics in soils. (See Chapter 5.)

REFERENCES

1. R. G. Butler, G. T. Orlab, and P. H. McGauhey, *J. Am. Water Works Assoc.* **46**, 97 (1954).

2. D. B. Aulenbach, T. B. Glavin, and J. A. R. Rojas, *Groundwater* **12,** 161 (1974).
3. G. Bitton, N. Lahav, and Y. Henis, *Plant and Soil* **40, 373 (1974).**
4. W. A. Ziebell, J. L. Anderson, J. Bouma, and E. McCoy, "Fecal Bacteria:Removal From Sewage by Soils," Amer. Soc. Agr. Engin. Paper #75-2579, Amer. Soc. Agr. Engin., St. Joseph, MI (1975).
5. H. Bouwer, J. C. Lance, and M. S. Riggs, *J. Water Pollut. Contr. Fed.* **46,** 844 (1974).
6. J. C. Lance and F. D. Whisler, *Hydrology and Water Resources in Arizona and the Southwest* **5,** 57 (1975).
7. H. Bouwer, R. C. Rice, J. C. Lance, and R. G. Gilbert. *J. Water Pollut. Contr. Fed.* **52,** 2457 (1980).
8. C. P. Gerba, C. Wallis, and J. L. Melnick. *J. Irrig. Drain, Div.* Am. Soc. Civil Engin. **101,** 57 (1975).
9. G. G. Robeck, W. A. Clarke, and K. A. Dostal, *J. Am. Water Works Assoc.* **54,** 1275 (1962).
10. W. A. Drewry and R. Eliassen, *J. Water Pollut. Contr. Fed.* **40,** R257 (1968).
11. R. H. F. Young and N. C. Burbank, Jr., *J. Am. Water Works Assoc.* **65,** 598 (1973).
12. G. F. Carlson, Jr., F. E. Woodward, D. F. Wentworth, and O. J. Sproul, *J. Water Pollut. Contr. Fed.* **40,** R89 (1968).
13. C. Wallis, M. Henderson, and J. L. Melnick, *Appl. Microbiol.* **2,** 476 (1972).
14. J. C. Lance, *Trans, Am. Soc. Agr. Engin.* **21,** 1114 (1978).
15. S. M. Duboise, B. E. Moore, and B. P. Sagik, *Appl. Environ. Microbiol.* **31,** 536 (1976).
16. E. F. Landry, J. V. Vaughn, McH. Z. Thomas, and C. A. Beckwith, *Appl. Environ. Microbiol.* **38,** 680 (1979).
17. J. C. Lance, C. P. Gerba, and J. L. Melnick, *Appl. Environ. Microbiol.* **32,** 520 (1976).
18. J. C. Lance and F. D. Whisler, *J. Environ. Qual.* **1,** 180 (1972).
19. De-S Wang, J. C. Lance, and C. P. Gerba, *Appl. Environ. Microbiol.* **39,** 662 (1980).
20. C. J. Hurst, C. P. Gerba, J. C. Lance, and R. C. Rice, *Appl. Environ. Microbiol.* **40,** 192 (1980).
21. C. P. Gerba and J. C. Lance, *Appl. Environ. Microbiol.* **36,** 247 (1978).
22. J. C. Lance and C. P. Gerba, *J. Environ. Qual.* **9,** 31 (1980).
23. G. L. Dugan, R. H. F. Young, L. S. Lau, P. C. Ekern, and P. C. S. Loh, *J. Water Pollut. Contr. Fed.* **47,** 2067 (1975).
24. E. L. Koerner and D. A. Haws, *Long-Term Effects of Land Application of Domestic Wastewater: Roswell, New Mexico, Slow Rate Irrigation Site,* EPA 600-279047, U.S. Environmental Protection Agency, 1979.
25. F. M. Wellings, A. L. Lewis, and C. W. Mountain, "Virus Survival Following Wastewater Spray Irrigation of Sandy Soils," In J. F. Malina, Jr. and B. P. Sagik, Eds., *Virus Survival in Water and Wastewater Systems,* Center for Research in Water Resources, Austin, TX, 1974, pp. 253–260.
26. R. G. Gilbert, R. C. Rice, H. Bouwer, C. P. Gerba, C. Wallis, and J. L. Melnick, *Science* **192,** 1004 (1976).
27. J. M. Vaughn and E. F. Landry, "The Occurrence of Human Enteroviruses in a Long Island Groundwater Aquifer Recharged with Tertiary Wastewater Effluent,"

in H. L. McKim, Ed., *State of Knowledge in Land Treatment of Wastewater*, U.S. Army Corps of Engineers, Cold Region Research and Engineering Lab, Hanover, NH, 1978, pp. 233–243.

28. S. A. Schaub and C. A. Sorber, *Appl. Environ. Microbiol.* **33,** 609 (1977).
29. E. L. Koerner and D. A. Haws, *Long-Term Effects of Land Application of Domestic Wastewater: Vineland, New Jersey, Rapid Infiltration Site*, U.S. Environmental Protection Agency, Ada, OK, 1979.
30. J. M. Vaughn, E. F. Landry, L. J. Baranosky, C. A. Beckwith, M. C. Dahl, and N. C. Delihas, *Appl. Environ. Microbiol.* **36,** 47 (1978).
31. J. C. Lance and C. P. Gerba, "Virus Removal with Land Filtration," in E. J. Middlebrooks, Ed., *Water Reuse*, Ann Arbor Sci. Publ., Inc., Ann Arbor, MI, 1982, pp. 305–324.
32. S. A. Schaub, K. F. Kenyon, B. Bledsoe, and R. E. Thomas, *Appl. Environ. Microbiol.* **39,** 127 (1980).
33. S. Damgaard-Larsen, K. O. Jensen, E. Lund, and B. Nissen, *Water Res.* **11,** 503 (1977).
34. G. Bitton, J. M. Davidson, and S. R. Farrah, *Water Air Soil Pollut.* **12,** 449 (1979).
35. C. J. Hurst, C. P. Gerba, and I. Cech, *Appl. Environ. Microbiol.* **40,** 1067 (1980).
36. W. N. Mack, Y. S. Lu, and D. B. Coohan, *Health Services Report* **87,** 271 (1972).
37. J. C. Lance, *J. Water Pollut. Contr. Fed.* **44,** 1352 (1972).
38. W. E. Sopper, "Crop Selection and Management Alternatives—Perennials," *Proc. Joint Conf. on Recycling Municipal Sludges and Effluents on Land*, Champaign, IL, 1973, pp. 143–154.
39. F. E. Broadbent and F. E. Clark, "Denitrification," W. V. Bartholomew and F. E. Clark, Eds., *Soil Nitrogen*, Am. Soc. Agron., Madison, WI, 1965, pp. 347–379.
40. G. S. Cooper and R. L. Smith, *Soil Sci. Soc. Amer. Proc.* **27,** 659 (1963).
41. M. Alexander, *Introduction to Soil Microbiology*, John Wiley and Sons, Inc., New York, 1950.
42. R. G. Gilbert, J. B. Robinson, and J. B. Miller, "The Microbiology and Nitrogen Transformations of Recharge Basin Used for Wastewater Renovation," *Proc. Internatl. Conf. on Land for Waste Management*, Ottawa, Ontario, Canada, 1973, pp. 87–96.
43. J. M. Bremner and K. Shaw, *J. Agr. Sci.* **51,** 40 (1958).
44. B. D. Meek and A. J. MacKenzie, *Soil Sci. Soc. Am. Proc.* **33,** 575 (1969).
45. R. C. Stefanson, *Australian J. Soil Res.* **10,** 183 (1973).
46. J. W. Woldendrop, *Plant and Soil* **17,** 267 (1962).
47. P. O. St. Amant and P. L. McCarty, *J. Am. Water Works Assoc.* **61,** 659 (1969).
48. J. C. Lance, F. D. Whisler, and R. C. Rice, *J. Environ. Qual.* **5,** 102 (1976).
49. W. V. Bartholomew, "Mineralization and Immobilization of Nitrogen in the Decomposition of Plants and Animal Residues," in W. V. Bartholomew and F. E. Clark, Eds., *Soil Nitrogen*, Am. Soc. Agron., Inc., Madison, WI, 1965, pp. 285–306.
50. F. E. Broadbent, W. D. Burge, and T. Nakashima, *Trans. 7th Intl. Congress of Soil Sci.*, Madison, WI, Vol. 2, 1970, pp. 509–516.
51. W. D. Burge and F. E. Broadbent, *Soil Sci. Soc. Am. Proc.* **25,** 199 (1961).
52. J. L. Young, *Soil Sci. Soc. Am. Proc.* **28,** 334 (1964).
53. U.S. Environmental Protection Agency, U.S. Army Corps of Engineers, and U.S.

Department of Agriculture, *Process Design Manual for Land Treatment of Municipal Wastewater*, EPA 625/1-77-008 (COE EM 100-1-501), 1977.

54. L. A. Richards, *Diagnosis and Improvement of Saline and Alkali Soils*, Agri. Handbook #60, U.S. Dept. of Agriculture, Washington, D.C. (1954).
55. S. S. Brar, R. H. Miller, and T. J. Logan. *J. Water Pollut. Contr. Fed.* **50**, 709 (1978).
56. D. R. Linden, W. E. Larson, and R. E. Larson. "Agricultural Practices Associated with Land Treatment of Wastewater," in H. L. McKim, Ed., *State of Knowledge in Land Treatment of Wastewater*, Vol. 1, U.S. Army Corps of Engineers Cold Region Research and Engineering Laboratory, Hanover, NH, 1978, pp. 313–322.
57. J. C. Lance, F. D. Whisler, and H. Bouwer, *J. Environ. Qual.* **2**, 345 (1973).
58. J. C. Lance and F. D. Whisler. *J. Water Pollut. Contr. Fed.* **48**, 346 (1976).
59. J. C. Lance, R. C. Rice, and R. G. Gilbert, *J. Water Pollut. Contr. Fed.* **52**, 381 (1980).
60. P. E. Levine, J. V. Olson, and R. W. Crites. "Nitrogen and Phosphorous Removal after 30 Years of Rapid Infiltration," in H. L. McKim, Ed., *State of Knowledge in Land Treatment of Wastewater*, Vol. 2, U.S. Army Corps of Engineers, Cold Region Research and Engineering Laboratory, Hanover, NH, 1978, pp. 17–26.
61. J. P. McPherson, *Progress in Water Technol.* **9**, 1035 (1978).
62. J. P. Law, R. E. Thomas, and L. H. Meyers, *J. Water Pollut. Contr. Fed.* **42**, 1621 (1970).
63. R. E. Thomas, "Spray Runoff to Treat Raw Domestic Wastewater," *Proc. Internat'l. Conf. on Land for Waste Management*, Ottawa, Ontario, Canada, October 1973, pp. 354–360.
64. Council for Agricultural Science and Technology, *Application of Sewage Sludge to Cropland: Appraisal of Potential Hazards of the Heavy Metals to Plants and Animals*, Report No. 64 (1976).
65. L. D. King, *J. Environ. Qual.* **2**, 356 (1973).
66. D. R. Keeney, K. W. Lee, and L. M. Walsh, "Guidelines for the Application of Wastewater Sludge to Agricultural Land in Wisconsin," Tech. Bull. 88, Dept. of Natural Resources, Madison, WI (1975).
67. P. F. Pratt, F. E. Broadbent, and J. B. Martin. *California Agri. J.* **27**, 10 (1973).
68. G. B. Willson, J. F. Parr, E. Epstein, P. B. March, R. L. Chaney, D. Colacicco, W. D. Burge, L. J. Sikora, C. F. Tester, and S. Hornick, *Manual for Composting Sewage Sludge by the Beltsville Aerated-Pile Method*, EPA-600/8-80-022, U.S. Department of Agriculture and U.S. Environmental Protection Agency, 1980.
69. C. W. Kao and R. W. Blanchar, *J. Environ. Qual.* **2**, 237 (1973).
70. R. H. Dowdy, C. E. Clapp, G. C. Martin, D. R. Linden, and W. E. Larson, *Proc. Conf. Vegetation Management for Municipal Wastewater Land Application* (1981).
71. L. T. Kardos and J. E. Hook, *J. Environ. Qual.* **5**, 87 (1976).
72. J. C. Lance, *J. Environ. Qual.* **6**, 279 (1977).
73. K. W. Brown, "The Fate of Sewage Effluent Heavy Metals in Land Wastewater Disposal Sites," in H. L. McKim, Ed., *State of Knowledge in Land Treatment of Wastewater*, Vol. 2, U.S. Army Corps of Engineers, Cold Region Research and Engineering Laboratory, Hanover, NH, 1978, pp. 301–314.
74. A. C. Chang and A. L. Page. *Proc. Sympos. Wastewater Reuse for Groundwater Recharge.* Office of Water Recycling—California State Water Resources Control Board, 1980, pp. 118–136.

75. J. R. Sims and F. T. Bingham, *Soil Sci. Soc. Am. J.* **32,** 369 (1968).

76. J. D. Rhoades, R. D. Ingvalson, and J. T. Hatcher, *Soil Sci. Soc. Am. Proc.* **34,** 871 (1979).

77. H. Bouwer and R. L. Chaney, "Land Treatment of Wastewater," in N. C. Brady, Ed., *Advances in Agronomy*, Vol. 26, Academic Press, New York, 1974, pp. 133–176.

78. H. Bouwer, *J. Irrig. Drainage Div.*, Am. Soc. Civil Engin. **95,** 153 (1969).

79. P. L. McCarty, B. E. Rittman, and M. Reinhard, *Proc. Sympos. Wastewater Reuse for Groundwater Recharge*, Office of Water Recycling—California State Water Resources Control Board, 1980, pp. 93–117.

80. E. J. Bouwer, P. L. McCarty, and J. C. Lance. *Water Res.* **15,** 151 (1981).

81. E. J. Bouwer, M. Reinhard, and P. L. McCarty, "Organic Contaminant Behavior during Rapid Infiltration of Secondary Wastewater at the Phoenix 23rd Avenue Project," Tech. Report No. 264, Dept. of Civil Engineering, Stanford University, CA, 1982.

82. M. B. Tomson, J. Dauchy, S. Hutchins, C. Curran, C. J. Cook, and C. H. Ward, *Water Res.* **15,** 1109 (1981).

83. D. H. Foster and R. S. Engelbrecht, "Microbial Hazards of Disposing of Wastewater on Soil," in W. E. Sopper and L. T. Kardos, Eds., *Recycling Treated Municipal Wastewater and Sludge through Forest and Cropland*, Pennsylvania State University Press, University Park, PA, 1954, pp. 247–270.

84. C. P. Gerba, C. Wallis, and J. L. Melnick, *J. Water Pollut. Contr. Fed.* **49,** 575 (1977).

85. U.S. Environmental Protection Agency, *Disinfection of Wastewater*, Task Force Report, EPA-430/9-75-012, Washington, D.C. (1976).

86. R. L. Culp, "Breakpoint Chlorination for Virus Inactivation," in J. F. Malina, Jr. and B. P. Sagik, Eds., *Virus Survival in Water and Wastewater Systems*, Center for Research Water Resources, Austin, TX, 1974, pp. 158–165.

87. E. M. Nupen, B. W. Bateman, and N. C. McKenney, "The Reduction of Virus by the Various Unit Processes Used in the Reclamation of Sewage to Potable Waters," in J. F. Malina, Jr. and B. P. Sagik, Eds., *Virus Survival in Water and Wastewater Systems*, Center for Research in Water Resources, Austin, TX, 1974, pp. 107–114.

88. O. J. Sproul, "Removal of Viruses by Treatment Processes," in G. Berg, H. L. Bodily, E. H. Lennette, J. L. Melnick, and T. G. Metcalf, Eds., *Virus in Water*, Am. Public Health Assoc., Washington, D.C., 1976, pp. 167–197.

89. C. P. Gerba, C. Wallis, and J. L. Melnick, *Environ. Sci. Technol.* **9,** 1122 (1975).

90. N. A. Clarke, R. E. Stevenson, S. L. Chang, and P. W. Kabler, *J. Am. Health* **51,** 1118 (1961).

91. N. Buras, *Water Res.* **8,** 19 (1974).

92. W. D. Burge, P. B. Marsh, and P. D. Millner, *Proc. 1977 National Conf. on Composting of Municipal Residues and Sludges*, 1977, pp. 128–134.

93. D. J. Lambert and H. E. McKim, "Deer Creek Lake, Inland Wastewater Treatment System," in R. C. Loehr, Ed., *Food, Fertilizers, and Agricultural Residues,* Ann Arbor Sci. Publ., Inc., Ann Arbor, MI, 1977, pp. 79–94.

10

MICROORGANISMS AS GROUNDWATER TRACERS

Charles P. Gerba

Department of Microbiology

and

Department of Nutrition and Food Science
University of Arizona, Tucson, Arizona

10.1. Introduction

The movement of bacteria and viruses into groundwater is influenced by many environmental factors (1) that are difficult to define completely. Thus, the ability to trace microbial movement in groundwater is essential in recognizing the potential for transmission of disease-causing microorganisms. Therefore, it is desirable to use tracer techniques to monitor the movement of microorganisms with percolating water through soil systems and groundwater when assessing new sites for land application of wastewater, septic-tank drainfields, and investigating sources of waterborne disease outbreaks.

Most groundwater tracers used for hydrological and geological studies are chemicals, most often fluorescein dyes or halogen salts. They are used to determine directions and velocities of subsurface flow. Chemical tracers, however, do not always reflect the movement of microorganisms into groundwater (2–5). Techniques are needed to predict the movement of suspected sources of groundwater pollution with highly distinctive tracer organisms that can be identified in the groundwater system. The criteria for selecting a suitable tracer organism for groundwater contamination are survival time and retention in soil water systems. Microbial tracers have the advantage of not being mutagenic (6), not having potential toxic effects, and having a finite lifetime.

Because of their size, the movement of yeast (3–7μm) and bacteria (1–2μm) is more limited in groundwater than viruses (0.02–0.4μm). Large reductions of bacteria can occur after percolation of sewage through a meter or less of sandy soil (see Chapter 9). Thus, when bacteria are used, reductions of 99 percent or greater should be anticipated. Different species of bacteria may also be removed at different rates. For example, we have observed in laboratory soil column experiments that coliform bacteria are removed more effectively than fecal streptococci (D. Wang and C. P. Gerba, unpublished). Virus removal by soil is believed to be due largely to adsorption. In general, viruses would be expected to travel longer distances in the subsurface if ionic concentrations are low and the pH of the groundwater is above neutral. Virus adsorption is also highly type and strain-dependent (7) and some viruses are poorly removed by soils. In recent laboratory and field tests, we have found *Escherichia coli* phage f2 to have the lowest removal of any animal or bacterial virus studied (C.P. Gerba, unpublished). Other viruses such as bacteriophage MS-2 and poliovirus adsorb very readily to most soils and their movement in the subsurface would be expected to be more limited.

Microorganisms such as *Escherichia coli* can be grown to high concentrations in overnight cultures in the laboratory. For example, an overnight culture of a one-liter volume of broth will yield approximately 10^{11}-10^{12} organisms. If possible, bacteria should be used from overnight cultures within 1--12 hr to avoid excessive bacterial die-off. One-hundred-ml volumes of water can be assayed for bacteria using standard membrane filtration techniques (8). Results are usually obtained within 24–48 hr. Stocks of *E. coli* phage f2 and MS-2 can be produced to yield 10^{11}–10^{12} phage/ml in crude cell lysates. One to two ml samples can be conveniently assayed using standard assay techniques (9). For greater sensitivity, methods are available for the concentration of phages from large volumes of water. With the use of these techniques, it is possible to detect as little as 2–3 phages in a 20-liter volume of water (10). Phage assay requires 12–24 hr, which may be a drawback in some applications, but for most studies would pose no difficulties when samples are collected over long periods of time; furthermore, samples can be frozen for long-term storage and assayed later.

An excellent, practical guide to the use of microorganisms as tracers of groundwater can be found in the *Water Tracers Cookbook* by Aley and Fletcher (11).

10.2. BACTERIA AS TRACERS

Bacteria are the most commonly used microbial tracers because of their ease of growth and detection. Coliform bacteria and, in particular, *Escherichia coli* (a fecal coliform) are currently extensively used for monitoring groundwater quality. Coliform bacteria are excreted in large numbers in the fecal wastes of man and other warm-blooded animals and thus can be used to monitor the movement of septic and sewage wastes in groundwater and can be differentiated easily from common soil microflora by their ability to grow on selective media at elevated temperatures. Several investigators (12–18) have looked at the movement of these indicators of fecal pollution at pit latrines, septic fields, and sewage disposal sites. Martin and Noonan (16) traced fecal coliforms at a land disposal site in New Zealand over a distance of 900 m. The bacteria moved at a rate of approximately 150 m/day. Based on this rate of movement and the measured survival time for the bacteria, fecal coliforms were estimated to be traceable for at least 2.5 km from their source. The disadvantage of using fecal organisms as tracers is the difficulty in differentiating between those derived from the suspected source and those from other sources. Therefore, some investigations have been conducted by utilizing fecal bacteria with specific characteristics such as a marker that allowed the tracer organisms to be distinguished from all background organisms in the system (5, 19, 20).

Hagedorn et al. (19) studied the potential use of antibiotic-resistant strains of *Escherichia coli* and *Streptococcus faecalis* to monitor movement of subsurface water flow in one soil series in western Oregon. These organisms are distinguished from naturally occurring organisms by their ability to grow on media containing antibiotics that suppress the growth of the naturally occurring organisms. They observed that these microorganisms could travel 500 cm in a septic drainfield under saturated conditions and could survive a 32-day sampling period. It was suggested that these organisms were suitable as indicators of microbial translocation and may be utilized satisfactorily as tracers. Similar experiments were conducted by Rahe et al. (20) using strains of antibiotic-resistant *E. coli* to study the transport of fecal organisms through two western Oregon hillslope soils. The strains of *E. coli* were found to survive in significant numbers in the soils over periods of at least 96 hr and they appeared to be satisfactory as tracers of subsurface water flow. Recently Rahe et al. (5) reported a comparison study between antibiotic-resistant *E. coli* and a fluorescein dye as tracers of ground water flow in a soil series in Oregon. The bacteria were recovered at the 45 cm depth from 15 m downslope 1 hr after inoculation. In addition, large numbers of *E. coli* survived during the 12-hr sampling periods. However, the failure to visually detect the fluorescein dye in groundwater samples raised serious concerns about the continued use of such dyes as groundwater tracers for microorganisms. The results indicated that consideration should be given to more frequent use of marked strains of bacteria as tracers since these organisms more closely represent subsurface movement of microorganisms in soil and groundwater.

More recently, a series of tracer experiments were conducted at Canterbury, New Zealand, using a number of different species of bacteria (21). The results indicated that *Bacillus stearothermophilus* and H_2S-producing strains of *E. coli* traveled over a distance of 920 m in the underlying groundwater system. Both species moved from the injection well at a rate of 200 m/day; however, *Bacillus stearothermophilus* was found to occur naturally in Canterbury soils and groundwater systems, so the use of this species as a groundwater tracer was considered to be limited to periods of low rainfall when numbers were low. Furthermore, since both *E. coli* (H_2S^+) and *Bacillus stearothermophilus* appeared to be present in raw and treated sewage effluent, they were found to be unsuitable as tracers in sewage-polluted water. Tracer experiments were also conducted by injecting a mixture of antibiotic-resistant *E. coli* strains into a well site and monitoring their movements in downstream wells (22). The rate of movement during this experiment was found to be 350 m/day. However, this high velocity was considered to be due to the higher than normal groundwater table present during this study rather than the type of tracer. Pyle (4) investigated aquifer contamination at Heretauga, New Zealand using H_2S-producing strains of *E. coli* together with rhodamine WT dye as groundwater tracers, It appeared that the movement of *E. coli* was faster than that of rhodamine dye. In addition, these bacteria were found to survive for at least 1 month in the groundwater.

Other species of bacteria previously indicated to have potential as tracer organisms include *Bacillus chromobacterium*, *Streptococcus*, and *Serratia sp*; the latter produces red-pigmented colonies on nutrient agar when incubated at 25–30°C. Fournelle et al. (23) studied the lateral movement of bacteria through shallow groundwater near Anchorage, Alaska by introducing various types of bacteria in a well and then sampling for the presence of these bacteria in a series of test wells. The first study was conducted using *Serratia marcescens*, *Chromobacterium violaceum*, and *Bacillus globigii*. However, none of these bacteria were found to be satisfactory as tracer organisms due to an inability to distinguish these organisms from those naturally present in the soil. In separate studies, *Streptococcus zymogenes* (*S liquefaciens*) was demonstrated successfully as a ground water tracer in terms of its long-term survival and its capability to move in groundwater; movement of as much as 15.2 m was observed in 70 days. Antibiotic-resistant strains of *Serratia marcescens* were used by Wimpenny et al. (24) as water tracers in a heavily polluted river. Although the numbers of this tracer organism recovered on antibiotic-containing media were at least tenfold less than on normal media, it was considered to be a reasonable organism for use as a water tracer because it was easily distinguished on agar plates.

Marti et al. (25) studied several bacterial tracers and found significant differences in their travel through the aquifer under study. *Bacillus pomilus* was removed after a travel distance of 45 m. The speed at which *Streptococcus fecalis*, *Serratia marcescens*, and a fluorescent dye travelled over a distance of 90 m was found to be 220–240 m/day. *Bacillus subtilis*, on the other hand, moved at a speed of 150 m/day.

Bacteria have the following advantages as groundwater tracers. They are easy to grow in large numbers and to assay. Further, through the use of various markers, they are usually distinguishable from other flora likely to be present in the groundwater and soil. However, care should be taken in the choice of organisms since they could be potential pathogens; even *E. coli* has enterotoxigenic strains that can cause disease in man. Some are capable of growth in the environment, thus producing erroneous results; they are large enough to be filtered out in certain soils; they can adsorb to a variety of surfaces removing them from circulation; their movement probably does not reflect the movement of other types of microorganisms like viruses. In addition, even with markers like antibiotic-resistance, it is often difficult to distinguish test organisms from those naturally present because markers can be lost during interaction with natural populations. It also would seem to be imprudent to inject antibiotic-resistant microorganisms into groundwater, because of the possible transfer of resistance to potential human pathogens especially when the water might be consumed later. The odds of this can be reduced greatly by the use of bacteria that do not carry the genetic information for antibiotic sensitivity on plasmids. Thus prudence should be used in the selection of bacterial species to be used as tracers and should be used only when experimental conditions dictate or where the movement of groundwater is traced back to a source of pollution where organisms are already present.

Serratia sp. has been used over the years as a tracer organism for water and air movement because its highly colored colonies on agar plates can be readily distinguished (24, 25). *Serratia sp.* is now recognized as a potential human pathogen (26) and its use as a tracer is not recommended (27, 28). The data on the use of bacteria as tracers is summarized in Table 10.1.

Table 10.1. Use of Bacteria as Tracers

	Maximum Distance of Movement (m)		
Tracer Microorganism	Horizontal	Vertical	Reference
Fecal coliforms	900	15–22	16
Bacillus stearothermophilus	920	18	22
Escherichia coli (H_2S^+)	920	18	22
E. coli (antibiotic resistant)	920	18	22
E. coli	125	5	4
Total coliforms	15	0.6	18
Fecal coliforms	15	0.6	18
Fecal streptococci	15	0.6	18
Pseudomonas sp.	15	0.6	18
E. coli (antibiotic-resistant)	15	1.1	20
E. coli (antibiotic-resistant)	5	0.75	35
E. coli	NR[a]	1.3	36
Streptococcus zymogenes	15	NR	23
E. coli	1	5	37
Fecal coliforms	5	NR	38
Bacillus subtilis	900	NR	25
Bacillus pumilus	<40	NR	25
Serratia marcescens	90	NR	25
Streptococcus faecalis	90	NR	25
Fecal streptococci	180	18.3	34

[a] NR = Not reported.

10.3. YEASTS AS TRACERS

Limited research has been done on the use of yeasts as groundwater tracers. Wimpenny et al. (24) used two pigmented yeasts as water markers in the river Taff, Great Britain. Both yeasts were detected in the middle of the river 2926 m below the dosing point in 60–80 min. The selective medium for yeasts described in this study eventually eliminated growth of the natural population of bacteria. However, the authors indicated that yeasts could be used satisfactorily as river markers only if the wild-type yeasts in the river were low or absent.

Wood and Ehrlich (29) performed both laboratory and field tests using a suspension of baker's yeast (*Saccharomyces cerevisiae*) to trace movement of microorganisms in artificially recharged groundwater. Laboratory studies showed that yeast cells were able to travel through fine sand at least 60 cm in 30 min indicating their potential use for tracing microbial movement in aquifer systems. In the field experiments where 35 lbs of yeast cells were injected into wells, yeast cells were found to penetrate more than 7 m into a sand and gravel aquifer in less than 48 hr. Results of the field tests also indicated that yeast cells moved through the aquifer faster than bromide and iodide, suggesting the chemical tracers did not represent the movement of yeast cells adequately. In evaluating the use of yeasts, it should be noted that the distance over which the yeast cells were followed in these experiments was short and their survival is dependent on nutrients present in the environment. They are also of a much larger size than bacteria and viruses.

10.4. VIRUSES AS TRACERS

Animal viruses such as hepatitis and the Norwalk agent are known to be transmitted by contaminated groundwater. Other animal viruses pathogenic to man have been isolated from groundwater. Because all human enteric viruses are capable of causing disease, they cannot be used safely as tracers. To overcome this problem, vaccine strains of poliovirus type 1 LSc have been used as tracers (Table 10.2). But the vaccine strain of poliovirus can be responsible for serious disease in man, although the odds are extremely low. Thus, caution must be observed carefully, even if the vaccine strain is used. Most animal enteric viruses are not known to infect man. Thus, bovine enterovirus type 1 has been used to trace the movement of viruses from septic-tank drainfields (30).

Bacteriophages were first evaluated as water tracers in a river by Wimpenny et al. (24). A lambdalike bacteriophage of *E. coli* K 12 was used and found to be an excellent tracer of water movement especially in polluted rivers. The advantages of using bacteriophages as water tracers include: (1) The phage is nonpathogenic to man and domestic animals; (2) it is specific for its host bacteria; (3) assay is simple and rapid; (4) it has good survival characteristics.

Type T4 bacteriophage, a tailed phage, was used by Fletcher and Meyers (31) to trace groundwater movement in the carbonate rock terrain of the Ozark

Table 10.2. Viruses Used as Tracers

Virus	Maximum Distance of Movement (m) Horizontal	Vertical	Reference
Coliphage T4	1600	NR[a]	31
Coliphage ϕ174, T4	900	18	33
Poliovirus, type 1, Coliphage f2	180	18.3	34
Type 2 phage of *Aerobacter aerogenes*	680	NR	32
Bovine Enterovirus, type 1	35	NR	30
Poliovirus, type 1, LSc	NR	NR	39
Poliovirus (vaccine strain)	<40	NR	25

[a]NR = Not reported.

region of southern Missouri. The phage was injected on the surface and was traced approximately 1600 m from the point of injection site, demonstrating a long potential migration distance for viruses in groundwater. The first arrival of the phages took approximately 16 hr.

Martin and Thomas (32) employed type 2 phage of *Aerobacter aerogenes* 243 to follow the movement of shallow groundwater in South Wales over a distance of up to 680 m. Ten liters of phage suspension were introduced through piezometers into sandstone strata. Phage was detected in their monitoring wells for 9 days at a distance of at least 680 m, yielding a velocity of 1.5–7.5 m/hr (36–180 m/day). They concluded that the use of phage was advantageous where nonpolluting substances were required. In light of recent evidence concerning mutagenicity of some tracer dyes (6) this may be especially significant.

Noonan and McNabb (33) used T_4 and ϕ X 174 phages to trace groundwater movement at a land disposal area in New Zealand. They injected the phages to a depth of 14 m below ground level and followed their movement over a distance of 920 m in 7 days. The coliphage T_4 was observed to travel 920 m to a monitoring well within approximately 96 hr. The rate of movement was about 300 m/day which was similar to that found by Sinton (22) using a bacterial tracer.

Coliphage f2 has been used as a tracer virus to examine its accumulation and movement in the soil and groundwater beneath a rapid-infiltration land application site (34). This study was conducted by continuous addition of stock f2 phage into the wastewater over a seven-day application period. Sewage effluent, soil, and groundwater samples were obtained prior to, during, and after dosing of the wastewater with f2 bacteriophage. The tracer virus was found to penetrate into groundwater at the same rate as the percolating primary effluent and was observed in a 18.3-m deep observation well beneath the land application site within 48 hr after application. Both pathogenic animal viruses indigenous to sewage and tracer f2 were detected in the groundwater at a horizontal distance of

183 m from the application site, indicating the potential use of f2 phage for tracing movement of pathogenic animal viruses.

REFERENCES

1. C. P. Gerba, C. Wallis, and J. L. Melnick, *J. Irrig. Drain Div. ASCE* **101,** 157 (1975).
2. K. J. Buchtela, J. Mairhofer, V. Maurin, T. Papadimitropoulos, and J. Zotl, *Natl. Speleol. Soc. Bull.* **30,** 55 (1968).
3. S. N. Davis, G. M. Thompson, H. W. Bentley, and G. Stiles, *Ground Water* **18,** 14 (1980).
4. B. H. Pyle, "Bacterial movement-experience at Aeretaugna," in *The Quality and Movement of Ground Water in Alluvial Aquifers of New Zealand,* Dept. of Agric. Microbiol. Tech. Publ. No. 2, Lincoln College, Canterbury, New Zealand, 1979, p. 105.
5. T. M. Rahe, C. Hagedorn, and E. L. McCoy, *Water Air Soil Pollut.* **11,** 93 (1979).
6. E. R. Nestmann, J. K. David, and J. Ellenton, *Water Res.* **14,** 901 (1980).
7. C. P. Gerba, S. M. Goyal, I. Cech, and G. F. Bogdan, *Environ. Sci. Techn.* **15,** 940 (1981).
8. American Public Health Assoc., *Standard Methods for the Examination of Water And Wastewater,* 15 Ed., APHA, Washington, D.C., 1980.
9. M. H. Adams, *Bacteriophages*, Interscience Publishers, New York, 1959.
10. S. M. Goyal, K. S. Zerda, and C. P. Gerba, *Appl. Environ. Microbiol.* **39,** 85 (1980).
11. T. Aley and M. W. Fletcher, *J. Mo. Speleological Survey* **16,** 1 (1976).
12. E. L. Caldwell, *J. Infect. Dis.* **61,** 264 (1937).
13. E. L. Caldwell, *J. Infect. Dis.* **61,** 270 (1937).
14. E. L. Caldwell, *J. Infect. Dis.* **62,** 225 (1938).
15. E. L. Caldwell and L. W. Parr, *J. Infect. Dis.* **61,** 148 (1937).
16. G. M. Martin and M. J. Noonan, Water and Soil Tech. Publ., No. 7, Dept. Agri. Microbiol., Lincoln College, Canterbury, New Zealand, 1977.
17. R. B. Reneau and E. E. Pettry, *J. Environ. Qual.* **4,** 41 (1975).
18. T. Viraraghavan, *Water Air Soil Pollut.* **9,** 355 (1978).
19. C. Hagedorn, D. T. Hansen, and G. H. Simonson, *J. Environ. Qual.* **7,** 55 (1978).
20. T. M. Rahe, C. Hagedorn, E. L. McCoy, and G. F. Kling, *J. Environ. Qual.* **7,** 487 (1978).
21. I. W. Sinton, Water and Soil Techn. Series No. 17, National Water and Soil Conservation Organ., Wellington, New Zealand (1980).
22. L. W. Sinton, "The use of bacteria as tracers of ground water movement," In N. J. Noonan, Ed., *The Quality and Movement of Groundwater in Alluvial Aquifers of New Zealand*, Dept. of Agricult. Microbiol., Tech. Publ., No. 2, Lincoln College, Canterbury, New Zealand, 1979, p. 93.
23. H. J. Fournelle, E. K. Pay, and W. B. Page, *Publ. Hlth. Rep.* **72,** 203 (1957).
24. J. W. T. Wimpenny, N. Cotton, and M. Statham, *Water Res.* **6,** 731 (1972).

25. F. Marti, G. D. Valle, M. Krech, R. A. Gees, E. Baumgrat, *Alimenta* **18,** 135 (1979).
26. W. H. Dobson, *Arch. Intern. Med.* **121,** 141 (1968).
27. J. T. Davis, E. Flotz, and W. S. Balkemore, *J. Am. Med Assoc.* **214,** 145 (1970).
28. S. A. Sattar, E. J. Synek, J. E. N. Westwood, and P. Neals, *Appl. Microbiol.* **23,** 1053 (1972).
29. W. W. Wood and G. E. Ehrlich, *Water* **16,** 398 (1978).
30. J. E. Scandura and M. D. Sobsey, *Abs. Ann. Mtg. Am. Soc. Microbiol.*, 1981. p. 175.
31. M. W. Fletcher and R. L. Meyers, *Abs. Ann. Mtg. Am. Soc. Microbiol.*, 1974, p. 52.
32. R. Martin and A. Thomas, *J. Hydr.* **23,** 73 (1974).
33. M. J. Noonan and J. F. McNabb, *Abs. Ann. Mtg. Am. Soc. Microbiol.*, 1979 p. 221.
34. S. A. Schaub and C. A. Sorber, *Appl. Environ. Microbiol.* **33,** 609 (1977).
35. E. L. McCoy and C. Hagedorn, *J. Environ. Qual.* **9,** 686 (1980).
36. P. J. Glantz and T. M. Jacks, *An Evaluation of the Use of Tracing Microbial Pollution of Water*, Penn. State Univ. Water Resources Publ. 1–67, 1967, p. 57.
37. B. M. Kudryavtseva, *J. Hyg., Epid., Microbiol.* **16,** 503 (1972).
38. N. Johnson and D. H. Urie, *Ground Water* **6,** 403 (1976).
39. S. L. Stramer and D. O. Cliver, *Abs. Ann. Mtg. Am. Soc. Microbiol.*, 1981, p. 178.

11

MICROBIOLOGICAL SAMPLING IN THE ASSESSMENT OF GROUNDWATER POLLUTION

James F. McNabb

United States Environmental Protection Agency
Robert S. Kerr Environmental Research Laboratory,
Ada, Oklahoma

Gail E. Mallard

U. S. Geological Survey, National Center, Reston, Virginia

11.1. INTRODUCTION

Groundwater makes up over 95 percent of the world's available freshwater reserves and is the main source of drinking water for a large percentage of the world's people. As a result of various human activities, particularly those involving disposal of wastes to the land, the quality of groundwater is being increasingly threatened by the continuing and expanding release of both chemical and biological pollutants. Because groundwater is so important it is vital that we understand the behavior of such pollutants in the subsurface and the processes governing this behavior. Knowledge of biological processes in the subsurface is especially important since biological activity may be a major factor in controlling the fate of pollutants in groundwater. However, relatively little is known about the microbiology of groundwater compared to what is known about the microbiology of surface waters.

This lack of knowledge about microorganisms within the subsurface hydrologic regime and their ecology may, in part, be attributed to the difficulties associated with sampling or observing environments that are anywhere from several inches to hundreds of feet below the land surface. Microorganisms in the subsurface cannot be directly observed and are accessible only through a limited number of points that are technically difficult, costly, and time-consuming to install. The microbiologist is faced not only with the problem of collecting a representative sample—no easy task in itself—but must also attempt to collect a microbially uncontaminated sample. The microbiologist's problems are further complicated by having to deal with complex and unfamiliar environments and technologies. In addition, microbiological investigations may be part of a larger study involving chemical and physical analyses and the sampling methods may not be suitable for microbial studies.

Yet, these problems must be faced in order to assess the impact of human activities on groundwater quality. Indirect evidence gained from laboratory studies, hydrogeological investigations, and chemical sampling can be invaluable, but eventually microbial sampling in the field is required. However, data gained from improper samples may be worse than no data at all! If these data are misleading, a considerable part of the microbiologist's time and budget will have been expended for naught.

One decision that must be made early in an investigation is whether to sample water or subsurface solids. The objective of the investigation and the available resources will usually dictate the type of sample that is required. For example, if the objective is to determine the presence or absence of some pathogen or indicator organism in a particular water supply, it is simple and relatively inexpensive to collect a sample at a water tap. However, if the objective is to define the horizontal and vertical distribution of some pathogen or indicator organism within the subsurface and/or determine the environmental factors affecting the transport and survival of the organism being studied, then soil cores, special monitoring wells, and/or special sampling devices and techniques may be necessary.

There are uncertainties that mandate caution in the extrapolation of information obtained from either sample type. Samples of subsurface solids are taken at only one point in three-dimensional space and the sampling process is destructive—that point cannot be sampled again. Water samples, collected from a network of either lysimeters or wells, can be collected at the same point over a long period of time and provide information about processes occurring in the subsurface because the water will have moved through some volume of solid material in its path to the sampling point. However, the microorganisms in water samples may be transient and may not accurately reflect, quantitatively or qualitatively, the microbial populations existing in the formation being sampled.

The purpose of this chapter is to acquaint microbiologists with the problems of sampling a remote and relatively inaccessible environment and to present some of the procedures used to obtain groundwater and subsurface solids for microbial analyses. Hopefully, the procedures described summarize the current state-of-the-art. However, it should be noted that the technology for the study of microorganisms in groundwater is a decade or more behind that of surface water. The recent surge of activity in this previously neglected area is resulting in a constant stream of improved methods and equipment.

11.2. SAMPLING THE SUBSURFACE: HYDROGEOLOGICAL CONSIDERATIONS

Hydrogeology is the science that addresses the interrelationship of geological materials and processes with the occurrence, movement, and chemistry of water. A large number of texts, manuals, and handbooks of hydrogeology have been published and the microbiologist will find many of them useful (1–5; see Chapter 2). However, because sampling the hydrologic regime may be the most critical and expensive part of a groundwater study and because hydrogeologic conditions are site-specific, the services of a hydrogeologist are needed in designing a field sampling program. The information that follows is, therefore, intended only to introduce some of the hydrogeological considerations that can affect a sampling program.

Subsurface organisms are essentially restricted to the spaces known as voids, interstices, pores, or pore space between the rock and soil particles of the earth's crust (Fig. 11.1). These interstices provide storage and, where interconnected, serve as the conduits for subsurface water. They differ widely in shape, arrangement, and aggregate volume, and range in size from large solution cavities through all gradations to minute pores. The aggregate volume of the interstices in sedimentary rocks is usually a significant portion of the total volume of these materials (Fig. 11.1).

Voids or interstices may or may not contain appreciable water and that water may be held tightly to rock and soil particles by adsorptive forces. The organisms within these interstices are either free floating in the interstitial water or are attached or adsorbed in some manner to the solid surfaces. Recent studies have shown that most microorganisms found in surface soils are strongly attached to the solid phase (7). Microorganisms found deeper in the subsurface are also believed to be attached to solid materials (8, 9). A consequence of this is that the organisms found in water samples from wells or lysimeters may not be representative of the total indigenous, subsurface microflora.

Free-floating, unattached, or nonadsorbed subsurface microorganisms are

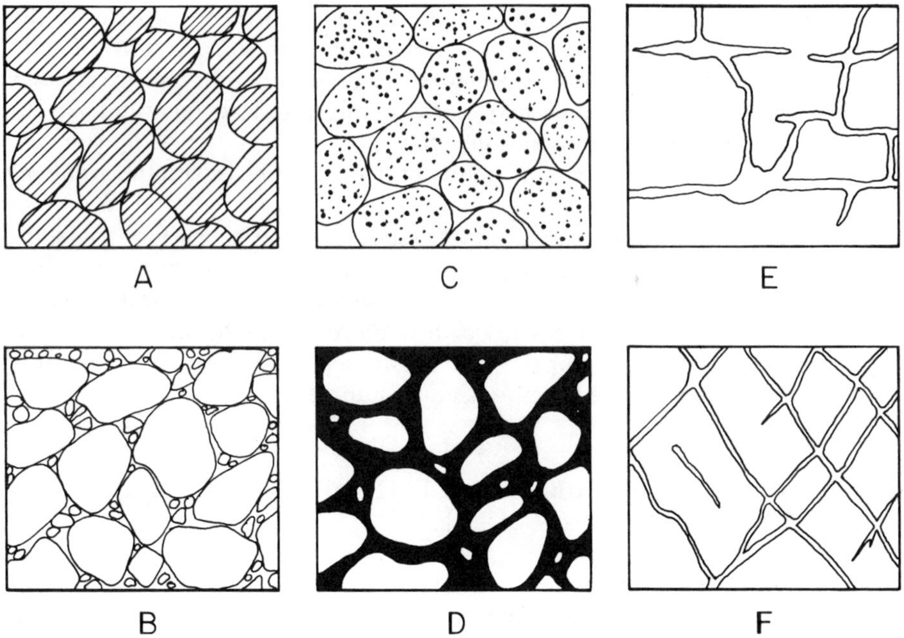

Figure 11.1. Diagram showing several types of rock interstices and the relation of rock texture to porosity (6). (A) Well-sorted sedimentary deposit having high porosity; (B) Poorly sorted sedimentary deposit having low porosity; (C) Well-sorted sedimentary deposit consisting of pebbles that are themselves porous, so that the deposit as a whole has a very high porosity; (D) Well-sorted sedimentary deposit whose porosity has been diminished by the deposition of mineral matter in the interstices; (E) Rock rendered porous by solution; (F) Rock rendered porous by fracturing.

transported by the water as it moves through the interconnected interstices. Microorganisms are unlikely to move on a macroscopic scale in direction opposite to the direction of groundwater flow. The distance a microorganism will be carried is dependent on many factors including its size relative to the size of the interstitial spaces, the extent of interconnected spaces, and the degree to which the microorganism is adsorbed to the solid material. These factors are discussed in detail in other chapters of this book.

Water samples, if properly collected, allow the investigator to determine the concentration of microorganisms and chemical constituents in the water at a particular point at a given time. Water samples can be taken from the same point over an extended period and, therefore, provide information about how the concentrations change with time. However, the proper collection of subsurface water for microbiological analyses is extremely difficult.

Subsurface water is generally distributed within the rock and soil interstices in two major zones: (1) the unsaturated zone (also referred to as the zone of aeration or vadose zone) which includes the zone of soil moisture, and (2) the zone of saturation (groundwater) as shown in Fig. 11.2.

The unsaturated zone is between the land surface and the water table. The

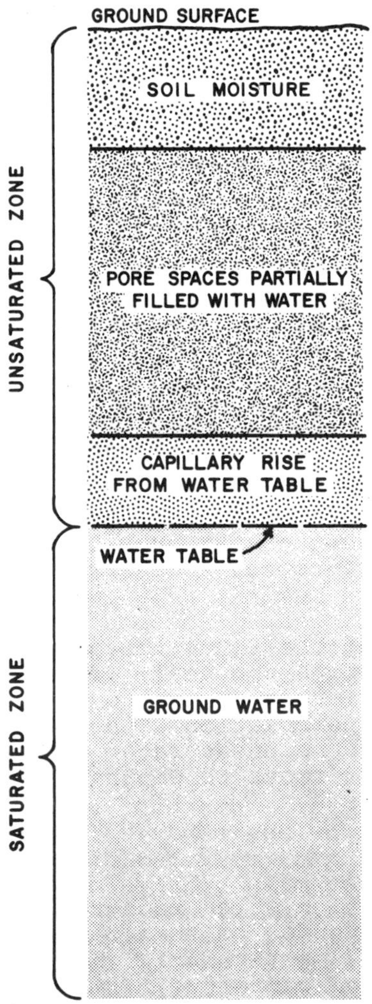

Figure 11.2. An illustration of the relationship between unsaturated and saturated zones (10).

interstices of this zone contain liquid water under less than atmospheric pressure and water vapor and other gases generally at atmospheric pressure.

Although several methods have been developed for sampling the chemical and physical characteristics of water within the unsaturated zone (11), microbiological data is generally obtained from cores. Because water in this zone exists under pressures less than atmospheric, water will not move into wells or other sampling devices that are exposed to atmospheric pressure. Sampling devices used in the unsaturated zone such as suction lysimeters normally preclude microbial sampling because these devices require the use of porous cups that either restrict the passage of microorganisms or become quickly contaminated after installation (12). However, some porous ceramic cups have been shown to be suitable for the passage of viruses (13) and specially designed gravity lysimeters (14) have been used successfully to recover bacteria and viruses from the unsaturated zone (15).

Water contained in the zone of saturation is *groundwater*. Groundwater is generally derived from precipitation and/or stream infiltration and the depth to the upper boundary of the zone of saturation is controlled by topography, climactic conditions, structure of the rocks, and the number, size, shape, and arrangement of their open spaces. Water-bearing formations that will transmit and yield significant quantities of water to wells and springs are called *aquifers*.

Within the zone of saturation, groundwater occurs under either watertable or artesian conditions. Under water table conditions, the groundwater is not confined. The upper surface of the saturated zone is called the *water table* and may be near the land surface or hundreds of feet below it. It is free to rise and fall with changes in the volume of the stored water caused by either pumping or seasonal variations in the rate of natural recharge. Such changes in the water table can be significant and must be considered when installing sampling wells or coring. Under artesian conditions the groundwater is confined between an upper and lower confining bed. If the artesian aquifer is penetrated by a well, the water rises in the well above the bottom of the upper confining bed. The potentiometric surface is the surface that represents the static head of ground water. As related to an aquifer, it is defined by the level to which water rises in tightly cased wells.

The zone immediately above the water table in which some or all of the interstices are filled with water that is under less than atmospheric pressure and that is continuous with the water below the water table is the *capillary fringe*. The water is held above the water table by surface tension. The capillary fringe is typically saturated to some distance above its base at the water table. Upward from the saturated part only progressively smaller pores are filled and the upper limit is indistinct. Its thickness is generally greater in fine-grained materials than in coarse-grained materials. Hence, it may extend several feet in clay and only a fraction of an inch in gravel. The capillary fringe may be a site of increased microbial activity relative to the adjacent unsaturated and saturated zones. Stewart et al. (16) studied the distribution of bacteria and pollutants in the soil profile beneath Colorado fields and corrals. They took cores at one-foot intervals and found that there was often a distinct increase in bacterial counts at or immediately above the water table.

Generally, the rate of groundwater movement is slow, usually measured in feet/day or feet/year, but the rate is dependent on the local hydrogeology and may be much faster. The slow rate of groundwater movement has at least two consequences for the investigator designing a sampling program. First, groundwater and whatever substances or organisms it contains are in an environment with relatively constant conditions (17). For example, temperatures in the earth's crust rarely fluctuate, except in the topmost layer of approximately 10 m (33 ft) which may be affected by seasonal temperature variations (18). Because the concentration of many constituents may change only minimally over a period of a few days to weeks, it usually is unnecessary to sample as often as required in a surface water study (19). However, seasonal variations in the number of microorganisms in groundwater have been documented (20-23). A second consequence of the slow rate of groundwater movement is that it may take months for groundwater to move from a source of contamination to the nearest well downgradient from it. Negative data in such a situation may not mean that the groundwater is not being contaminated—it may only mean that the wells being sampled have not yet been affected.

Movement of subsurface water within the unsaturated zone is essentially vertical; whereas the movement of groundwater in the zone of saturation can be either lateral or vertical as the water moves from a high potentiometric head to a lower head. Groundwater flow patterns can be developed from water level contours; however, the actual movement of a contaminant plume may be very complex. For example, in unconsolidated materials such as gravel and sand where the water occurs in interconnected void spaces between particles, the shape of the plume will be controlled by abrupt changes in permeability.

Groundwater flow patterns are even predictable in aquifers with fractures or solution channels. Obviously, the problem in siting monitoring wells in such geologic units is to intercept fractures or solution channels that are hydraulically connected to the source of contamination. It is possible in many formations of this type to drill a well and miss the pollutant plume and move only a few feet away and drill another that intercepts the plume.

11.3. OBTAINING A WATER SAMPLE

Generally groundwater is obtained by a well and some sort of device to bring water to the surface. In some locations, groundwater appears naturally at the surface as a spring or seep where the water table intersects the ground surface. Springs may provide meaningful information concerning the biological situation in their vicinity, but microorganisms found in spring waters may not accurately reflect the subsurface population (9).

A well is a hole in the ground that may be lined or cased with some material to keep it from caving in. For thousands of years, man dug wells with hand tools and lined them with rocks. Hand-dug wells still exist. However, today wells are generally drilled with machines and lined (cased) with plastic or metal pipe.

There are several important considerations in using wells to obtain groundwater samples. These include the location of wells within an aquifer, well construction and completion, and procedures used to sample a well. Why are these considerations important to the microbiologist? Each step in the process of locating a sampling point, gaining access to the groundwater, and securing a sample will greatly impact the reliability of the microbiologist's data. Ideally, the sample will be representative of the water in the aquifer and will not be contaminated with surface organisms. Yet, not only is it difficult to collect a representative sample, it is extremely difficult to collect an uncontaminated one.

11.3.1. Well Location

In order to correctly interpret microbiological or chemical data obtained from groundwater samples, the investigator must know something about the location of the well within the aquifer. This is necessary because an underlying assumption in the interpretation of the data is that free-floating microorganisms travel with groundwater. Therefore, two conditions must be met if one is interested in either detecting microorganisms moving from a waste-disposal site or defining a plume of groundwater contaminated by some chemical that travels with groundwater. First, wells to be sampled must be sited and screened to a depth that will intercept a particular flow path. Second, the investigator must have some standard of comparison in order to recognize a change in water quality. This second requirement is generally satisfied by using data on the quality of water obtained from one or more wells located upgradient or at a site believed to be unaffected by the source of contamination being investigated.

There are some basic guidelines for locating wells. However, it is usually advisable to obtain the services of a hydrogeologist. The expense will save money and effort in the long run. Prior to initiating any field work, all existing geologic and hydrologic data should be collected, compiled, and interpreted. Procedures for doing this are detailed in several manuals (24–26). Data that may be available include geologic maps, cross sections, and an array of water-well data including location, date drilled, depth, name of driller, water level and date, well completion methods, use of well, geophysical data, formation samples, aquifer test(s), and water quality data.

Static water level measurements can be used to make a preliminary determination of gross direction of flow. Water level data can be obtained from wells and boreholes or from surface water manifestations of the groundwater systems such as springs, lakes, and streams. Note, however, that nearby pumping wells or other artificial discharges or recharges may alter the natural hydraulic gradient.

The vertical portion of the saturated zone that is actually sampled is dependent upon where the casing is screened (i.e., has openings) for water to enter the well from the formation. A well may be screened to allow the entry of water along the entire length of the casing or it may be screened to withdraw water only at a specific depth. Knowledge of the total depth of a well and where it is screened is important in interpreting data on the quality of water obtained from that well. For example, if a well is screened through two or more discrete water-bearing layers, water obtained from that well will be a mixture and will not reflect the quality of either zone tapped.

To overcome the problems caused by using wells screened over a large portion of their depth, many monitoring manuals recommend screening at discrete depths (24–26). However, a contaminant can migrate along fairly restricted pathways and may not enter depth-discrete wells.

11.3.2. Methods of Drilling, Constructing, and Completing Wells

The following discussion is not intended as an exhaustive discussion of well drilling; procedures involved are outlined in detail in a number of well-drilling manuals (26–29). Instead, the purpose of this discussion is to alert the microbiologist to the fact that a well can be grossly contaminated during its construction because well drilling is by nature a nonaseptic procedure.

11.3.2.1. Methods of Drilling Wells

A well may be drilled by several different methods—mud and air rotary, cable tool, auger, and jetting. The principles of operation, advantages, and disadvantages of the more common types of drilling techniques suitable for constructing groundwater wells are discussed in detail by Scalf et al. (26).

Water-supply wells in the United States are commonly drilled by the mud rotary method (30). Mud rotary drilling rigs have a number of advantages for drilling water-supply wells, but are not very desirable for drilling wells to be used for microbiological sampling. With a mud rotary system a drilling fluid is pumped down the inside of the drill pipe, and then returned to the surface through the annulus between the drill pipe and the borehole wall. The drilling fluid may be clear water, water mixed with bentonite or a nonbiodegradable synthetic polymer, or water mixed with a biodegradable organic polymer.

Drilling fluids often contain various sorts of additives and may be a source of bacterial contamination (9, 31) or cause other microbial problems (32, 33). A full-scale field and laboratory investigation conducted in the Soviet Union in 1948 and 1955 found that drilling muds were definitely a potential source of contamination (31). These muds contained a variety of bacteria, including sulfate-reducers and hydrocarbon utilizers. To determine the range of penetration of bacteria from drilling muds into various types of cores, a series of experiments were carried out. Circulating drilling fluids were seeded with a culture of *Bacterium prodigiosum* (*Serratia marcescens*). The seeded drilling fluid was introduced and circulated in the drill holes for 15–20 minutes before pulling the cores. The cores were then carefully examined layer by layer for penetration by the test organisms. As would be expected, it was found that more porous rocks offered a greater possibility for the core to be contaminated by the drilling fluid. Test bacteria penetrated throughout cores of sand and sand layers, but barely penetrated hard, compact rocks such as limestone.

The air rotary drilling method operates in the same manner as mud rotary drilling except that air rather than drilling fluids is circulated. Some rotary rigs are equipped to operate either with mud or air as the drilling fluid. Air rotary rigs are available throughout much of the United States and are well-suited for many groundwater quality programs. Because no drilling fluids are used, contamination of the formation is not a severe problem. However, when more than one water-bearing zone is encountered and where the hydrostatic pressures are different, flow between these zones will occur within the wellbore after drilling is completed and before the hole can be properly cased and different zones sealed off. This may lead to contamination problems.

Some rotary rigs can be modified to use a device called a casing hammer. The method of drilling is the same as with the air rotary drilling method except that when caving formations are encountered, the casing hammer drives a casing down to prevent the hole from caving. This drilling method is generally excellent for constructing wells in unconsolidated formations. The use of casing minimizes flow into the hole from upper water-bearing layers and thus minimizes contamination. However, the use of the air rotary method with casing hammers is much more expensive than other drilling methods and is not in common use.

The cable tool drilling method uses a heavy, solid-steel, chisel-type drill bit suspended on a steel cable, which when raised and dropped chisels or pounds a hole through the soils and rock. When drilling through the unsaturated zone, some water must be added to the hole. The cuttings are suspended in the water and then bailed out periodically. Adding water during drilling increases the possibility of contamination of the borehole and the formation.

The use of augers for drilling wells and coring has a great many advantages because drilling fluids are not used and contamination problems are minimized (Fig. 11.3). Auger flights consist of either solid or hollow stems. Solid-stem augers must be removed from the borehole before a core can be taken or well casing and screen can be installed. This may be a problem in saturated, soft materials because the hole may cave-in when the auger is removed. Hollow-stem augers have a sufficiently large inside diameter to permit coring or the placing of well casing and screens down inside the augers before they are removed.

The auger drilling method is generally fast and inexpensive to operate in unconsolidated formations. Augering cannot be used in hard rock and the depth that can be reached is relatively limited. Hollow-stem augers with well screens in the lead section of the augers are available commercially. They allow the collection of water samples at any desired depth below the water table during the drilling operation without removing the augers or setting a screen and casing (34).

The bucket auger consists of a relatively large bucket with a cutting edge at the bottom that is slowly rotated by the drill stem. When the bucket fills, it is brought to the surface and emptied. This method is good for constructing shallow wells just into the water table in unconsolidated formations. No drilling fluids are required when either drilling above or below the saturated zone in noncaving formations. However, the hole diameter is large and requires careful backfilling and sealing to prevent movement of the water down the outside of the well casing. To prevent caving below the water table it is necessary to continuously add water which would increase the chances of contamination during drilling.

Jetting consists of pumping water or drilling mud down through a small diameter (1½–2-in.) standard pipe that may be fitted with a chisel bit or a special

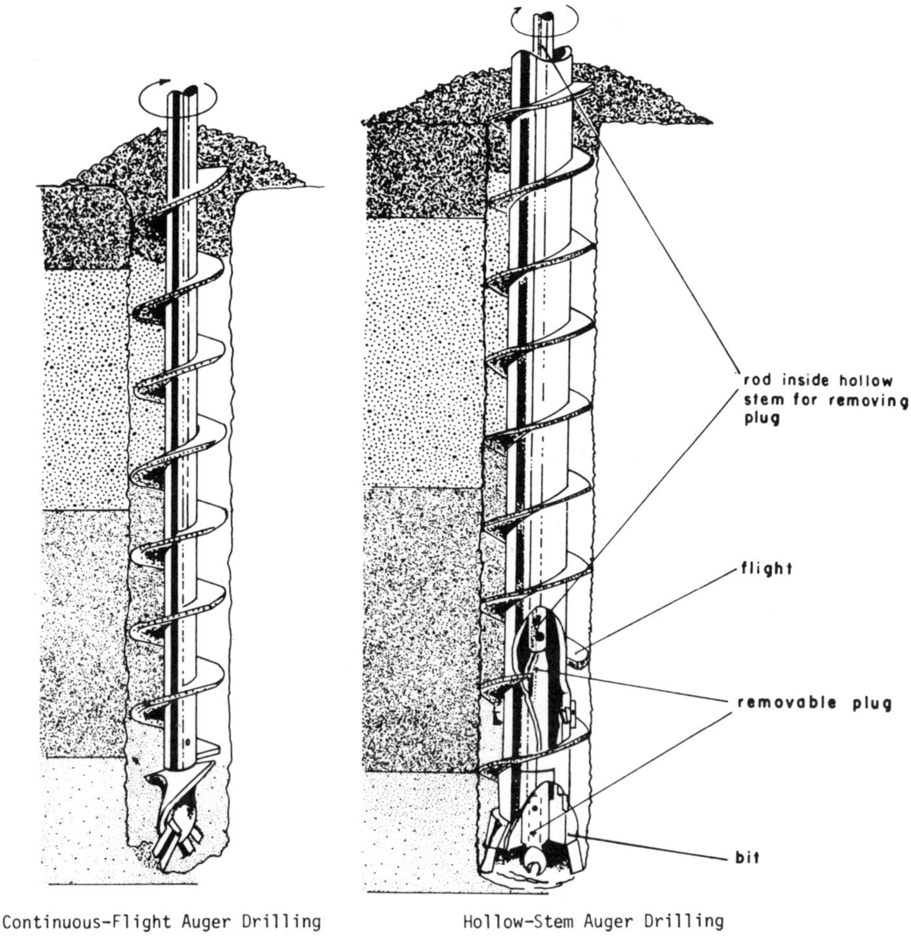

Figure 11.3. Auger drilling: The continuous-flight auger bores into the soil and rotates the cuttings upward along the flights. In order to core, the auger must be removed when the desired depth is reached. The hollow-stem, continuous-flight auger bores into soft soils and carries the cuttings upward along the flights. When the desired depth is reached, the plug is removed from the bit and withdrawn from inside the hollow stem. A core barrel can then be inserted to the bottom of the hollow stem.

jetting screen. Jetting is inexpensive, but only feasible in very soft, shallow formations. The diameter of the pipe limits the type of samplers that can be used. The major problem for the microbiologist is that a large amount of foreign water or drilling mud is introduced above and into the formation to be sampled, and it is difficult to place seals so that only one zone is sampled.

11.3.2.2. Well Construction and Completion

After drilling is completed, the borehole should be cased, screened, sealed (grouted), and completed. Well construction and completion measures are usually not determined by the microbiologist, unless the wells are to be used strictly for a microbiological investigation. The microbiologist should be aware when sampling preexisting wells that proper measures may not have been used.

Well casings should be constructed of materials that have the least potential for affecting the sample. Production well casings are usually steel or polyvinyl chloride (PVC) pipe. Other materials such as polystyrene, ceramic tile, stainless steel, and even teflon have been used for monitoring wells. The best type of casing for microbiological investigations is not known. PVC or other plastic casings may affect the growth of bacteria in wells due to leaching of or adsorption of chemicals (35). Steel pipe, on the other hand, may become a site for active microbial corrosion and/or incrustation (28).

Care should be given to the preparation of the casing and well screens prior to installation. These materials should either be washed with a detergent and rinsed thoroughly with clean water or steam cleaned. Washing with a disinfectant is not always practical if the well is to be used to obtain water for chemical analyses.

After the casing is installed, a sand or gravel pack may be placed in the annular space between the casing and the wall of the wellbore to minimize the flow of fine materials into the well and to increase the yield of the well. The gravel pack should be placed opposite the screened area and should not extend throughout the entire length of the casing. If the well is to be used as a drinking-water well, the gravel or sand pack should be disinfected before it is installed (29), but this probably is not always done.

The remainder of the annular space outside the casing should be sealed (grouted) with an impervious material to prevent water movement down the annular space between aquifers and/or the surface. Cement-based materials are the most commonly used grouting materials, although bentonite clay or well cuttings are sometimes used, One serious problem with preexisting wells is that the investigator may not know whether they are properly sealed, It is especially important that wells are sealed at the surface with an adequate ground seal to prevent passage of water from the surface down the annular space of the wall. Improper grouting of wells is a major cause of microbial contamination of wells used for drinking water in the United States (36). After the well is completed, the top should be capped and secured to prevent anyone from introducing materials into the well that might contaminate the water.

11.3.3. Collecting Samples From Wells: General Considerations

Under the best of circumstances, a microbiologist will have access to the groundwater system through a well that has been carefully located, properly constructed, and completed. A considerable challenge still remains, however, if the investigator is to collect a water sample that is representative of the groundwater at that location and that is uncontaminated by the sampling process.

There are at least two practical problems that should be considered prior to sample collection. First, the water standing in the well casing and in proximity to the well is unlikely to be representative of the groundwater at a distance from that well. Second, most of the equipment commonly used to obtain water samples cannot be sterilized by heat. The following discussion will present some of the methods commonly used to minimize these problems. However, a third, more theoretical problem exists, which must also be faced by the microbiologist—no one has yet defined the relationship between microorganisms carried with groundwater and those attached to subsurface solids. Whether or not this is a major problem will probably depend upon the purpose of the investigation. In any case, the problem is intrinsic to the system. The investigator should be aware of the fact that data obtained from even the most carefully collected groundwater sample may need to be interpreted with caution.

11.3.3.1. Well Effects

Wells provide numerous habitats for microbial colonization. How does this affect a groundwater sampling program? To some extent the answer to that question depends upon the purpose of the investigation. For example, an individual studying the transport of viruses or bacterial pathogens from a waste-disposal site may be less concerned about the bacteria growing on the side of the well casing than an individual studying the "normal flora" of the aquifer. Obviously, it is beyond the scope of this discussion to consider every possible circumstance in which a groundwater sample might be collected. However, a brief overview of the environments within a well and in the immediate hydrogeologic system tapped by a well is appropriate because it will illustrate some of the factors that must be considered in obtaining a groundwater sample.

Figure 11.4 illustrates a generalized cross section of an observation well screened in unconsolidated material. If the well is not pumped regularly, it will have a column of stagnant water standing in it. This water is exposed to oxygen at the surface, is in contact with the casing material, is not in direct contact with the aquifer material, and is different in quality from water at some distance from the well (37–40). This water and the film on its surface is also likely to have its own unique population of microorganisms. In addition, there may be film of bacterial growth on the surface of the casing and the well screen. All of these microbial populations are probably more characteristic of the well than the aquifer. Another distinct population may exist in the aquifer immediately adjacent to the well. This area may have been contaminated during drilling and redox and nutrient conditions may be different here than in the aquifer some distance from the well.

Public-supply, industrial, or irrigation wells that are pumped continuously may provide a representative sample of water in the aquifer. However, these wells are often either unavailable or are inappropriate for microbiological sampling. For example, they may be screened over a large interval or withdraw water from more than one aquifer. If observation wells are used, the well should be pumped to replace the standing water with fresh aquifer water prior to sample collection. General guidelines have been established for groundwater samples collected for chemical analyses. These include removing 1 to 10 times the volume of water in

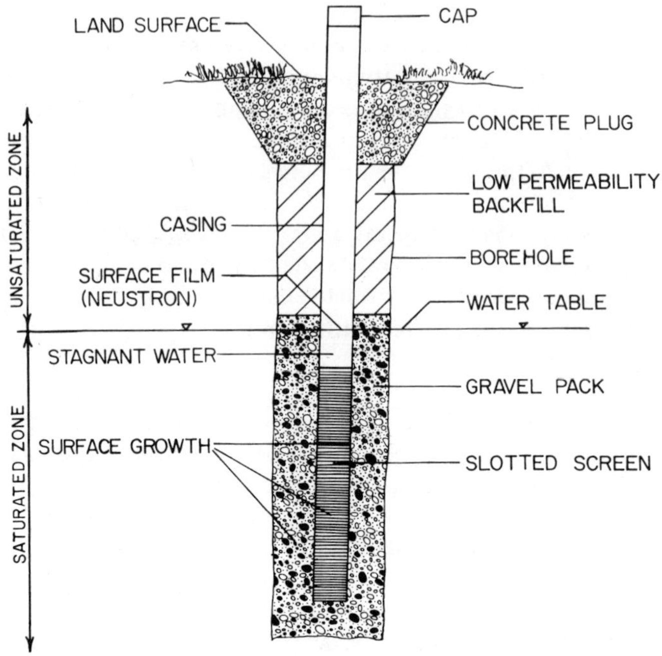

Figure 11.4. Cross section of a representative observation well illustrating different habitats for microbial colonization.

the well casing prior to collection (25, 26, 40) or monitoring certain parameters such as temperature, pH, or specific conductance as the well is pumped and collecting the water sample when these characteristics stabilize (41). Whether or not either of these procedures is valid for microbiological samples is not known. At present, there is no general rule for the number of times that water in the casing must be removed before collecting samples for microbiological analysis. Hirsch and Rades-Rohkohl found in studies in Germany that an initially higher number of bacteria found in water samples dropped to a constant level after approximately three times the volume of the water in the wells had been removed (42).

11.3.3.2. Well Sampling Equipment

A variety of equipment is available to withdraw water from wells. In general, the choice of the equipment to be used in a sampling program will be dictated by factors such as the size of the well, depth to water, and the desired pumping rate. Any equipment or procedures used to obtain a groundwater sample should not alter or contaminate that sample. Ideally, sampling equipment should be completely inert; economical to manufacture; easily cleaned, sterilized, and reused; able to operate at remote sites in the absence of external power sources; and capable of delivering continuous but variable flow rates for well flushing and sample collection.

Numerous types of pumps exist and many models of each type are available from various manufacturers. Also, new and improved designs are being introduced continuously. Instead of attempting to catalogue all of these sampling devices or providing exhaustive explanations of the principles and operating characteristics of each, the brief discussion that follows is intended to introduce the range of alternatives available for the individual interested in obtaining a sample of groundwater. More extensive discussions of pumps in general (26, 27, 43–45) and of pumps designed for small-diameter wells or special purposes (46–49) are available in a number of handbooks, manuals, and other technical publications.

1. *Bailers* are one of the oldest and are certainly the simplest means of obtaining a groundwater sample. They may consist of either a weighted bottle or a length of pipe capped at one end that can be lowered and raised by hand. Bailers are very versatile and highly portable because they do not require an external power source. Bailers can be constructed of a variety of materials which are sterilizable and can be almost any size although generally they are small. The main problem with bailers is they are not very effective for flushing the casing prior to sampling.

2. *Suction-lift pumps* are a class of pumps that lift water to the surface by creating a partial vacuum inside a tube extending down into the water in a well. Atmospheric pressure forces the water up the tube. Therefore, the practical suction lift with such pumps is about 8 meters or less depending on altitude and the pump capacity. Any type of pump that creates a vacuum can be used for microbiological sampling by placing a thick-walled collector flask in-line before the pump (Fig. 11.5).

3. *Shallow-well centrifugal pumps* are a type of suction-lift pump that generally have both the pump and the drive motor located at land surface (45). These pumps can have relatively high pumping rates but can only be used when

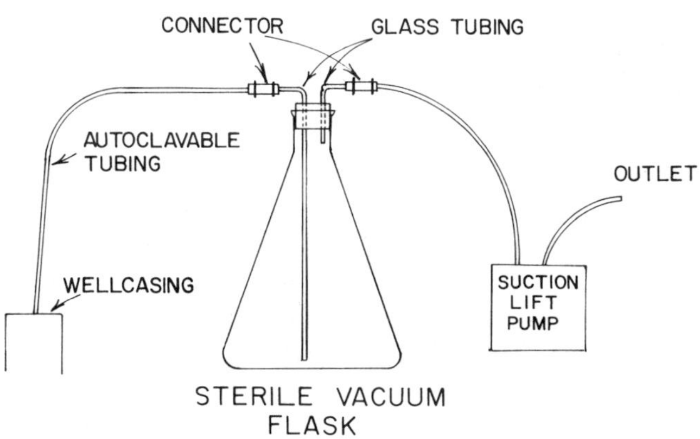

Figure 11.5. System for microbiological sampling of wells using a suction-lift pump (26). After well is flushed, tubing is clamped off and initial flask is replaced with sterile flask for collecting sample. (Used by permission of the National Water Well Association.)

the depth to water is within suction lift (about 20 ft. at sea level). A foot valve on the end of the sampling tube is sometimes required. The tubing can be sterilized and a sterile suction-flask can be placed ahead of the pump.

4. *Peristaltic pumps* are a type of suction-lift pump that squeeze water through flexible tubing ahead of the contact points on a rotor. Pumping rates are fairly low and depth to water cannot exceed suction lift (about 20 ft). One advantage of this type of pump is that the water flows through tubing which can be sterilized and does not contact the pump itself.

5. *Diaphragm pumps* are suction-lift pumps that can be hand-operated. They are readily available, relatively portable, and inexpensive. A sterile suction flask can be put ahead of the pump to catch the sample (26).

6. *Submersible pumps* are basically centrifugal pumps with a small motor inside a water-tight casing. The entire assembly is submerged in the well with the discharge pipe and power cable emerging at land surface. Pump characteristics are variable; all pumps can deliver water from depth greater than 25 ft. Submersible pumps are generally not sterilizable by autoclaving, but they can be disinfected.

7. *Air-lift pumps* lift water by ejecting compressed gas into the water. Gas bubbles mix with the water resulting in a lower water density inside the discharge pipe that allows the water to be driven to the surface. A possible drawback to the use of an air-lift pump is that alteration of some water quality parameters may result from the mixing of gas and water (40).

8. The *air-squeeze pump* was first designed by Middelburg (47) who described its operation as follows:

> When the pump is submerged, water in the well enters a collapsible bladder confined inside a rigid housing. After the bladder has filled, air pressure is applied to the annular space between the rigid housing and the bladder. The bladder is squeezed, forcing the water upward in the direction of least resistance. Two check valves, one located above the pump and the other located at the intake to the pump, control the direction of flow.

Although these pumps have relatively low pumping rates, their advantage is that they can be made small enough to fit into many small-diameter wells. Some air-squeeze pumps probably can be sterilized by autoclaving and all can be disinfected.

9. *Gas-driven piston pumps* have double-acting pistons operated by compresssed gas. The driving gas enters and exhausts from the gas chambers between the two pistons and the intermediate connector that joins them. Built-in check valves at each end of the pump allow water to enter the cylinders on the suction stroke and to be expelled to the surface on the pressure stroke. Present designs are generally constructed of stainless steel, brass, and PVC. Pumping rates vary with the pumping head but pumping rates of 2.5–8 gallons/hour have been noted at 100 ft of pumping head. These pumps require no electrical power source and can be used at depths in excess of 400 ft, but are relatively expensive.

10. A submersible, small diameter *gas-lift* pump that can be autoclaved has been developed by Tomson et al. (49). The two-stage pump is made of glass and

teflon for trace organics sampling. Because gas (usually nitrogen) is used to drive the water to the surface, the gas must be filtered to remove microbial contaminants. Stripping of CO_2 and volatile compounds can occur when the water is lifted to the surface.

11.3.3.3. Well Sampling Procedures

Depending upon local hydrologic conditions, the type of equipment available, and the purpose of the investigation, several methods are available for collecting groundwater samples. Each has advantages and disadvantages. It should, however, be recognized that in many cases the microbiologist has very few choices and that none of them are ideal.

If the sample is to be collected from an existing well with an in-place pump, the precautions that the investigator can take to ensure a noncontaminated sample are limited. Ridgeway and Olson (50) have demonstrated that the surfaces of drinking-water distribution systems are colonized by bacteria. Samples should, therefore, be obtained from outlets as close as possible to the pump and leaky or faulty spigots or spigots that contains screens or aeration devices should be avoided, if possible (51). A steady flowing stream of water at moderate pressure is desirable in order to prevent dislodging particles in the faucet or water line. Pumping should continue for some time prior to sample collection.

If a well does not have an existing in-place pump, samples can be obtained by using either a portable surface or submersible pump or by using a bailer. Each method presents special problems in obtaining an uncontaminated sample.

The main problem in using a sterilized bailer is obtaining a representative sample of water in the aquifer. Pumping or bailing the well before sampling in order to exchange the water in the wellbore for fresh formation water is difficult because such presampling activities must be carried out in such a way as to not contaminate the bailer. Care must also be taken so that bailers do not contaminate the sample with any scum on the surface of the water in the well. This is usually accomplished by using a weighted, sterilized sample bottle suspended by a nylon rope and lowering the bottle rapidly to the bottom of the well. Alternatively, the well can be evacuated with a pump and the actual sample collected with a sterilized bailer.

The use of portable pumps provides a way to pump out a well before sampling and thus provide a more representative sample, but this technique presents a potential source of contamination if the pumping apparatus cannot be sterilized beforehand. The method of disinfection will depend on what other samples are to be taken from the well because the use of disinfectants may not be feasible if the well is also sampled for chemical analyses. If disinfection is not ruled out by other considerations, one method (52) of sanitizing a submersible pump system is submerging the pump and any portion of the pump tubing that will be in contact with the well water into a disinfectant solution and circulating the disinfectant through the pump and tubing. The most widely used method of disinfection (because of its simplicity) is chlorination. After disinfection the

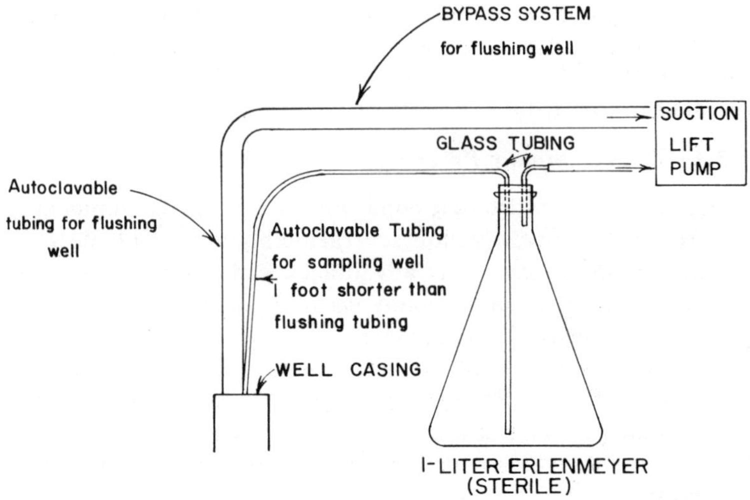

Figure 11.6. System for microbiological sampling of wells using a suction-lift pump (26, 53). (Used by permission of the National Water Well Association.)

pump should be carefully placed in the well and then operated for a period of time in order to thoroughly rinse the disinfectant from the pump system.

If the pump cannot be disinfected, then the pump and tubing should be carefully handled to avoid gross surface contamination. The well should be pumped until the pump and tubing have been rinsed several times before taking a sample. It may be desirable to remove the pump after flushing out the well and take the sample with a sterile bailer.

In wells where the water level is less than 20 to 30 ft below the land surface, a surface vacuum-pumping system such as that developed by Noonan and McNabb (53) can be used for flushing out the well and withdrawing a sample. An apparatus for this is depicted in Fig. 11.6. This apparatus consists of two lengths of tubing that can be sterilized by autoclaving and a portable vacuum system. The two tubing lengths that are attached side-by-side to each other are sterilized in the laboratory in large covered containers. In the field, they are lowered into a well using sterile gloves, attached to a vacuum system, and the stagnant water in the well is removed with the bypass system. Then the well is sampled by drawing water into a sterile vacuum flask on the intake side of the pump. Large volume sampling for viruses or pathogenic bacteria can be accomplished by substituting filters or columns with various absorbents in place of the vacuum flask. Standing water is prevented from entering the sampling tubing before the well is flushed by making the sampling tube a few feet shorter than the flushing tubing and turning on the pump to the flushing system as the tubing is put into the well.

11.3.3.4. Sample Handling

At the land surface, the water sample should be handled with the usual microbiological procedures; these do not need to be detailed here. However, it is appropriate to mention that any investigator interested in studying the "normal microflora" of an aquifer should be aware of the chemical and physical conditions within that geologic unit. For example, precautions taken during sample collection to avoid or minimize the introduction of oxygen have been described by Olson et al. (54), Elkan and Horvath (55), and Godsy and Ehrlich (56).

11.4. SAMPLING SUBSURFACE SOLIDS: GENERAL CONSIDERATIONS

A variety of procedures and equipment have been developed by civil engineers, geologists, and soil scientists for collecting undisturbed soil and rock materials for physical or chemical analyses (57, 58), but relatively little attention has been paid to collecting undisturbed, uncontaminated subsurface solids for microbiological analyses. In some cases, tools as simple as a shovel or backhoe can be used to dig a pitch or trench to the desired depth. The investigator can then aseptically cut away the exposed surface of the wall with a sterile tool and take samples from the wall through the fresh face. This type of sampling is effective, but costly, time-consuming, and limited to relatively shallow depths. It is also destructive to the site, in the sense that the site is ruined for further sampling due to the introduction of surface organisms when the hole is filled and because the normal profile and soil structure are destroyed.

Coring procedures can also be used to collect solid materials. This is commonly done by pushing a thin metal tube into the soil or rock, removing the tube with the sample intact, and then processing and analyzing the sample as desired. The problems in satisfactorily completing each step increase with depth, the extent of saturation, and the hardness of the material being sampled.

Increased depth requires the investigator to drill or auger a hole to the desired depth before sampling. The investigator must then pull the drill pipe or auger from the hole, attach a coring device, and take a sample. In loose, unconsolidated sands or in the saturated zone, caving of the borehole occurs and maintaining an open borehole while collecting a core sample is difficult. Use of a hollow-stem auger can sometimes mitigate this problem but in extreme situations, sand and water may flow up the interior of the auger stem before the core sample can be taken. Drilling methods that require drilling fluids should not be used when core samples are to be taken because of the increased potential for sample contamination (26). Sampling the saturated zone increases problems of contamination because the coring tube must travel through water in the bore hole and this water will almost surely contaminate at least the outside of the core sample.

Care should be taken during retrieval of the coring device to minimize loss of the sample. This is a serious problem when sampling loose materials such as sand. A sampling technique that may be used in some circumstances is that described by Danielpol (59) in which he drove a pipe into the material he wished to sample, poured liquid nitrogen down the pipe until the material near the pipe was frozen to the pipe, and then pulled the pipe out with the frozen material adhered to it. This procedure was used in shallow, saturated gravels.

The general problems associated with coring have been discussed by Gilmore (58) who points out that "the outside surface of a core obtained with a soil tube must be considered contaminated by particles and organisms from the overlying soil, because these are unavoidably carried along by the wiping action of the cutting point and the inner walls of the tube." The extent to which the outside surface of a core is contaminated is dependent on many factors, the major one being whether water or drilling fluids were present in the borehole when the sample was taken.

Gilmore (58) concluded that the best way to obtain an uncontaminated core sample is to core in as clean a manner as possible and then subsample the core through a fresh, cleanly fractured surface using a sterile sampling device. This can be done most effectively by removing a subsample from the center of the core and avoiding contact with the outer 1–2 cm of the core.

It should be recognized that a core sample is one point in an extremely heterogenous, three-dimensional system. Most coring procedures are so costly and time-consuming that relatively few cores are ever taken in any field investigation (60, 61). It is realistically impossible to acquire enough samples to satisfy statistical requirements for surface soil sampling (62).

11.4.1. Coring Procedures

If sampling is to be done more than a meter or so beneath the surface, the first step is to drill a hole to the desired sampling depth. This can be done with a continuous-flight or hollow-stem auger, rotary core drill, or other drilling method. Scalf et al. (26) recommend the use of augers, preferably the hollow-stem type, to minimize the introduction of surface contaminants.

The types of coring equipment available are described in detail by Acker (57). They include solid-tube samplers, split-tube samplers (with or without liners), the Shelby thin-wall tube sampler, the stationary piston sampler, and the Denison sampler or core barrel. There are advantages and disadvantages associated with all of these samplers. Theoretically, all could be used to obtain an uncontaminated sample in most materials other than hard rock.

Scalf et al. (26) and Wilson et al. (60, 61) have used the thin-wall tube sampler illustrated in Fig. 11.7. This device consists principally of an adapter threaded to fit a standard drill rod and a replaceable thin wall seamless steel tube. The tube sizes normally used are 70–100 mm O.D. although sizes from 50–120 mm are available. In some subsurface materials, drag on the inner wall of the tube may cause considerable compression of the sample. To reduce this distortion and to aid in keeping the sample in the tube during withdrawal from the borehole, a drive shoe can be put on the bottom of the tube that has an opening with an inside diameter slightly less than the core barrel. This reduces the drag of the

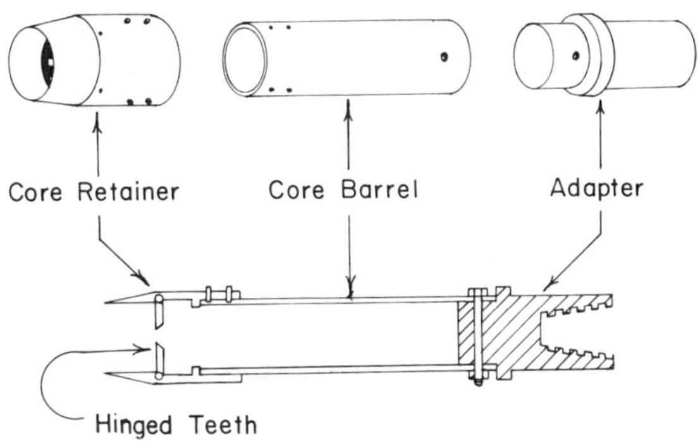

Figure 11.7. Core-sampling device used by Wilson et al. (60, 61, 63) to obtain microbiological samples of subsurface material.

core sample on the inner wall of the tube as the core enters. The drive shoe can also be fitted with a sample retaining device that consists of spring-loaded teeth that open as the sample enters the tube and then close as the sample is withdrawn, thus preventing the sample from falling out of the barrel.

At depths beyond the practical capability of augers, it is much more difficult to collect uncontaminated samples for microbiological analysis. For example, air drilling with a casing hammer offers the potential for opening a borehole at these depths without drilling fluids. However, air drilling may disturb the bottom of the borehole. In cases where the drilling method requires the use of a drilling fluid, the stationary piston sampler may be the best device to obtain solid materials. Its use is described by Acker (57) and by Scalf et al. (26). Considerable additional work is needed to evaluate the extent of core contamination with this sampler and to develop optimum methods for avoiding contamination whenever drilling fluid is used.

11.4.2. Core Processing Procedures

After a core sample is obtained, care must be taken not to disturb or contaminate the sample. Regardless of the analyses to be performed, processing should be done as soon as possible. If cores cannot be processed in the field, it is sometimes possible to retain the sample in the core barrel, which is then placed in a sterile container and transported to the laboratory. This procedure works sufficiently well in many cases, but is not satisfactory if the core materials shift or mix within the core barrel. If mixing is a problem, one possible solution is to rapid freeze the core barrel and its contents with liquid nitrogen until the core could be processed. This procedure, however, may affect the number of microorganisms recovered.

In some cases the core may fit so tightly in the core barrel that removing the core may be difficult. Scalf et al. (26) recommended a single piece coring tube and the use of a hydraulic extruding device when sampling subsurface solids for

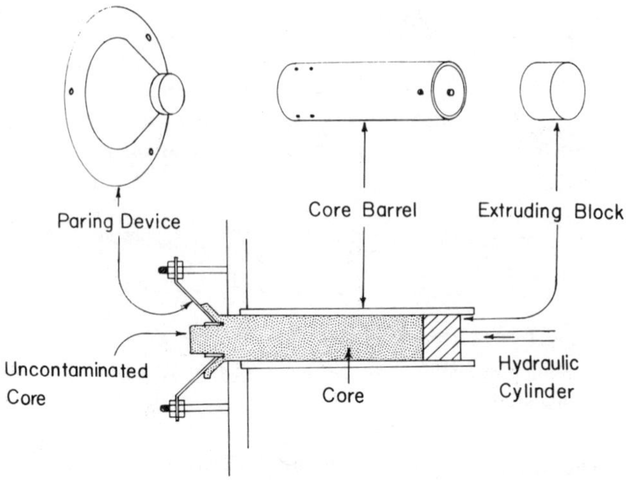

Figure 11.8. Device used by Wilson et al. (60, 61, 63) to process core samples. Sterile paring device is used to remove potentially contaminated outer part of core.

organic chemicals and microorganisms. As soon as the core is obtained, the drive shoe, if used, is removed and the sample tube is placed into an extruding device that uses a hydraulic jack to force the core from the core barrel (Fig. 11.8). As the core sample is forced out of the tube, the first 5–8 cm are broken off with a sterile scalpel and discarded or used for analyses of chemical or physical parameters. The center of the core is then subsampled to obtain material for microbial analysis by either pushing a sterile tube into the core for a subcore or the core is pushed through a device that cuts away the outer layer of core material that is presumed to be contaminated (63). The procedure can be done under either a bacteriological hood or in an anaerobic glove box.

11.4.3. Completion of Sampling Site

The sampling site should be returned to as near presampling conditions as possible and steps taken to prevent the site from being a hazard. Also it is important to prevent a pit, trench, or borehole from allowing entry of pollutants from the surface into the groundwater. The hole may be backfilled with the material taken from the hole originally or it may be filled with concrete or some highly impermeable material such as bentonite. The borehole could be further used as a monitoring well by completing it with proper procedures (26).

11.5. DISCUSSION AND SUMMARY

In many microbiological studies of ground water and subsurface environments, little attention has been paid to securing good samples. Many papers in the literature give little information regarding the sampling methods and the characteristics of the wells. Further, there is a tendency for investigators to compare data from different types of samples. Keith et al. (64) have addressed the problem of variability in ground water quality data and concluded that variables such as different types of well construction, pumping rates, and sample collection procedures can affect chemical data. It is reasonable to assume that these factors might also cause variability of microbiological data. Much more information is needed, however, to determine how these factors affect microbiological data.

Geohydrologic conditions are of overriding importance in designing and conducting subsurface microbiological sampling programs. Both the type of equipment used and the location of sampling points are dependent on site-specific conditons. The type of sample collected, whether solid or water, is, however, generally determined by the purpose of the study. Groundwater samples may be ideal in studies of the movement of pathogens or indicator organisms from waste-disposal sites or in investigations using microorganisms as tracers. In contrast, for studies of microorganisms indigenous to an aquifer, core samples probably provide more information. In either case, the microbiologist will face a considerable challenge in collecting representative and uncontaminated samples of either subsurface solids or water. Microbiological data obtained from either type of sample should be interpreted with caution. Samples of subsurface solids provide information about one point in three-dimensional space and can be obtained only once because the sampling process is inherently disruptive. Water samples can be obtained from the same well repeatedly; however, this information may not adequately reflect conditions in the aquifer.

At present, there are probably more questions than answers for an individual studying microorganisms in the subsurface. The recent emphasis on groundwater quality and the increasing number of microbiologists interested in the subsurface hydrologic regime is likely to result in a rapid increase in the information available.

REFERENCES

1. D. K. Todd, *Groundwater Hydrology*, 2nd ed., John Wiley & Sons, New York, 1980.
2. R. A. Freeze and J. A. Cherry, *Groundwater*, Prentice–Hall, Englewood Cliffs, N.J. 1982.
3. C. W. Fetter, Jr., *Applied Hydrogeology*, Charles E. Merrill, Columbus, Ohio, 1980.
4. R. C. Heath and F. W. Trainer, *Introduction to Ground Water Hydrology*, National Water Well Association, Worthington, Ohio, 1982.
5. S. W. Lohman et al., "Definitions of Selected Ground-Water Terms—Revisions and Conceptual Refinements," U.S. Geological Survey, Water-Supply Paper 1988, Washington, D.C., 1972.

6. O. E. Meinzer, *The Occurrence of Ground Water in the United States, with a Discussion of Principles*, U.S. Geological Survey, Water-Supply Paper 489, Washington, D.C., 1923.

7. D. L. Balkwill, T. E. Rucinsky, and L. E. Casida, *Antonie van Leeuwenhoek J. Microbiol. Serol.* **43,** 73 (1977).

8. J. Marxen, *Verh. Internat. Verein. Limnol.* **21,** 1371 (1981).

9. J. B. Davis, *Petroleum Microbiology*, Elsevier–North Holland, New York, 1967.

10. *Waste Disposal Practices and Their Effects on Ground Water*, U.S. Environmental Protection Agency, Report No. EPA-570/9-77-001, Washington, D.C., 1977.

11. L. G. Wilson, *Monitoring in the Vadose Zone*, U.S. Environmental Protection Agency Report No. EPA-600/7-80-134, Washington, D.C., 1980.

12. F. B. Dazzo and D. F. Rothwell, *Appl Microbiol.* **27,** 1172 (1974).

13. D. Wang, J. C. Lance, and C. P. Gerba, *Appl. Environ. Microbiol.* **39,** 662, (1980).

14. R. C. Prill, E. T. Oaksford, and J. E. Potorti, "A Facility Designed to Monitor the Unsaturated Zone During Infiltration of Tertiary Treated Sewage on Long Island, N.Y.," U.S. Geological Survey, Water-Resources Investigation 79-48, Washington, D.C., 1979.

15. J. M. Vaughn, E. F. Landry, C. A. Beckwith, and M. Z. Thomas, *Appl. Environ. Microbiol.* **41,** 139 (1981).

16. B. A. Stewart, F. G. Viets, Jr., G. L. Hutchinson, W. D. Kemper, F. E. Clark, M. L. Fairbourn, and F. Strauch, "Distribution of Nitrates and Other Water Pollutants under Fields and Corrals in the Middle South Platte Valley of Colorado," Agricultural Research Service Report 41-134, U.S. Department of Agriculture, Washington, D.C., 1967.

17. W. J. Dunlap and J. F. McNabb, *Subsurface Biological Activity in Relation to Ground Water Pollution*, U.S. Environmental Protection Agency, Report No. EPA-660/2-73-014, Washington, D.C., 1973.

18. S. I. Kuznetsov, M. V. Ivanov, and N. N. Lyalikova, *Introduction to Geological Microbiology*, McGraw-Hill, New York, 1963.

19. J. D. Hem, *Study and Interpretation of the Chemical Characteristics of Natural Water*, U.S. Geological Survey, Water-Supply Paper 1473, Washington, D.C., 1970.

20. K. S. Hanks, D. E. Wallace, and H. A. Schreiber, *Bacteriological Ground Water Quality Characteristics of the Walnut Gulch Experimental Watershed*, U.S. Dept. Agricult., Conservation Research Report No. 28, Washinton, D.C., 1981.

21. K. G. Lamka, M. W. LeChevallier, and R. J. Seidler, *Appl. Environ. Microbiol.* **39,** 734 (1980).

22. C. Ritter and W. J. Hausler, *American J. Public Health* **51,** 1347 (1961).

23. F. M. Wellings, A. L. Lewis, C. W. Mountain, and L. V. Pierce, *Appl. Environ. Microbiol.* **29,** 751 (1975).

24. Procedures Manual for Ground Water Monitoring at Solid Waste Disposal Facilities, U.S. Environmental Protection Agency, Report No. EPA/530/SW-611, Washington, D.C., 1977.

25. S. W. Sisk, *NEIC Manual for Groundwater/Subsurface Investigations at Hazardous Waste Sites*, U.S. Environmental Protection Agency, Report No. EPA-330/9-81-002, Washington, D.C., 1981.

26. M. R. Scalf, J. F. McNabb, W. J. Dunlap, R. L. Cosby, and J. Fryberger, *Manual of Ground Water Sampling Procedures*, National Water Well Association, Worthington, OH, 1981.

27. U.P. Gibson and R. D. Singer, *Water Well Manual*, Premier Press, Berkeley, CA, 1971.

28. M. D. Campbell and J. H. Lehr, *Water Well Technology*, McGraw-Hill, New York, 1973.

29. *Manual for Water Well Construction Practices*, U.S. Environmental Protection Agency, Report No. EPA-570/9-75-001, Washington, D.C., 1975.

30. Anonymous, *Water Well J.* **36,**(5), 61 (1982).

31. Z. S. Smirnova, *Microbiology* (a translation of *Mikrobiologiya*) **26,** 717, (1957).

32. M. R. Richard, *Water Well J.* **33,**(4), 66 (1979).

33. M. R. Richard, *Water Well J.* **33,**(5), 50 (1979).

34. R. Minning, in D. Neilsen, Ed., *Proceedings of the 2nd National Symposium on Aquifer Restoration and Ground Water Monitoring*, National Water Well Association, Worthington, OH, 1982.

35. G. D. Miller, in D. Neilsen, Ed., *Proceedings of the 2nd National Symposium on Aquifer Restoration and Ground Water Monitoring*, National Water Well Association, Worthington, OH, 1982.

36. G. F. Craun, *Ground Water* **17**(2), 183 (1979).

37. W. K. Summers and L. A. Brandvold, *Ground Water* **5**(1), 9 (1967).

38. K. D. Schmidt, *Ground Water*, **15**(2), 130 (1977).

39. J. M. Marsh and J. W. Lloyd, *Ground Water*, **18**(4), 366 (1980).

40. R. M. Schuller, J. P. Gibb, and R. A. Griffin, *Ground Water Monitoring Review* **1**(1), 42 (1981).

41. W. W. Wood, *Techniques of Water Resources Investigations*, Book 1, Ch. D-2, U.S. Geological Survey, Washington, D.C., 1974.

42. P. Hirsch and E. Rades-Rohkohl, In *Developments in Industrial Microbiology*, Vol. 24, Soc. Industrial Microb., Arlington, VA, 1983.

43. *Gound Water and Wells*, Edward E. Johnson, Inc., Saint Paul, MN, 1966.

44. *Ground Water*, American Water Works Association, AWWA Manual M21, New York, 1973

45. Interagency Advisory Committee on Water Data, *National Handbook of Recommended Methods for Water-Data Acquisition*, U.S. Geological Survey Office of Water Data Coordination, Reston, VA, 1977.

46. F. C. Koopman, *Downhole Pumps for Water Sampling in Small-Diameter Wells*, U.S. Geological Survey, Open File Report 79-1264, Reston, VA, 1979.

47. R. F. Middelburg, In *National Conference on Quality Assurance of Environmental Measurements*, Information Transfer, Silver Spring, MD, 1979, p. 97.

48. D. C. Signor, *Gas-Driven Pump for Ground-Water Samples*, U.S. Geological Survey, Water Resources Investigation 78-72, Reston, VA, 1978.

49. M. B. Tomson, S. Hutchins, J. M. King, and C. H. Ward, *Ground Water* **18,** 444 (1980).

50. H. F. Ridgway and B. H. Olson, *Appl. Environ. Microbiol.* **41,** 274 (1981).

51. E. E. Geldreich, *Handbook for Evaluating Water Bacteriological Laboratories*, 2nd

ed., U.S. Environmental Protection Agency, Report No. EPA-670/9-75-006, Cincinnati, Ohio, 1975.

52. B. G. Katz and G. E. Mallard, in W. J. Cooper, Ed., *Chemistry in Water Reuse*, Vol. 1, Ann Arbor Science, Ann Arbor, MI, 1981, p. 165.

53. M. J. Noonan and J. F. McNabb, in M. J. Noonan, Ed., *The Quality and Movement of Groundwater in Alluvial Aquifers of New Zealand*, Lincoln College Department of Agric. Microbiol., Tech. Pub. No. 2, Canterbury, New Zealand, 1979.

54. G. J. Olson, W. S. Dockins, G. A. McFeters, and W. P. Iverson, *Geomicrobiol.* **2,** 327 (1981).

55. G. Elkan and E. Horvath, *The Role of Microorganisms in the Decomposition of Deep Well Injected Liquid Industrial Wastes*, National Technical Information Service, PB-268 646, Springfield, VA, 1977.

56. E. M. Godsy and G. G. Ehrlich, *J. Research U.S. Geological Survey* **6,** 829, (1978).

57. W. L. Acker, III, *Basic Procedures for Soil Sampling and Core Drilling*, Acker Drill, Co., Scranton, PA, 1974.

58. A. E. Gilmore, *Soil Science* **87,** 95 (1959).

59. D. L. Danielpol, *Der Einfluss Organischer Verschmutzung auf das Grundwasser-Okosystem der Donau in Raum Weinund Niederosterreich*, Limnologisches Institut, Mondsee, Austria, 1982.

60. J. T. Wilson, J. F. McNabb, and B. H. Wilson, in *Developments in Industrial Microbiology*, Vol. 24, *Soc. Indust. Microbiol.*, Arlington, VA, 1983.

61. J. T. Wilson, M. J. Noonan, and J. F. McNabb, in C. H. Ward, Ed., *Proceedings of First International Conference on Ground Water Quality Research*, John Wiley, New York, 1983.

62. C. A. Black, Ed., *Methods of Soil Analysis*, Vol. 1, American Society of Agronomy, Madison, WI, 1965.

63. J. T. Wilson and M. J. Noonan, in G. Bitton and C. P. Gerba, Ed., *Groundwater Pollution Microbiology*, John Wiley, New York, 1983.

64. S. J. Keith, L. G. Wilson, and H. R. Fitch, in D. Nielsen, Ed., *Proceedings of the 2nd National Symposium on Aquifer Restoration and Ground Water Monitoring*, National Water Well Association, Worthington, OH, 1982.

12

GROUNDWATER CONTAMINATION: DATA ANALYSIS AND MODELING

I. Cech and R. Harrist

School of Public Health
The University of Texas, Houston, Texas

12.1. INTRODUCTION

This chapter discusses the use of statistics in various stages of microbiological experimentation and research and illustrates the application of statistical strategy with several examples drawn from recent work in aquatic microbiology. The presentation is not intended to be exhaustive, but it is intended to be indicative of the valuable role of statistical methods in investigations of questions that are typically encountered by aquatic microbiologists. Detailed information on most of the statistical methods can be found in standard textbooks. We recommend Remington and Schork (1) for an introductory exposition and Snedecor and Cochran (2) for intermediate readers.

In a broad sense, statistical methods are used whenever specific observations or measurements are made and the properties of a larger population of such observations are inferred from this sample. The discipline of statistics seeks to formalize this procedure and reveal its essential elements. This allows the experimenter or observer to collect information in the most efficient manner and to draw valid generalizations. In particular, it is recognized that estimates or decisions made in this manner are subject to error. Much of the effort involved in statistical analysis is directed toward calculating the probability of error. This consideration also dictates many of the constraints on the analysis and interpretation of the data.

Ideally, any experimental or observational study should be planned in detail, including the statistical analysis, before the data are collected. This is not the usual situation, however; in many studies, the conditions necessary for careful and precise analysis are not met. Even when a detailed plan has been worked out, compromises often must be made, for example, strict random sampling may be impossible or impractical or a sample may be lost. This means that a researcher must be flexible and resourceful in his analysis and conservative in his interpretation of the results.

An outline of the evolution of a study with the involvement of statistical considerations emphasized could be as follows:

1. Formulate a question of interest. (State a research hypothesis.)
2. Decide what measurements can and should be made to answer the question. (Plan the study or experiment; this includes deciding on the number of observations or sample size.)
3. Collect the data according to the above plan.
4. Prepare a written summary of the data. (Descriptive statistics and statistical analysis are used here—assistance of a computer for data processing could be helpful at this stage.)
5. Interpret the statistical results to reach an answer to the original question. Possible sources of error or bias should be carefully acknowledged. (Statistical estimation or tests of hypothesis with calculation of confidence intervals of error probabilities should be presented.)

As can be seen, statistical issues play a part from Step 2 onward, because of the inferential nature of the process. Whenever we reason from the specific information obtained from the sample (Step 3) to a general conclusion (Step 5), we must account for the possibility that our conclusions are in error. In order to be assured that the probability of error is small or that our estimates are acceptably close to the mark, the error probabilities or confidence intervals must be calculated and presented as part of the conclusion.

Step 1, the careful formulation of a research question is a natural and indispensable beginning point for any study. Surprisingly, studies are sometimes begun with no clear purpose in mind. This may be due in part to the inclination of energetic investigators to begin some visible activity, but such efforts are likely to be wasted unless preceded by disciplined contemplation of the real question to be investigated. In order for the question to be interesting in the sense indicated here, the answer should make a substantial scientific or practical difference.

The research hypothesis is the motivating factor for the study. Once it is clearly stated, the next steps follow logically. The study must be planned, beginning with the decision about which measurements or observations are to be made to answer the research question. Careful work requires that we formulate a statistical model for the measurements to be made, that is, a formal statement of the distribution of the observations to be expected if the hypothesis (null hypothesis) is true or if it is false.

In experimental studies, this model can be used (along with other information) to determine the number of observations required to answer the research question with a given degree of precision. This is sometimes difficult to calculate exactly, but suitable estimates can usually be found. This can be an especially important step if the observations are costly or difficult to obtain and can sometimes point the way to significant savings in time and money. The planning of the experiment is one of the crucial tasks of the study. This specifies in detail the numbers of observations to be made and the conditions under which they are to be made. This, in turn, determines what statistical summaries and analyses are appropriate.

Efforts in planning are fruitless unless Step 3, the collection of data, is carried out in strict accord with the plan. If stated observational or experimental conditions are not met, the data collected will be compromised and the statistical summaries and tests to follow may be invalidated. It is almost inevitable that there will be some variation from the experimental design. Usually minor variations can be accommodated, but it is the researcher's duty to minimize such inconsistencies and to carefully report those that do occur.

When data collection is completed, the next step is to summarize and describe what has been observed. At this point it is important that the investigators become thoroughly familiar with the data; many errors and inconsistencies are discovered because the numbers "just don't look right." Descriptive devices such as tables, graphs, charts, histograms, and statistical summaries such as means, medians, standard deviations, and ranges can be produced at this point. Descriptive studies (i.e., those that seek to describe only the data at hand) may culminate at this step, foregoing the generalization to larger populations.

The fifth step is to infer from the sample data the characteristics of the population from which it was drawn. Such inferences take the form of estimates (e.g., the average concentration of a particular organism in a certain body of water at a given time) or tests of hypotheses (e.g., that recent rainfall has no effect on the concentration of these organisms.) The uncertainty (variation) in the estimate can be reported by calculation of a confidence interval and the uncertainty in the case of a test hypothesis is properly reported as the probability of a false positive (rejection of the null hypothesis when it is true—a Type I error or sometimes a "p-value") or a false negative (failure to reject the null hypothesis when it is false—Type II error.) Finally, the statistical results must be interpreted in terms of the original question and related to other knowledge in the field.

The examples that follow are presented to illustrate these principles as applied to recent research activities in aquatic microbiology. In each instance, statistical issues and methodology are seen as an integral and natural part of the scientific inquiry, not as arbitrary addenda. Most of the examples have been drawn from recent research projects. They sometimes fall short of the ideal study described above, particularly with regard to explanation of sample selection.

12.2. ADSORPTION OF VIRUS TO SOIL: STRATEGY FOR ANALYSIS; MULTIPLE LINEAR REGRESSION

This example is drawn from studies by Gerba et al. (3) conducted in an effort to expand our understanding of processes controlling virus adsorption to soil. An understanding of factors controlling virus migration through subsurface material is crucial for the effective and safe operation of septic systems and wastewater land treatments. Earlier studies on virus adsorption and removal by various soils conducted by the same laboratory at Baylor College of Medicine, Department of Virology (4), failed to reveal a consistent predictive pattern of adsorption related to any particular property of soil. Part of the reason for this could have been the complexity of the original research scheme. It became apparent that in the search for patterns, it might be advantageous to group factors influencing virus adsorption as well as viruses themselves with respect to their adsorptive patterns. Thus, the purpose of the study was to statistically reexamine earlier data by grouping variables of interest into a number of broad logical categories.

Information on adsorption of viruses in soil was available from nine experiments with 16 viruses and with nine different soils. In each study, the mean adsorption rates for each of the 16 viruses was measured and 15 measurements describing various physical and chemical properties of the soil were recorded.

In order to summarize this rather formidable array of numbers, it was decided to arrange the viruses into relatively homogeneous groups and to consider the relationship of the average adsorption for the group to the properties of the soil. In the original study, a version of factor analysis was used to suggest such groups (Table 12.1). Viruses were grouped into two general categories with a mean adsorption of 44 percent for virus Group I and 78 percent for Group II. Bacteriophage f2, because its 16 percent adsorption was apparently outside either of these groups, was placed in a third category and considered separately in the following analysis.

Within each virus group, the adsorption rates were averaged to form a group adsorption rate, that is, the 16 response variables were reduced to 3, giving the obvious advantage of greater simplicity. One potential difficulty is the possibility that a given virus might be "misclassified" because of random variation in its mean adsorption rate. Another potential difficulty is that of characterizing the

Table 12.1. Descriptive Statistical Summary of Virus Adsorption to Nine Soil Types, by Individual Virus Types and by Their Groups (as Determied by Factor Analysis)[a]

	Adsorption (Percent)			
Virus	Mean	Standard Deviation	Minimum	Maximum
Echovirus I (Farouk)	39.19	35.00	11.00	99.70
Echovirus 1 (V212)	45.08	31.80	1.00	99.70
Echovirus I (V239)	6.89	29.98	5.00	98.00
Echovirus I (V248)	45.89	37.62	0.10	99.90
Coxsackievirus B4 (V216)	38.14	29.66	2.00	99.30
Coxsackievirus (V240)	55.76	36.36	0.10	98.70
MS-2	45.34	26.58	16.16	99.91
ΦX174	34.51	44.91	0.01	98.98
Group I	43.86	33.25	0.01	99.98
Poliovirus 1 (LSc)	85.26	21.60	42.00	99.90
Echovirus 7 (Wallace)	68.13	40.06	8.00	99.90
Coxsackievirus B3 (Nancy)	81.07	22.48	35.00	99.61
T2	72.12	31.69	8.40	99.99
T4	83.12	32.52	0.01	99.70
Group II	77.94	29.88	0.01	99.99
f2 (Group III)	16.44	32.39	0.01	99.20
All Viruses	56.97		0.01	99.99

[a]*Source:* Gerba et al. (3). Used by permission of the American Chemical Society.

groups formed by inspection of the response variables; that is, deciding what the viruses in a given group have in common other than similar adsorption rates.

To evaluate the effect of specific characteristics of various soils on virus adsorption, these characteristics were subdivided into three sets. The first set contained variables describing general physical and chemical characteristics of the soils. Included in this set were pH, percent organic matter, conductivity, cation-exhange capacity, and soil surface area. The second and third sets contained variables describing specific chemical elements in the soils (viz., set 2: concentration of magnesium, calcium, phosphorus, aluminum, and iron; set 3: concentration of exchangable magnesium, calcium, phosphorus, aluminum, and iron). The adsorption was regarded as a function of the physical and chemical characteristics of the soils. The linear dependence of adsorption upon each group of explanatory variables was investigated (for each virus group) by multiple linear regression.

The results of the regression analysis are summarized in Tables 12.2 through 12.4. These must be interpreted with caution since the choice of experimental conditions (i.e., the virus groups) was made on the basis of their response (i.e., their observed adsorption rates). This makes the distribution of the test statistics to be calculated dependent upon this choice. This dependence is not taken into account; hence, the nominal probability levels associated with ensuing statistical tests may not be accurate and should be used only as a suggestion of the relative importance of the influence of the explanatory variables upon adsorptive rates. To illustrate this point, suppose that a given virus type that was assigned to Group I received its (low) score due to random variation, that is, it actually belongs to group II. This would result in an inappropriate analysis and would tend to obscure any consistent relationships that might be present.

It would be appropriate to preface the analysis with the qualifying statement, "Given that the viruses were divided into the three groups in the manner described, the following relationships were seen." This acknowledges the possibility that if the entire experiment were repeated the groupings might be different, and thus the conclusions concerning the relationships of adsorption of viruses to soil characteristics might differ.

The results are interesting, however, for the development of hypotheses for future experiments. They suggest that particular soil characteristics do influence adsorption to a different degree, depending on the virus group. For example, in virus Group I, of all the soils variables studied, the most important appeared to be pH. This single factor, which was found to be negatively correlated with adsorption of viruses in Group I, "explained" about 83 percent of the variation. (Table 12.2.) Another variable of importance to Group I from the same set of soil characteristics appears to be the content of organic matter. For this variable, the correlation was also negative. From other sets of soil variables, the only element found to contribute significantly to the regression model for virus Group I was exchangable iron (Table 12.2). The direction of association for this variable was positive.

The regression analysis of adsorption on soil characteristics for viruses in Group II showed that none of the characteristics were significant predictors of viral adsorption (Table 12.3). Inasmuch as bacterial phage f2 was not included in either virus Group I or II, it was presented by itself (Table 12.4). For this virus, the concentration in soil of exchangable aluminum was found to be a significant predictor of f2 adsorption. The direction of this association was positive.

The data suggest that Group II viruses adsorbed very well to most soils, sludges, and marine sediments, while Group I viruses adsorbed to a lesser degree and were affected more by the pH and the organic content of the soil. Coliphage f2 adsorbed the least to all of the soils, and may actually represent a third group.

The authors concluded that differences in adsorption between various types and strains of viruses probably resulted from variability in the configuration of proteins on the surface of the virus. This, in turn, should affect the electrostatic potential beween virus and soil, and thus influence the degree of interaction between the two particles. Perhaps the net charge on Group I viruses is more easily influenced over the range of pH values and organic matter concentrations found in nature than Group II viruses. Groundwater, in the opinion of the investigators, probably would be more susceptible of contamination by viruses

Table 12.2. Stepwise Multiple Regression of Mean Virus Adsorption in Group I on Soil Characteristics[a]

Dependent Variable	Significance of Individual Variables	Multiple R	R^2	R^2 Change	Simple R	Overall F	Significance of Regression
First set of soil variables							
pH	0.004	0.911	0.8300	0.8210	−0.9110	24.404	0.004
Organics	0.032	0.976	0.9525	0.1226	−0.4920	40.132	0.002
Ion-exchange capacity	0.094	0.991	0.9839	0.0314	−0.0435	61.259	0.003
Surface area	0.149	0.998	0.9955	0.0116	−0.1130	112.196	0.009
Conductivity	0.754	0.998	0.9962	0.0006	−0.2662	52.332	0.105
Second set of soil variables							
Total phosphorus	0.138	0.619	0.3830	0.3830	−0.6189	3.100	0.138
Total calcium	0.672	0.643	0.4135	0.0305	−0.4581	1.400	0.344
Total aluminum	0.834	0.650	0.4234	0.0010	−0.3377	0.730	0.597
Total iron	0.764	0.675	0.4555	0.0321	−0.2024	0.420	0.793
Third set of soil variables							
Exchangeable aluminum	0.056	0.743	0.5515	0.5515	0.7427	6.150	0.056
Exchangeable iron	0.025	0.943	0.8902	0.3386	0.5755	16.216	0.012
Exchangeable magnesium	0.359	0.960	0.9209	0.0307	−0.1645	11.649	0.037
Exchangeable calcium	0.347	0.977	0.9546	0.0337	−0.1608	10.523	0.089
Exchangeable phosphorus	0.835	0.978	0.9576	0.0029	−0.4409	4.517	0.342

[a]*Source:* Gerba et al. (3). Used by permission of the American Chemical Society.

Table 12.3. Stepwise Multiple Regression of Mean Virus Adsorption in Group II on Soil Characteristics[a]

Dependent Variable	Significance of Individual Variables	Multiple R	R^2	R^2 Change	Simple R	Overall F	Significance of Regression
First set of soil variables							
Ion-exchange capacity	0.106	0.661	0.437	0.437	0.661	3.887	0.106
Conductivity	0.214	0.797	0.636	0.198	0.552	3.493	0.133
Surface area	0.064	0.950	0.903	0.267	0.592	9.333	0.050
Organic matter	0.485	0.964	0.929	0.026	0.386	6.532	0.137
pH	0.413	0.987	0.947	0.045	−0.044	7.509	0.270
Second set of soil variables							
Total magnesium	0.133	0.625	0.391	0.391	0.625	3.213	0.133
Total calcium	0.146	0.814	0.663	0.272	0.373	3.943	0.113
Total phosphorus	0.293	0.884	0.781	0.118	0.250	3.573	0.162
Total iron	0.185	0.962	0.927	0.145	0.602	6.324	0.141
Total aluminum	0.824	0.965	0.932	0.005	0.617	2.749	0.427
Third set of soil variables							
Exchangeable magnesium	0.126	0.634	0.402	0.402	0.634	3.365	0.126
Exchangeable aluminum	0.196	0.792	0.627	0.224	0.365	3.357	0.139
Exchangeable calcium	0.316	0.865	0.748	0.121	0.517	2.967	0.198
Exchangeable phosphorus	0.597	0.888	0.789	0.041	0.518	1.869	0.378
Exchangeable iron	0.683	0.915	0.837	0.048	0.238	1.028	0.631

[a]*Source*: Gerba et al. (3). Used by permission of the American Chemical Society.

Table 12.4. Stepwise Multiple Regression of f2 Virus Adsorption on Soil Characteristics[a]

Dependent Variable	Significance of Individual Variables	Multiple R	R^2	R^2 Change	Simple R	Overall F	Significance of Regression
First set of soil variables							
pH	0.215	0.535	0.287	0.287	−0.535	2.009	0.215
Conductivity	0.729	0.557	0.310	0.024	−0.091	0.900	0.476
Surface area	0.567	0.666	0.443	0.128	0.116	0.398	0.803
Organic matter	0.893	0.678	0.459	0.016	−0.044	0.170	0.940
Second set of soil variables							
Total phosphorus	0.262	0.492	0.242	0.242	−0.491	1.594	0.262
Total aluminum	0.418	0.608	0.270	0.128	−0.085	1.175	0.397
Total iron	0.209	0.812	0.659	0.289	−0.140	1.934	0.301
Total magnesium	0.658	0.836	0.699	0.040	−0.157	1.162	0.511
Total calcium	0.196	0.986	0.972	0.273	−0.188	7.055	0.278
Third set of soil variables							
Exchangeable aluminum	0.001	0.957	0.916	0.916	0.957	54.602	0.001
Exchangeable iron	0.103	0.980	0.960	0.044	−0.218	48.289	0.002
Exchangeable phosphorus	0.136	0.991	0.983	0.023	−0.352	58.452	0.004
Exchangeable calcium	0.511	0.993	0.987	0.004	−0.250	38.578	0.025

[a] *Source*: Gerba et al. (3). Used by permission of the American Chemical Society.

with properties similar to those of Group I. The overall implication of this study is that adsorption–elution of Group I viruses is more sensitive to certain soil characteristics than Group II. These differences should be taken into account in both field and laboratory studies of virus adsorption and migration through soil.

12.3. CONTAMINATION OF RURAL WATER SUPPLIES: CONTINGENCY TABLES; SYNAGRAPHIC MAPPING

This example illustrates the use of statistical analysis and synagraphic computer mapping in assessing groundwater quality and patterns of contamination of rural water supplies. The example is drawn from a study at the University of Texas School of Public Health at Houston (5, 6). The study was done in part of rural East Texas, Houston County, with the purpose of obtaining information on causes of rural well contamination and developing a survey strategy for rural areas where water is derived mainly from individual sources. Such private rural supplies are not subjected to routine water quality evaluations and thus, relevant and up-to-date information on groundwater quality for rural users is usually not available.

The shortage of information on rural groundwater quality is particularly noticeable with respect to contaminants of potential health significance. In this particular survey, attention was focused on assessing the origin and extent of nitrate and bacterial contamination. Concern about nitrate in well water dates back to 1945 when Comly first reported infant methemoglobinemia resulting from ingestion of nitrate-enriched water (7). Considerable concern over nitrates in drinking water has been revived in the last two decades mainly because compounds formed by the products of nitrate reduction and secondary amines have been found to be carcinogenic in laboratory animal experiments (8, 9). The impact of nitrosamines on human health is unclear and remains among scientific research priorities.

The actual origin of excessive nitrates in shallow wells is itself not well known. A possibility was suggested in the 1950s that nitrates in groundwater are ubiquitous and not man-related, but of perfectly natural origin, being indigenous in Tertiary and Quaternary sands (10). Since information on nitrates in rural wells is sporadic and available primarily by chance rather than by purposeful design (and most of the data are over ten years old), appraisal of the causes and extent of the nitrate problem cannot be readily made.

The hypothesis of natural origin of nitrates in rural wells was challenged in the study described here. The sampling of rural domestic wells no deeper than 350 feet was conducted in the summer of 1976 after a comparitively rainy period and was repeated a month later, after a period of scarce rainfall. Fifty-three privately owned shallow wells were sampled. Sampling was designed to consider and distinguish natural geographic variability in nitrate concentration due to differences in water-bearing strata, soil characteristics, amount of antecedent rainfall, and fluctuations of water level from variations caused by human activity.

There are eight different sands underlying the study area, of geological age from Eocene to Recent. To account for possible differences due to geology, water samples were taken from all these sands except for the *Weches Greensand* and *Reklaw*, which are not known to yield water to wells, and from the *Jackson sand*, which was physically inaccessible (Fig. 12.1). An effort was made to distribute the sampling sites uniformly within these areas.

At each sampled site, data on all potential sources of well contamination as well as on the structural characteristics of wells were collected. Location of septic tanks, barnyards, feedlots, refuse disposals, grazing fields, gardens, croplands, cemeteries, and forest lands in relation to well sites was recorded. Other information collected at well sites was type and age of well, depth, water use, and methods of water lift. Special attention was given to the condition of the wells and the quality of their protection against contamination. Field documentation included a sketch of the well sites with appropriate man-made features indicated. A record of previous land uses was also made.

Samples were analyzed directly in the field by cadmium reduction methods (11). Tests were also performed for the presence of fecal coliforms and fecal streptococci in the water by the membrane filtration method (12). The methodological approach in this study was basically a comparison of natural geologic tendency in distribution in nitrates in this area with the patterns induced by man's activity. Results of the survey were evaluated using two and three-dimensional synagraphic computer mapping [*SYMAP* and *SYMVU* (13)]. The strength and significance of the association between nitrate levels in well water and the specific factors related to land use or the characteristics of the wells themselves were measured using the chi-square test for association in 2×2 contingency tables (1). Specific questions to be answered using statistical analyses were formulated as follows:

1. Is there an association between the distance of wells from field or gardens where nitrogenous fertilizers were applied and concentration of nitrates in wells?
2. Is there an association between the distance of wells from the nearest source of animal wastes (feedlots, barnyards) and concentration of nitrates?
3. Is there an association between the distance of wells from the domestic waste-disposal site (septic tank, cesspool, privy) and the concentration of nitrates?
4. Is there an association between the type of wells (drilled or dug) and the concentration of nitrates?
5. Is there an association between the depth of wells and the concentration of nitrates?
6. Is there an association between the concentration of nitrates and levels of fecal bacteria in well water (fecal coliforms, fecal streptococci, and the ratio of FC/FS)?

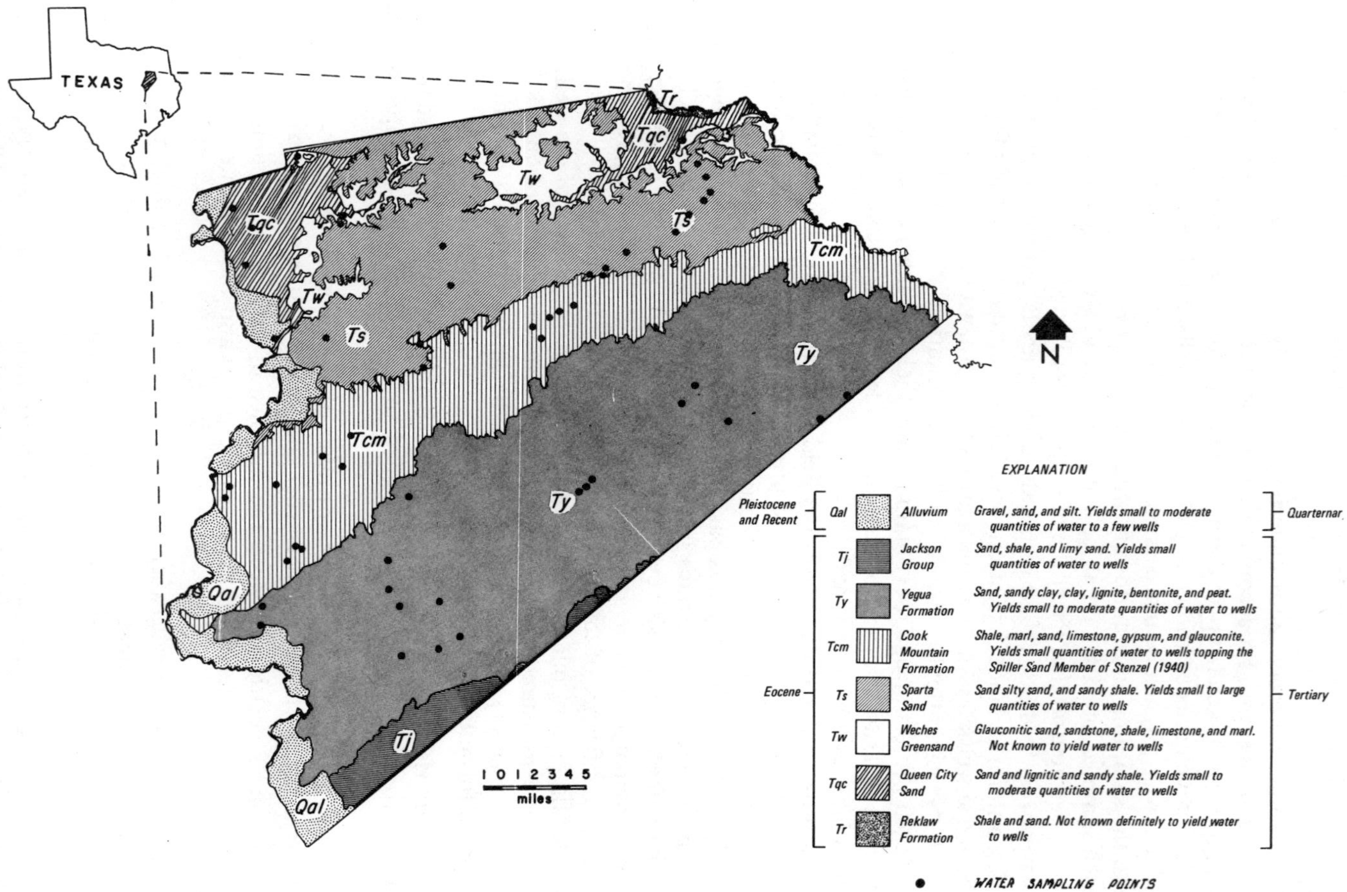

Figure 12.1. Schematic map of the well-sampling network with respect to water-bearing formations (in Houston County, Texas). (1 mile = 0.06 km). Used by permission of Pergamon Press.

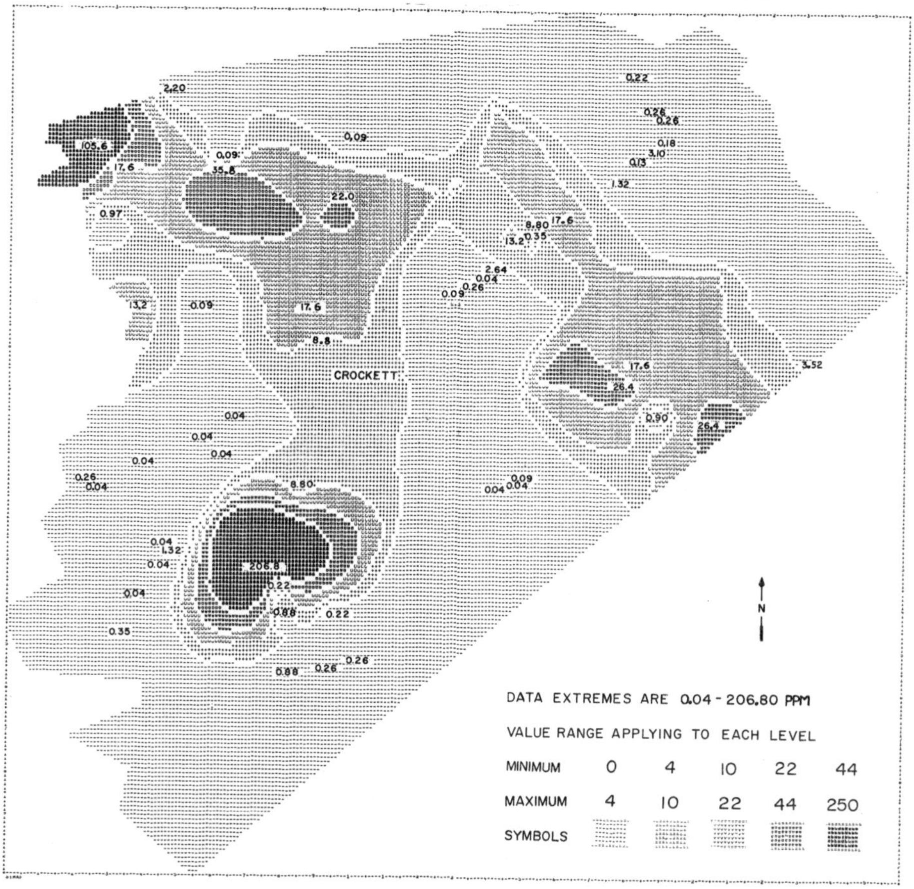

Figure 12.2. Concentrations of nitrates in the shallow wells of Houston County, Texas, June 1976. (Scale 1:390,300) (5). Used by permission of Pergamon Press.

Figure 12.2 exemplifies the use of SYMAP in synthesizing the information on nitrate concentration obtained during the first survey in June of 1976. This computer-generated map consists of contour lines connecting points with the same (interpolated) numerical values. The natural tendency in nitrate concentration is easily detectable from such a display; the prevailing concentration in the study area appears to be very low—from 0–4 parts per million (ppm). Somewhat higher values, 4–~20 ppm, predominated in the central part of the county in the area of comparatively higher population density. Overall, however, the concentration of nitrates throughout the study area was quite uniform with the exception of localized spots with concentrations of roughly 100–200 ppm, which clearly contrasted with the background. Such spots of high nitrate concentrations were observed in the southcentral and northwest parts of the county, irrespective of the character of the geologic deposits or other geographic factors. The isolated character of these spots of high nitrates become even more obvious

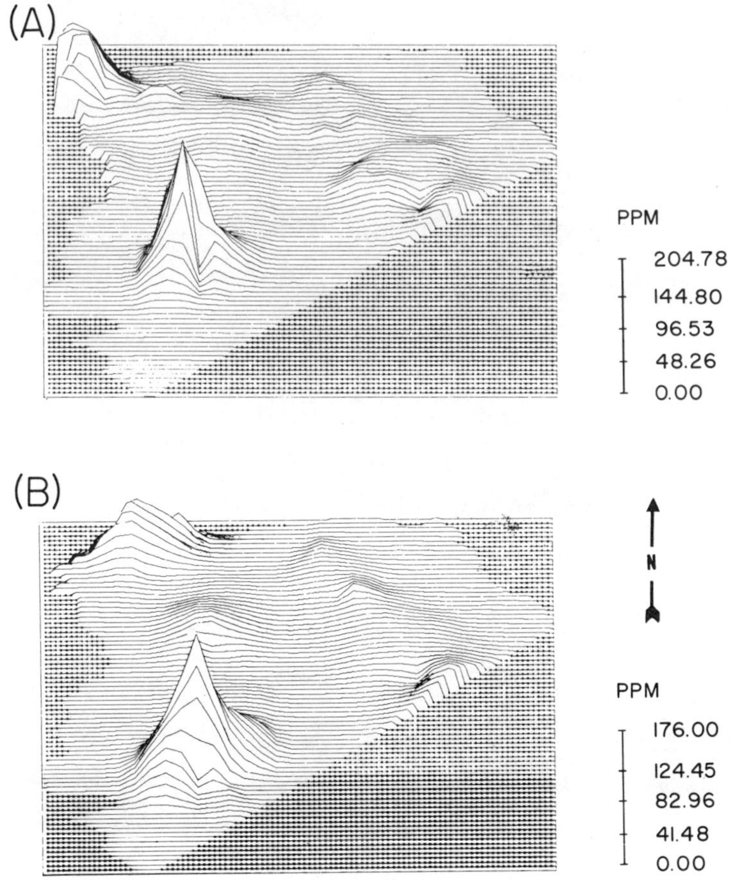

Figure 12.3. Nitrate concentrations in shallow wells after a rainy period, June 1976; (A) in comparison with (B) nitrates a month later after a period of low rainfall. Tilt is 45°; horizontal scale is 1:621,330; vertical scale is given to the right of each map (6) Used by permission of the American Water Resources Association.

from the three-dimensional SYMVU display (Fig. 12.3). The question, thus, was whether or not there were specific man-related factors or features that could have caused the elevated nitrates.

Since it was noted from some earlier surveys and especially from inspection of geographic tendency on synagraphic maps that the typical nitrate concentration in uncontaminated ground and surface water in this area was low, from nearly 0–4 ppm, it was suspected that levels higher than this would be indicative of the presence of nitrates from other sources, for example, fertilizers, animal, or human wastes. It was also noted that most older, non-cased wells were less than 15 m (about 50 ft) in depth. Because of their age, construction, and depth, these wells appeared to be the most susceptible to contamination from surface sources. A test for association in a 2×2 contingency table was used to examine the relationship between the concentration of nitrates and well depth. Each of the above stated questions could be, and indeed, was studied in the same statistical manner. A 2 × 2 contingency table was formed as shown in Table 12.5(*a*).

Conceptually, the cells of the table may be thought of as the four possible categories to which each representative well can be assigned. Under the assumption that the row and column probabilities are independent, the expected number of wells falling in each of the cells can be calculated and displayed as in Table 12.5(*b*). The reasonableness of the assumption of independence (the null hypothesis) can be tested by comparing Table 12.5(*a*) and Table 12.5(*b*).

One measure of the "distance" between the table of expected numbers and the actual observations is the well-known statistic $s = \Sigma$ (observed − expected)^2 ÷ (expected), where the sum is to be taken over all cells of the table. This is also called a "goodness of fit" statistic since it measures how well the hypothesized model fits the actual data. When the null hypothesis is true and the sample size is sufficiently large, this statistic approximately follows a (central) chi-square distribution with one degree of freedom (χ_1^2), so the probability of occurrence of any given magnitude of s can be calculated. For example, the probability that s will exceed 3.84 is about 0.05 and the probability that it will exceed 6.63 is about 0.01. When the null hypothesis is false, that is, the cell probabilities cannot be calculated from the marginal probabilities in the manner indicated, the test statistic s has a noncentral χ_1^2 distribution that tends to be located to the right of the central χ_1^2 distribution. This means that the probability that s will fall in the selected region above, say, 3.84, is increased. This probability is called the power of test and measures the ability of the test procedure to determine when the null hypothesis is false. Further discussion and details of calculation can be found in (14).

The acceptability of the approximation of the true probabilities by those calculated from the χ_1^2 distribution depends upon the number of wells in the sample, the relative sizes of the marginal totals, and, of course, the degree of accuracy required. Because of the complexity of the relationships of these factors, general rules are difficult to formulate, but one rule of thumb is to require that the expected number of observations in each cell be at least five (1). This criterion is satisfied in Table 12.5(*b*), so we know that the probabilities quoted in the example are fairly accurate. If the data had been such that one or more of the cell expectations was small, the approximation described here would be inappropriate. In some such cases, Fishers Exact Method could be applied to test a similar hypothesis (2).

Table 12.5. 2 × 2 Contingency Tables

		(*a*) Observations Depth			(*b*) Expectations Depth		
		<15m	⩾15m	Total	<15m	⩾15m	Total
Nitrates	<4 ppm	9	29	38	15.1	22.9	38
	⩾4 ppm	12	3	15	5.9	9.1	15
Total		21	32	53	21	32	53

In this study we have described the 2 × 2 contingency table analysis that was used in combination with synagraphic mapping. Computer packages for such mapping, known as SYMAP and SYMVU, have been developed by the Harvard University Laboratory for Computer Graphics (13). The synagraphic mapping treats each value in a geographic area as a centroid in a grid pattern. The surface that fits this pattern best is determined by least-squares approximation. Any two isopleths delimit the area with continuous variation within an established interval of values. Application of this technique made it possible to synthesize the sampling data on nitrates and bacteria in well water of the study area (Figs. 12.2–12.5). What was learned using a computer mapping and 2 × 2 contingency statistical analyses can be summarized as follows:

1. No relationship was found between the concentration of nitrate in the well water and the distances from the wells to areas where fertilizers were applied. Almost all of the wells were located in proximity to fields, lawns, or gardens where fertilizer was reported to have been applied in spring, and there was considerable precipitation afterwards until the time when the first (June) survey was taken. A month later, the second survey revealed essentially the same nitrate distribution as the first survey (compare Figs. 12.3A and 12.3B).

2. The overall relationships between the levels of nitrate in water and the distances to wells from the sources of animal waste were not found to be strong or statistically significant. However, in places where wells (especially dug wells) were located within 5 m of such sources, nitrate concentrations were between 20 and 44 mg/liter.

3. Distances from human waste facilities to wells was demonstrated to be a significant factor ($p < 0.05$). In some locations the distance between well and privy was no more than 5 m and the nitrate concentration was up to 250 mg/liter.

4. The relationship between the levels of nitrate in water and the type of well (such as drilled vs. dug well) and with the depth of well, was found to be strong ($p < 0.01$).

The overall implications are that the principal determinants in nitrate contamination were depth of well, type of well, and relative distances from sites of organic waste disposal. Determination of the densities of fecal coliform and fecal streptococci in rural wells (Figure 12.4(A) and Fig. 12.4(B)) showed that almost all of the dug wells were positive for one or both indicators. Even some wells drilled to depths of 80 m were contaminated. The higher nitrate levels, however, did not always correlate with bacterial density. Distributions of the fecal coliform/fecal streptococci (FC/FS) ratios are shown in Fig. 12.5. Ratios such as those have been used to estimate contamination due to animal and human sources. Areas that had the highest nitrate levels also had FC/FS values higher than 2.0. Those higher ratios were observed in the areas where contamination by human wastes was suspected. The tests did not show the association between nitrate concentrations and bacteria ratios to be significant statistically ($p <$ 0.05). The problem is apparently in the unknown factor of how long the bacteria were in the well water.

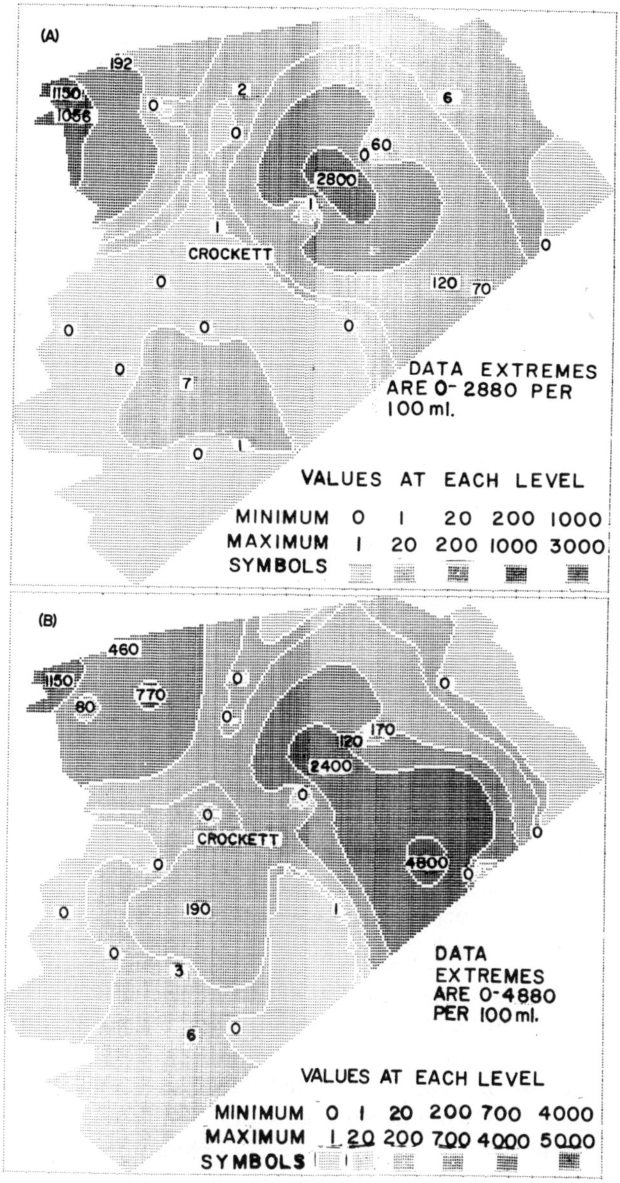

Figure 12.4. Fecal bacteria in the shallow wells of Houston County, Texas, July 1976. (A) Fecal coliforms; (B) Fecal streptococci. (Scale 1:557,840.) (6). Used by permission of the American Water Resources Association.

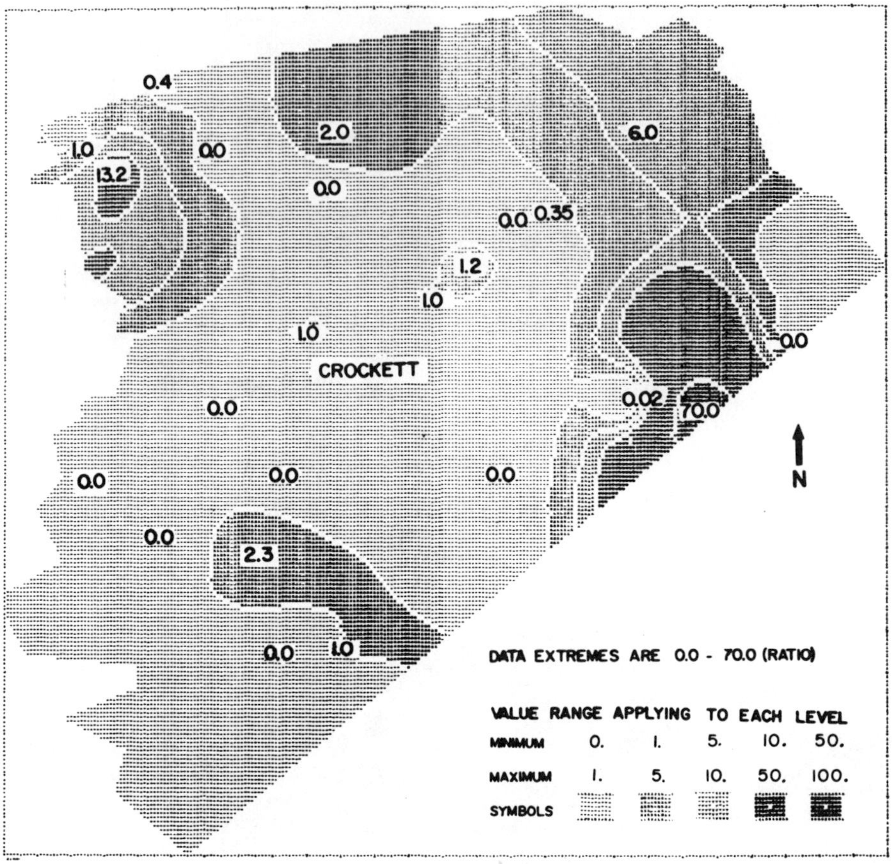

Figure 12.5. Ratio of fecal coliforms to fecal streptococci, July 1976. (Scale 1:390,300.) (6). Used by permission of the American Water Resources Association.

This study found, in general, that even though wells may be constructed in accordance with existing guidelines for locations and protection of water source, septic tanks or privies built afterward sometimes are placed without regard to these wells. Those improperly located facilities apparently create a direct recycling of drained fluid from the septic tanks to wells. In the present investigations, the focus was on the sources of contamination of groundwater by potentially toxic nitrate. Data suggested that water from the Tertiary and Quaternary sands in this area does not naturally contain high nitrates; this does not support the hypothesis stated by earlier investigators. In can be safely said that the ambient nitrate concentration in the water coming from the strata at a depth of 350 ft or more (about 115 m) is in the low range, from 0–4 ppm. The sources of nitrates higher than the indicated background are likely to be of animal, and more importantly, of human origin. The most important single cause of nitrate contamination is improper location of septic tanks in relation to wells.

12.4. VIRUS SURVIVAL IN SOIL: LINEAR REGRESSION; ANALYSIS OF VARIANCE

The land application of untreated or partially treated sewage brings into focus a question concerning virus survival in soil. It is important to evaluate the dependence of variations in virus inactivation rates upon specific environmental conditions in which land application of waste water might occur. Several different research hypotheses arise from this context. The following are examples:

1. What is the effect of different soil types on the survival of viruses?
2. What is the combined effect of soil type and environment on the survival rate of different viruses?
3. What are the effects of variation in environmental conditions, for example, temperature, moisture, anaerobic or aerobic environment on virus survival?

A crucial decision for the experiments is the choice of an appropriate response variable to characterize the survival of the organisms under various conditions.

In order to measure survival, laboratory experiments might be conducted in which several vials with soil are wetted with fluid containing viruses of known concentrations, N_0; the vials are then incubated under controlled conditions intended to simulate the environmental conditions of interest. Each vial can be sampled on succeeding days to determine the concentration of organisms, N_t. This can be continued for a predetermined period of time or until $N_t = 0$, as exemplified on Fig. 12.6(15).

In order to describe this survival pattern, it is desirable to find a mathematical model dependent upon a single parameter that summarizes the survival capability of organisms. In can be shown that if a population of nonreproducing organisms is exposed to a constant risk (probability) of death, the survival ratio may be expressed in the form $N_t/N_0 = e^{-kt}$ where t denotes elapsed time, e is the base of natural logarithms, and k is the desired parameter.

In the experiment at hand, it would then follow that

$$\log_{10} (N_t/N_0) = -kt \log_{10} e = k't$$

that is, the logarithm of the survival ratio is a constant multiple of elapsed time. Then if we plot $\log_{10} (N_t/N_0)$ against t, we would have a line passing through the origin. Thus, linear regression of $\log_{10} (N_t/N_0)$ upon T_t is appropriate and the regression coefficient (slope) can be taken as the desired summary of the survival of virus in that particular sample. Such a response variable was used in studies of virus survival in soil by Hurst and co-workers (15).

Another possible choice for a response variable would be, for example, the time required for the virus to reach a given (lower) concentration. This would have the disadvantage that the limit may not be reached during the experiment, so that there would be no appropriate numerical observation.

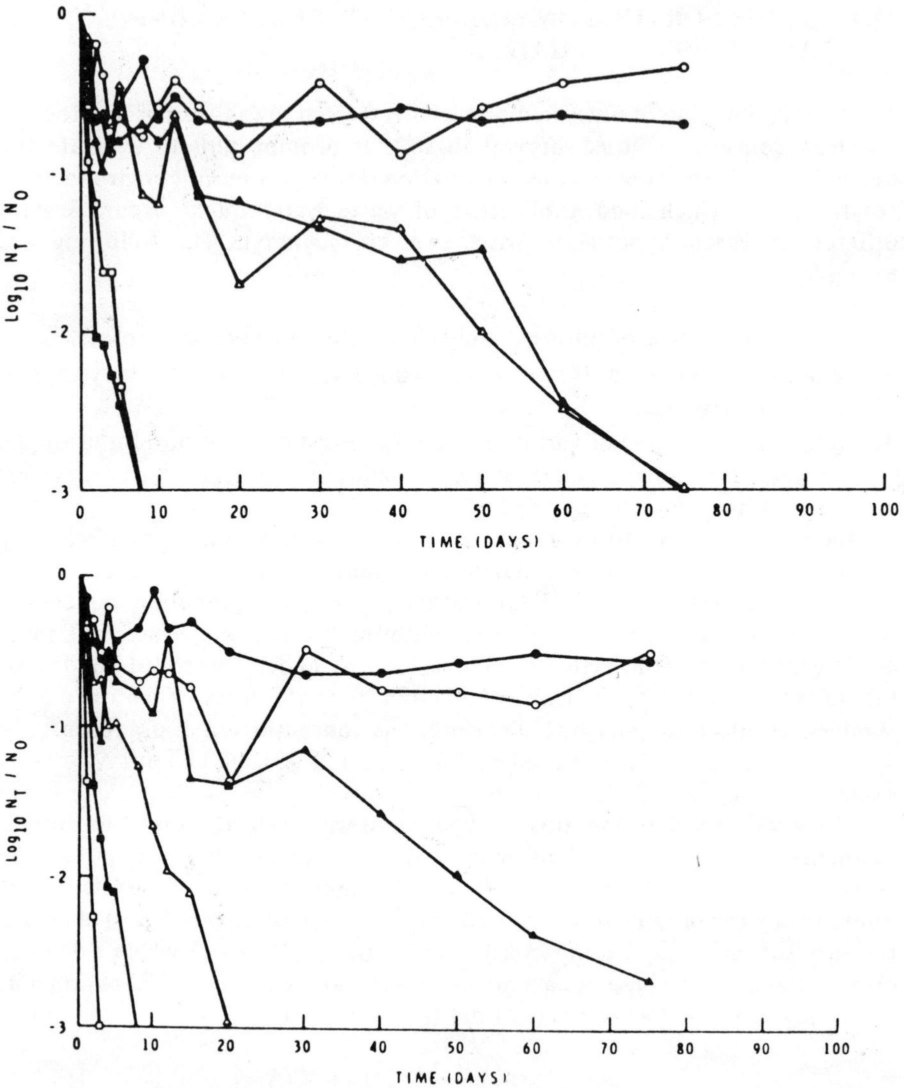

Figure 12.6. Poliovirus 1 survival in FM soil wetted with distilled water under sterile (top) and nonsterile (bottom) conditions. Symbols ○, 1° C, aerobic; ●, 1° C, anaerobic: Δ, 23° C, aerobic: ▲, 23°C, anaerobic: □ 37°C, aerobic; ■, 37°C, anaerobic (15). Used by permission of the American Society for Microbiology.

Once the response variable has been chosen, an experiment can be designed to answer the preceeding questions.

In order to illustrate this point we can adapt one of the experimental situations described in Ref. 15. Let the experimental conditions be restricted to soil moisture of 15 percent, sewage concentration of 50 percent, temperature of 23° C, and incubation of samples under nonsterile conditions. With this simplification, one may wish to compare virus inactivation rates in nine different soil types (i.e.,

Vernon, Clarita, Windhorst, Chigley, Pomello, Anthony, FM, Rubicond, and unclassified). One approach would be to use a one-way analysis of variance with nine "treatments" (soils). In order to obtain an estimate of within-soil error, it would be necessary to have at least two and preferably several observations within each treatment. It is most efficient to have the same number of observations n for each treatment.

An appropriate model would be written as:

$$y_{ij} = \mu + \alpha_i + \epsilon_{ij}$$

where $y_{ij} = \log_{10} N_t/N_0$ for the jth sample of the ith soil; μ = overall mean log survival slope; α_i = differential effect of the ith soil; ϵ_{ij} = random error associated with the jth sample of the ith soil. The ϵ_{ij} are independent observations from a normal distribution with mean zero and variance σ^2.

The null hypothesis is that there is no difference in the effect of the soils, that is, that $\alpha_i = 0$, $i = 1,2, \ldots ,9$. The alternative could be any negation of the null; there are many choices since we could have $\alpha_i \neq 0$ in many different ways.

In order to calculate the required sample size n, it is necessary to specify the level and the power of the test and an alternative hypothesis of interest. The choice of level and power is arbitrary, but should be made in view of the consequences (cost) of the associated errors.

The choice of a representative alternative hypothesis of interest is more difficult, since it involves many possible patterns of negation. The details will be omitted, but it can be shown that the power of the test is related to $\lambda = \Sigma\alpha_i^2$, the noncentrality parameter of a noncentral F-distribution. Winer (14) can be consulted for more details and the statistical tables needed to calculate the required sample size. The calculations necessary for the analysis can be displayed in a standard analysis of variance table:*

Source	DF	MS	F
Soils	8	MSS	MSS/MSE
Error	$9(n-1)$	MSE	
Total	$9n-1$		

The test of the original null hypothesis consists of comparing the F-ratio, MSS/MSE, to the chosen critical point from the F-table with $[8, 9(n-1)]$ degrees of freedom.

If significant differences are found, the location of the differences can be explored by any of several multiple comparison procedures, for example, the Newman–Keuls procedure [Winer (14)]. This will help the investigator see whether in some specific soil type the virus inactivation rate is significantly lower or significantly higher than in other soil types.

*The abbreviations are: DF = degrees of freedom and MSS = mean sum of squares. Details and formulas for calculation can be found in many introductory statistic texts, including Ref. (14). In this illustration it is assumed that there are $n \geqslant 2$ observations obtained for each soil type.

Turning to the second question, we see that two factors, soil type and virus type, must be considered simultaneously. The response variable is taken to be the estimated slope found by regressing $\log_{10}(N_t/N_0)$ on t as before. A 50 percent sewage solution, 15 percent moisture, 23°C temperature, and nonsterile conditions were used for all soil samples. A suitable design can be formulated by "crossing" the seven virus types with the nine soil types to form a 7 × 9 factorial arrangement in which each virus and each soil type occur once in each of the 63 cells. In this context, it is usual to say that there are two factors or main effects (virus and soil). The first of these (virus) has seven levels and the second (soil) has nine. This would allow for investigation of differences due to the main effects, viruses and soil types, but would be insensitive to any interaction of the two factors. To allow for this possibility, the experiment can be "replicated" by taking $n \geqslant 2$ samples per cell.

The model appropriate for this experimental situation would be:

$$y_{ijk} = \mu + \alpha + \beta_j + \alpha\beta_{ij} + \epsilon_{ijk}$$

where y_{ijk} = estimated survival slope for kth sample of the ith virus type in the jth soil type; μ = overall mean survival slope; α_j = effect of ith virus type; β_j = effect of jth soil type; $\alpha\beta_{ij}$ = interactive effect of ith virus and jth soil. The ϵ_{ijk} are independent, identically distributed random variables, each following a normal distribution with mean zero and variance σ^2.

It is helpful to specify the model as given in the above equation, because it makes clear the underlying assumptions upon which the ensuing analysis is predicated. Although the procedures to be suggested are fairly "robust" with respect to most of these assumptions (i.e., minor violations of the assumptions can be tolerated), gross violations can invalidate the calculation of probabilities upon which the final decisions about presence or absence of given effects must rest.

The equation above expresses the response y_{ijk} as the sum of an overall mean μ, main effects α_i and β_j, an interaction term $\alpha\beta_{ij}$, and a random error term ϵ_{ijk}; hence, this is sometimes called an additive effects model. The two-way interaction term β_{ij} models the differential effect of virus type upon survival time when observed in soil j, or, equivalently, the differential effect of soil j upon the survival time of virus i. The hypothesis that this interactive effect is zero can be explicitly tested.

The analysis of variance table would be of the form:

Source	DF	MS	*F*
Virus Type	6	MSV	MSV/MSE
Soil type	8	MSS	MSS/MSE
Virus × Soil	48	MSVS	MSVS/MSE
Error	$63(n-1)$	MSE	
Total	$63n-1$		

Since each virus and each soil type was selected for its own intrinsic interest, these two factors can be considered "fixed effects." We can formally test the three hypotheses:

$H_{01}: \alpha\beta_{ij} = 0$; there is no effect due to interaction of virus and soil.

$H_{02}: \alpha_i = 0$; there is no effect due to virus type.

$H_{03}: \beta_j = 0$; there is no effect due to soil type.

It is usual practice to test the interaction term first; if no effect is found, then one could proceed to test the virus effect and soil effect.

In order to decide on the number of replicates n to be used, we could consider the effect of principal interest or alternatively, the one with the fewest levels, and carry out a sample size calculation as if this were the only test to be done in the manner described for the one-way analysis in the preceding example. Since the degrees of freedom for the error term increase rapidly with n [DF $= 63(n-1)$], it is likely that few replicates will be required to carry out tests with chosen magnitudes of level and power.

Once the test or tests have been performed, the Duncan or Newman–Keuls multiple range tests can be used to determine the likely location of any differences. If it is desired to compare particular individual levels or groups of levels, the method of contrasts may be used (14).

The results of the calculations can be appropriately displayed as tables of means with marginal means (the same tables generated for preliminary analysis could be used) or as plots of means or both. It is helpful to accompany such data summaries with an indication of the variability in the means. Since means are usually compared by calculation of the differences, it is usually best to report the standard deviation of the difference of treatment means.

A third question of interest might concern the effect of environmental conditions upon virus survival in soil. Such a question was posed, for example, in the above-mentioned study by Hurst and co-workers (15). In this study, 48 samples of FM soil were innoculated with known amounts of poliovirus and an echovirus, and then incubated under varying environmental conditions. If it is presumed that survival of the poliovirus and echovirus are independent, the survival of each could be measured and interpreted independently of the other.

Factor	Levels
Temperature	1, 23, 37° C
Initial concentration of sewage	0, 25, 50, 100 percent
Atmosphere	Aerobic, anaerobic
Other soil organisms	Absent (sterile) Present (nonsterile)

The complete factorial design formed by "crossing" these factors would require $3 \times 4 \times 2 \times 2$ or 48 samples of soil for each replication. An appropriate model would be of the form: log (survival slope) = mean + main effects + two-way interactions + three-way interactions + four-way interactions + error.

If all of the indicated terms are used, there are so many parameters in this model (48 of them) that it would fit the data obtained from one replication exactly, so that there is no error, no estimate of error, and no summarization of the data. However, there appears to be no a priori (biological) reasons to think that any of the three- or four-way interactions are present, and it may be that some of the two-way interactions can be eliminated as well. If, for example, biological considerations suggest that it is unlikely that the interactions of temperature and initial concentration of sewage and sterility of experiment exist, these terms could be eliminated from the analysis of variance model. The variations due to the unmodeled terms will be designated as random error.

Now we have the following model:

$$y_{ijkl} = \mu + \alpha_i + \beta_j + \gamma_k + \delta_l + \alpha\gamma_{ik} + \alpha\delta_{il} + \gamma\alpha_{kl} + \epsilon_{ijkl}$$

where y_{ijkl} = estimated survival slope for the ith temperature, jth initial concentration, kth atmosphere, and lth sterility condition; μ = overall mean; α_i = effect of the ith temperature; β_j = effect of jth initial concentration; γ_k = effect of kth type of atmosphere; δ_l = effect of other soil organisms (sterility); $\alpha\gamma_{ik}$ = interactive effect of temperature and atmosphere; $\alpha\delta_{il}$ = interactive effect of temperature and other organisms; $\gamma\delta_{kl}$ = interactive effect of atmosphere and other organisms; ϵ_{ijkl} = random error. It is assumed that the ϵ's are normally and independently distributed with mean zero and common variance σ^2. It is also assumed that each of the effects is "fixed," that is, that the levels of each factor are chosed because of their intrinsic meaning and not as a representative of some larger population of possible levels. (The assumption of "random"effects entails the use of a slightly different analysis, described in Ref. 14.) The associated analysis of variance table would be:

Source	DF	MS	*F*-Ratio
Temperature	2	MST	MST/MSE
Concentration	3	MSC	MSC/MSE
Atmosphere	1	MSA	MSA/MSE
Sterility	1	MSS	MSS/MSE
T × A	2	MSTA	MSTA/MSE
T × S	2	MSTS	MSTS/MSE
A × S	1	MSAS	MSAS/MSE
Error	35	MSE	
Total	47		

We can now test the hypothesis of "no effect" for each of the interactions and main effects listed, beginning with the interactions and working upward. If significant differences are found for any effect, the most likely location of these differences can be determined by the Newman–Keuls procedure (14). The means for main effects and for interactions can be presented in tables or graphs and should be accompanied by the standard deviation of the differences of the means compared.

12.5 BACTERIAL INDICATORS AND OCCURRENCE OF ENTEROVIRUSES: MULTIPLE LINEAR REGRESSION

The following example is intended to demonstrate the application of multiple linear regression analyses in the study of a very important question in aquatic microbiology, that is, the representativeness of currently used bacterial indicators to reflect the occurrence of viruses in aquatic environments. Although this example uses data available from recent studies in marine environments, the question being addressed is also of great significance to groundwater microbiologists and the same methodology would be applicable in groundwater studies.

Several investigations conducted along the Texas Gulf Coast (16–18) showed that enteroviruses were detected in water that was judged acceptable for recreation by bacteriological standards and in water judged acceptable for shellfish harvesting. These and similar findings, as well as reports of the outbreaks of shellfish-associated infectious hepatitus resulting from consumption of oysters supposedly harvested from water that met bacterial standards (19), prompted substantial controversy over the adequacy of current bacteriological standards and the use of bacterial indicators in general to reflect the hazard of viral infection. The discrepancies are thought to be due to differences in survival time between viruses and bacteria and to the low number of viruses necessary to initiate infections as compared to bacterial pathogens.

In order to shed additional light on this question, several major field studies have been conducted in recent years by the Department of Virology and Epidemiology at Baylor College of Medicine on the occurrence of viruses in Galveston Bay of the Texas Gulf Coast. The research question of interest was whether coliform bacterial counts are useful as indicators of virus concentration in water.

In order to investigate this question, the concentration of coliform bacteria and viruses were determined in several water samples taken under varying environmental conditions. In this instance, the desired sample information was taken from studies producing a total of 150 suitable samples from 25 different sites. The data were collected from a wide range of marine environments, including both heavily polluted and nonpolluted waters as judged by bacteriological indicators. The combined data represent one of the largest samples of this kind from marine water ever studied. The details of the investigations can be found in Gerba et al. (16).

As a first step in analyzing the results of these investigations, all of the data collected were arranged in a systematic tabular form along with marginal sums and averages for easy visual inspection. This allowed the investigators to form an initial impression concerning trends, ranges, and relationships and to check for errors or omissions in the raw data.

Next, histograms of each of the measured variables were constructed and reviewed for obvious outliers or strangely shaped distributions. Scattergrams of virus counts with each of the other variables gave a visual impression of pairwise relationships. This step proved to be very helpful in the search for the best mathematical approximation of the complex process studied. For all the variables (except for rain index, which was a discrete variable), the bivariate relationships with the viral count as an independent variable were approximated by a least-squares fit of a linear model. The review of bivariate scattergrams was supplemented by a matrix of correlation coefficients (Table 12.6) to indicate the strength and direction (sign) of pairwise linear relationships.

The variables of interest were defined as follows:

y = concentration of viruses in water

x_1 = concentration of total coliform bacteria in water

x_2 = concentration of total coliform bacteria in bottom sediments

x_3 = concentration of fecal coliforms in water

x_4 = concentration of fecal coliforms in bottom sediments

x_5 = rainfall in previous 24 hours; 0 = no, 1 = yes

x_6 = salinity

x_7 = pH

x_8 = turbidity

After appropriate adjustment and corrections of data and modifications of procedures suggested by these preliminary steps were made, the formal analysis was begun. In this study the associations between the number of viruses found in water, the number of bacteria in water or in bottom sediments, and several other characteristics of aquatic environment listed above were investigated by methods of multiple linear regression.

It was assumed that an appropriate model is:

$$y = \beta_0 + \beta_1 + \beta_1 x_1 + \beta_2 x_2 + \ldots + \beta_8 x_8 + \epsilon$$

where the β's are coefficients reflecting the relative influence of the x's as predictor variables and ϵ is a random error. It was further assumed that the errors are normally and independently distributed with mean zero and common variance σ^2. Note that the model assumes the existence of a linear dependence of y upon the x's (e.g., quadratic, logarithmic, cyclic, or interactive terms are omitted) and that there are no other, unmodeled variables. Since many biological relationships are nonlinear, other transformations were considered in the course of the study, but the linear model is a reasonable beginning since many of the nonlinear relationships can be approximated fairly well over a limited range by a linear function.

Table 12.6. Product-Moment Correlation Matrix of All Measured Variables[a,b]

Variable	pH	Salinity	Turbidity	Rain	TCW	TCS	FCW	FCS	Virus
pH	1.0000								
Salinity	0.1149	1.0000							
Turbidity	−0.0173	−0.1368	1.0000						
Rain	−0.2400	−0.1399	−0.0835	1.0000					
TCW[b]	−0.5461[c]	−0.4902[c]	−0.0064	0.5066[c]	1.0000				
TCS[b]	−0.2367	−0.3532[c]	−0.1165	0.1793	0.4112[c]	1.0000			
FCW[b]	−0.3476	−0.2584[c]	−0.0572	0.41442[c]	0.6822[c]	−0.0152	1.0000		
FCS[b]	−0.0974	−0.2194[d]	−0.0947	0.0831	0.1441	0.1627	0.1460	1.0000	
Virus[b]	−0.1159	0.0992	−0.0176	0.2830	0.2476[d]	0.0834	0.1508	0.0155	1.0000

[a] Adapted from Gerba et al. (16). Used by permission of the American Public Health Association.
[b] TCW, TCS, FCW, FCS = Number/100 ml of total coliforms in water, total coliforms in sediment, fecal coliforms in water, and fecal coliforms in sediment, respectively.
[c] Significant ($p < 0.01$).
[d] Significant ($p < 0.05$).

In order to estimate the unknown parameters, least-squares regression was used. The relative contribution of each variable was determined by the forward stepwise selection procedure (20). In this procedure, the univariate regression upon each of the x's was calculated and the variable with the largest associated R^2 (proportion of explained variance) was selected to enter the model first, then the residuals were regressed upon the remaining explanatory variables and the process continued until all variables were included.

The variables and their associated R^2's are presented in Table 12.7 in the order of their entry into the model. The "multiple R" in this table is the correlation of the predicted values generated by the estimated model (including variables entered up to that point) with the virus concentration. This reflects the strength of the linear association of y with $\beta_0 + \beta_1 x_1$, $\beta_0 + \beta_1 x_1 + \beta_2 x_2$, etc. We expect this correlation to be positive and increase as additional variables are entered. In this example, the maximum reached is .4, indicating a relatively weak relationship.

The "R-square" entry is the proportion of total variance (of the y's about the sample mean of the y's) explained by the linear model including variables entered thus far. Here we see that the full model explains only about 17 percent of the variation, confirming our earlier impression that the model is not doing very well. The "R-square change" measures the explanatory power of each variable in addition to those previously entered.

Since the estimates of the β's are functions of the observed y's, these estimates themselves are random variables. Under the distributional assumptions we have made, the ratio of estimates to their estimated standard deviation follows a t distribution when the hypothesis $\beta_1 = 0$ is true. Ths ratio was calculated for each coefficient and the probability that a given number would be exceeded in absolute value by a random variable following the t-distribution is displayed under "significance of each variable." Stated more succinctly, this is the probability that the ratio would be as large or larger in absolute value as that actually observed when the true value of corresponding coefficient is, in fact, zero. This probability is commonly called a "p-value" and when it is "small" (e.g., .05 or less), we reject the hypothesis that $\beta_1 = 0$ and say that the associated variable has significant predictive power in this model. In this study, it appears that the coefficients of rainfall, salinity, and TCW (concentration of total coliforms in water) are the only ones which can be judged to differ from zero.

The "overall F" is the ratio of the variance explained by the model (divided by its degrees of freedom) to the total variance about the mean of the y's (divided by its degrees of freedom). Under the hypothesis that the β's are all zero, this statistic follows an F distribution and we can calculate the probability that such a random variable will exceed the calculated value of this ratio. We see in this example that all versions of this model have some explanatory power.

As seen in Table 12.7, Column 6, the variable for which regression coefficients were found to have p-values less than .05 were: (1) rainfall, (2) salinity, and (3) the number of total coliforms in water. The first factor, the presence or absence of

Table 12.7. Multiple Regression of Viruses in Water and Other Quality Indicators[a,b]

Coefficient	Variable	Multiple R	R-Square	R-Square Change	Significance of Each Variable	Overall F	Overall Significance of the Model
β_1	Rainfall	.30240	.09145	.09145	.001	10.76953	.001
β_2	Salinity	.35365	.12507	.03363	.046	7.57634	.001
β_3	TCW[b]	.40492	.16396	.03889	.029	6.86385	.000
β_4	FCW[b]	.41113	.16903	.00507	.428	5.28860	.001
β_5	pH	.41350	.17098	.00196	.623	4.24872	.001
β_6	Turbidity	.41619	.17321	.00223	.601	3.56145	.003
β_7	FCS[b]	.41828	.17496	.00175	.645	3.05971	.006
β_8	TCS[b]	.41848	.17512	.00017	.887	2.65379	.011

[a] *Source*: Gerba et al. (16). Used by permission of the American Public Health Association.
[b] TCW, FCW, FCS, TCS = Number/100 ml of total coliforms in water, fecal coliforms in sediment, and total coliforms in sediment, respectively.

rain, accounted for approximately 9 percent of variance in the number of viruses detected in water. The addition of two other variables—salinity and the concentration of total coliforms in water—added 3 and 4 percent, respectively, to the R^2 value and increased the amount of explained variance to about 16 percent. Thus, an association between viruses and some environmental factors exist, however small, and there is less than a 5 percent probability that it was found by chance.

The addition of the five remaining variables yielded an increase in the R^2 over the three-variable model of only one percent. The product-moment correlation matrix (Table 12.6) shows that there was some interdependence between variables, that is, the concentrations of total and fecal coliforms in water and sediment were significantly correlated with water salinity and, to some extent, with the presence or absence of rain and with pH.

What we have learned so far seems to indicate that the rainfall, salinity, and TCW taken together have some positive assocation with the number of viruses in water, but the association was quite weak. Thus, any one of these factors, even in combination with others is not likely to provide an adequate basis for predicting the concentration of viruses in water.

It should be noted that the validity of the tests just discussed depends upon the reasonableness of the assumptions: *viz*, that the relationship is of the linear form described earlier and that the errors are normally and independently distributed with homogeneous variance. It is appropriate to investigate the reasonableness of these assumptions by inspection of the residuals, that is, the differences between the observed and predicted counts. These differences are, in effect, a sample of the ϵ's. A histogram of the residuals can be compared (visually or analytically) to a normal distribution with mean zero to check the assumption that the ϵ's follow such a distribution. The next step is to plot the residuals against the predicted values and to determine if there is a trend. Presence of any trend indicates a violation of the assumption of homogeneous variance. (More details on residual analysis can be found in Ref. 20).

The preceding multiple linear regression analysis showed that the number of viruses detected in water was significantly correlated with the number of total coliforms in water. However, the amount of variation in the numbers of viruses accounted for by this factor (4 percent) was not large enough to make it a good predictor. Other bacterial indicators studied were not found to be important predictors either.

In this example, the combined results of several studies on the occurrence of enteroviruses in marine water were analyzed. Viruses were detected in 43 percent of the samples from recreational waters that were considered acceptable based on standards for total coliforms and in 44 percent of the water samples that were judged acceptable based on fecal coliform standards. Viruses were detected in waters that met standards of acceptability for shellfish harvesting. The failure to associate the occurrence of viruses in marine water with coliform bacterial indicators and the many instances of finding viruses in water that met current bacteriological standards led the authors to conclude that current bacterial standards do not reflect the occurrence of enteroviruses in marine waters.

12.6 FINDING THE SOURCE AND ROUTE OF WELL CONTAMINATION: TIME-SERIES ANALYSIS

In the following example, spatial graphics and statistical analysis are combined to assess the route of travel of pollutants underground. This example is drawn from a study by Birden and Cech in 1981 (21). Water-quality data used in this example were collected at the State of Connecticut Department of Health; the statistical evaluation of groundwater quality information, which is the main focus of the present discussion, was preformed at the University of Texas School of Public Health in Houston.

A routine inspection of a nursery school facility located on Birch Road in the city of South Windsor, Connecticut, carried out in September of 1977 as part of a license renewal procedure, showed that water used at this facility had a total coliform count of 50 colonies per 10 ml and slightly elevated sodium and nitrate concentrations. The water was from a cased well located on the school property. The well, casing, and plumbing were consequently disinfected with chlorine and flushed on four different occasions, but the total and fecal coliform counts rapidly returned to predisinfection levels each time. One of the samples also showed a detergent concentration of 0.43 mg/liter.

Water samples were then taken from residences on Birch Road, and over the next three months, water quality was examined on the adjacent downslope streets parallel and perpendicular to the street in question (Fig. 12.7). A total of 66 wells were examined, of which 19 were found to be nonpotable supplies as judged by the State Health Department water-quality guidelines.

In an effort to determine a possible source and route by which bacterial contamination was entering wells, fluorescent dye was put into the public sewer on Birch Road through a toilet at a residence on the north side of the street and into the storm drain through a catch basin on the north side of Birch Road (Fig. 12.7). A few days later, the well at residence 1 was sampled again. The tap water had the distinct color of the dye used. Analysis of the water showed a coliform count of 50 colonies/100 ml (TC), a fecal coliform count of 16/100 ml, and fluorescein concentration of 0.170 mg/liter.

Subsequently, much effort was put into locating the source of this problem. The public sewer was examined with cameras drawn through it for the possibility of infiltration and a pressure test of joints in the sewer line was done. No defects in the main lines or laterals were found. Further, the Environmental Health Services examined operating septic systems in the area using dye tracer. According to town records, only one house upslope from residence A was not connected to the public sewer (residence 2 in Figure 12.7). Dye was injected through the toilet of this house. Within approximately one hour, dye was observed to reappear in the catch basin. A sample of well water from house A also contained dye.

A direct connection between the septic system at residence 2 and the storm drain was found. This constituted a violation of the State Health Code (22). The town of South Windsor Health Department ordered this residence to be connected to the sanitary sewer.

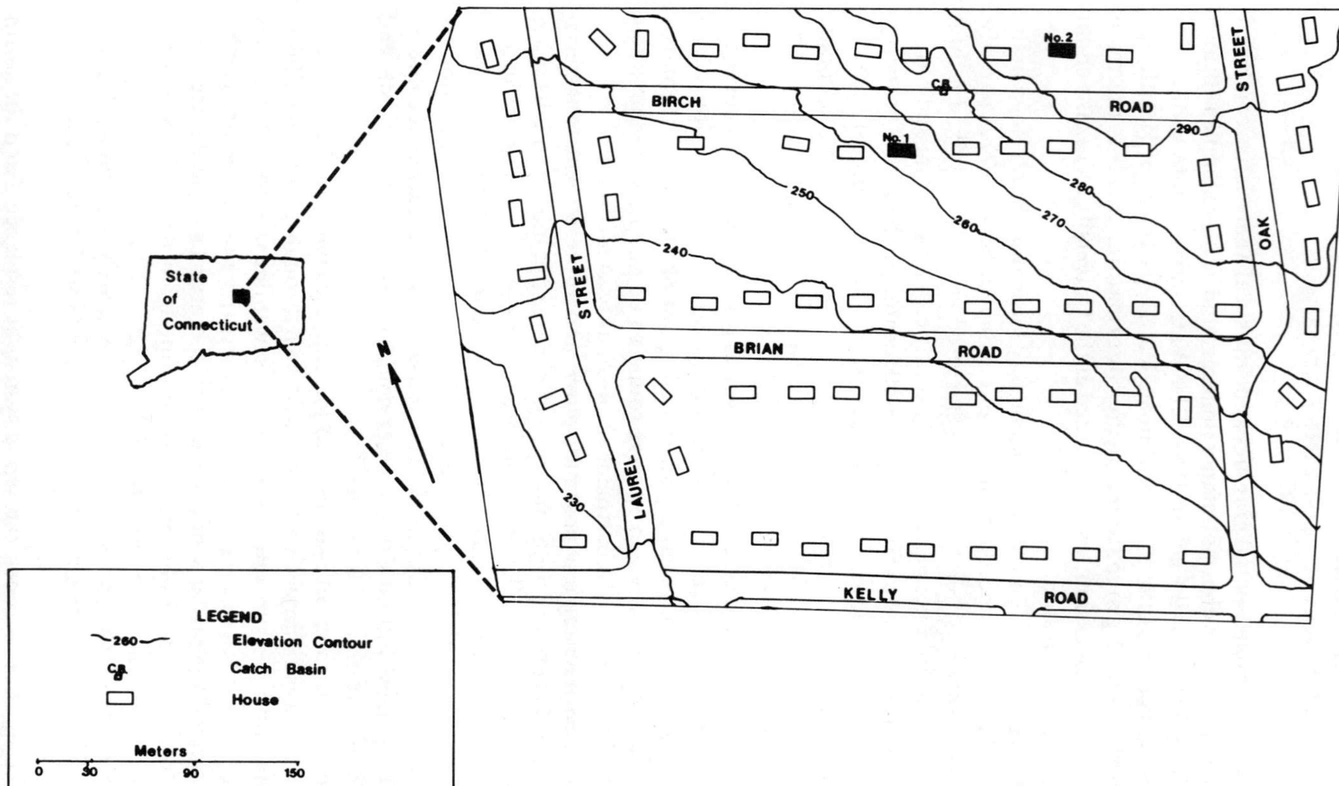

Figure 12.7. Schematic map of the study area (South Windsor, Connecticut). The household where a drinking-water well contamination was first discovered is residence 1. The household where a septic system was found to be directly connected to the storm drain system is residence 2 (21). Used by permission of Pergamon Press.

When this was done, the septic odor in the catch basin across from residence 1 dissipated. An extensive visual search, along with dye tests in the rest of the neighborhood, produced no evidence of other such connections to the storm drain system, although two septic systems were identified where leaching fields had failed and which were discharging to the surface.

It was speculated that bacteria found in the wells at house 1 were channeled by the storm drainage system and were flowing through fissures in the ledge rock (possibly caused by blasting when the public sewer was installed), thus entering the underlying water-bearing strata. Additional factors in support of this hypothesis were:

1. With the exception of two systems, all subsurface sewage disposal systems still in use appeared to be functioning normally.
2. Most residences in the area had been connected to the public sewer which appeared, judged by earlier tests, to be tightly constructed with good seals at the joints.
3. Wells in the study area that were contaminated had top seals in order and contamination from the surface by this route was ruled out.

If bacterial contamination from the storm drain system was entering the wells through fissured bedrock, then protection of the wells from infiltration of contaminants by the same route would most likely be an impossible task. Thus, it was recommended to extend mains to provide public water service for homes on Birch Road and Brian Road. This recommendation was met with controversy, and for the next several months, much debate followed with town meetings with the mayor, city council, and town planner of South Windsor. The residents of the area formed a coalition to study the problem.

A decision to extend water mains to Birch Road, Brian Road, and Laurel Road was subsequently adopted by the town council. The fact was, however, that at this point, information substantiating the explanation offered of the route by which bacteria entered the wells in this area was not sufficient and detailed analysis of spatial relationships of well water quality was needed to reaffirm or refute the fractured rock supposition.

At this point a set of 72 observations of total coliforms and fecal coliforms at residences was available, as shown in Fig. 12.7. A supplemental set of samples for nitrate concentrations was also available. To investigate the spatial patterns in the distribution of coliforms and nitrates and the interrelationships between points in space represented by individual households, statistical techniques originally developed for analysis of time series were adapted to the spatial data. The statistical approach selected in this case combines the elements of lagged correlation, spectral analysis, and synagraphic computer mapping. The general methodology of autocorrelation and spectral analysis was adapted from the work of Blackman and Tukey (23), Jenkins and Watts (24), and Box and Jenkins (25). Data processing was done using the computer package TIME written at the University of Texas School of Public Health.

A time series is a sequence of observations made on successive occasions, as, for example, temperature at a certain location measured at noon each day for a year or the closing price of a certain stock on each day of a year. A spatial series of observations consists of a sequence of measurements ordered by location rather than time, as in the present sample where houses are ordered linearly along the streets. Extensive and sophisticated statistical techniques have been developed for analysis of time series, but these methods are equally applicable to spatial series.

Two basic ideas from time-series analysis were used. The first of these was that certain features of a single series can be seen by calculating the autocorrelations and the second was that the relationship of two or more series can be investigated through the cross-correlations of the series.

The first-order autocorrelation is the ordinary correlation of each observation with its immediate predecessor. This is also called the autocorrelation of lag one. The second-order autocorrelation is the correlation of each observation with second predecessor, etc. For a series of n equally spaced observations, it is common to calculate a maximum of $\sqrt{n} + 10$ autocorrelations. These autocorrelations and their associated lags constitute the sample autocorrelation function (ACF). For a series of completely random observations, the plot of the ACF will fall rapidly from 1, which is the autocorrelation of each observation with itself at lag zero, to 0 as the lag increases, showing that there is no information in remote observations that could be used to predict the current observation. Other types of patterns in the plot of the ACF, which is also called the correlogram, are indicative of an association of remote observations with the current observation.

In the present example, such sample ACF's were calculated for the houses along each of the five east-west rows shown in Figure 12.7, *viz.*, Birch Road north, Birch Road south, Brian Road north, Brian Road south, and Kelly Road. Data collected from each of these five rows of houses were treated as five separate series with observations equally spaced. Figure 12.8 shows the correlograms of bacterial counts for each individual series (streets). If the series is produced by a random process, the autocorrelation should drop rapidly to zero with the introduction of progressively larger lags between the series and its delayed image.

As can be seen, some of the figures, such as that for Birch Road south, demonstrate a slight degree of serial association, suggesting there might be some interdependence between values (bacterial levels) in each of these series. In the context of the study, this would mean that bacteria may have some lateral spread on each street. However, in order to test the statistical significance of the lagged autocorrelation estimates and thus, to resolve the question of lateral spread, one must take this analysis a step further and examine the spectra of these series.

The spectrum of series of observations (again, time series in the classical application, spatial series in this case) is a Fourier transform of the autocorrelation function (23). It shows the variance at each of a number of frequency bands as a proportion or contribution to the total variance in the overall data representing the studied process. Spectra of each series of bacterial data are presented in Fig. 12.9. The unit of frequency on these graphs is the inverse of distance between houses. All spectral estimates lie within two standard deviations of the theoreti-

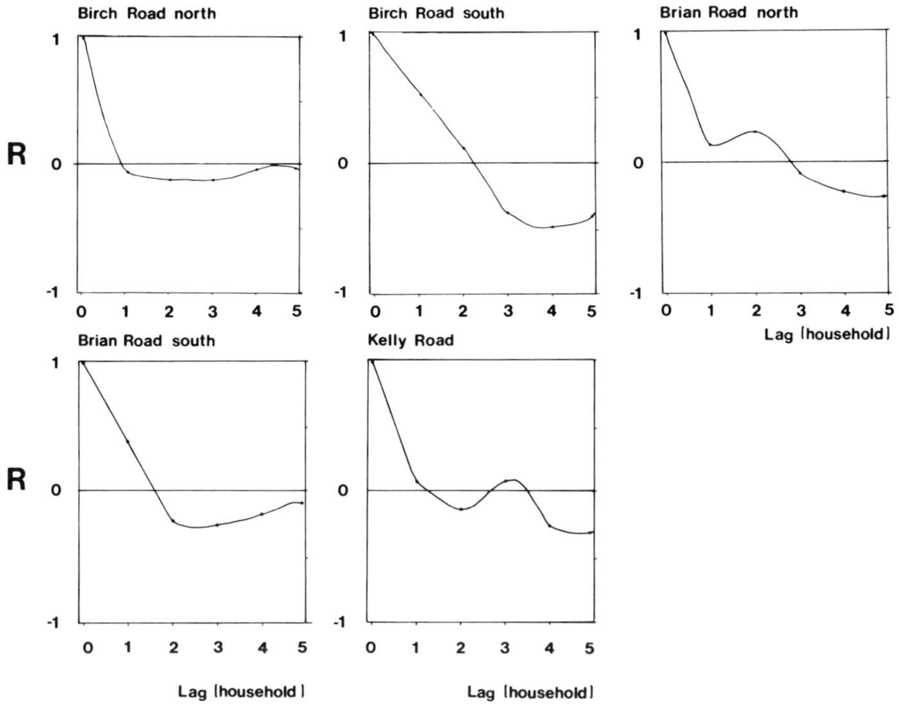

Figure 12.8. Curves of autocorrelation (for lags from 0 to 5 households) for series of total coliform bacteria in wells along individual streets in a section of South Windsor, Connecticut (21). Used by permission of Pergamon Press.

cal reference spectra for a purely random series, which means that there is no strong and statistically significant persistence or trend toward spread of bacterial contamination in the east–west or west–east direction on any of these streets. Nitrate spectra showed similar random tendencies.

The idea of autocorrelation can be modified to deal with two series observed simultaneously as coliform count and nitrate concentration were in this example. The ordinary correlation between paired observations (i.e., coliform and nitrate for the same house) is called the cross-correlation of lag zero. The correlation of an observation with its neighbor to the east is the cross-correlation of lag one, the correlation with the next neighbor is the cross-correlation of lag two, etc.

The cross-correlations and lags form the sample cross-correlation function (CCF). When plotted, this function can reveal some features of the spatial relation of the two series. For two completely unrelated series, the CCF would be approximately zero for all lags, with only random fluctuations from zero. In order to avoid spurious cross-correlations, any trends in the two series should be removed before calculations of the sample CCF. This is called filtering and is described in Chatfield (26).

Figure 12.10 shows the cross-correlation of total coliform bacteria for Birch Road south with the other streets. These curves, called cross-correlograms, are

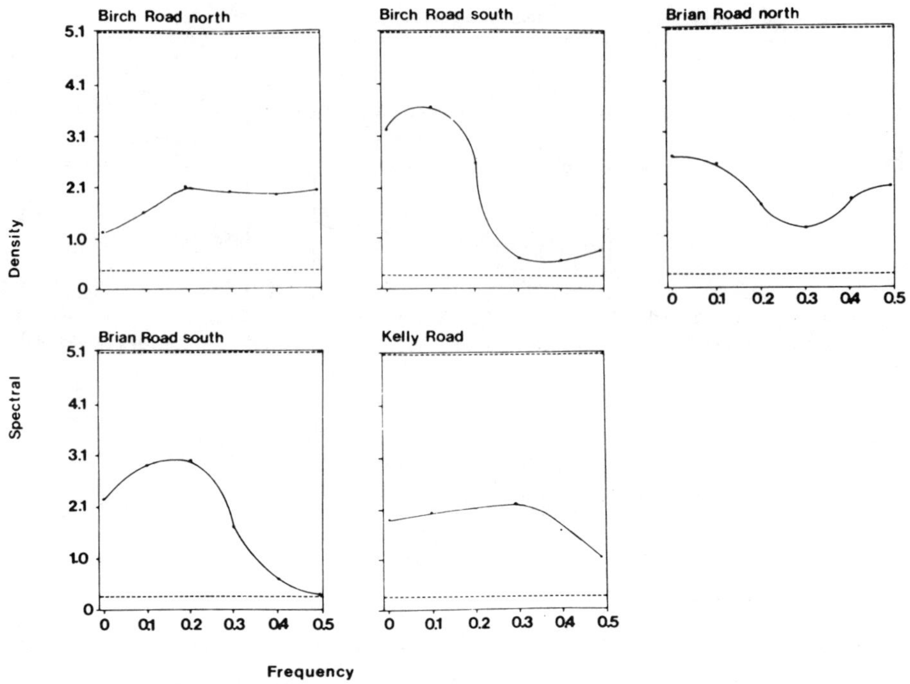

Figure 12.9. Spectral analysis of bacteria data series in wells along the individual streets in a section of South Windsor, Connecticut. Dotted lines delineate the 95 percent confidence interval for theoretical random spectra (21). Used by permission of Pergamon Press.

formed by plotting cross-correlations against lags (both positive and negative) equal to one to five households west and east from the respective households on the south side of Birch Road.

The approximate probability that the deviation of these cross-correlations from zero exceeds a given magnitude can be calculated by use of the fact that they are approximately normally distributed. The horizontal dashed lines in Figure 12.10 indicate the region in which about 95 percent of the cross-correlations will lie if the two series are unrelated.

The top graph, which is the cross-correlation for the north side of Birch Road, shows a significant correlation, $R = 0.7$, at a positive lag equal to three households, or in a direction northeast of the base street. The north side of Brian Road (the second graph down) shows a significant correlation, $R=0.7$, at a shift by one household to the west (again with respect to the south side of Birch Road). The south side of Brian Road shows a significant correlation, $R=0.85$, at a shift by two households to the west. Finally, for Kelly Road the significant correlation, $R = 0.75$, was observed at no shift (or zero lag).

Importantly, no significant correlation was observed at lags other than the above indicated and for the situation where there was no shift (north–south) only the coefficient of correlation for Kelly Road was significantly different from

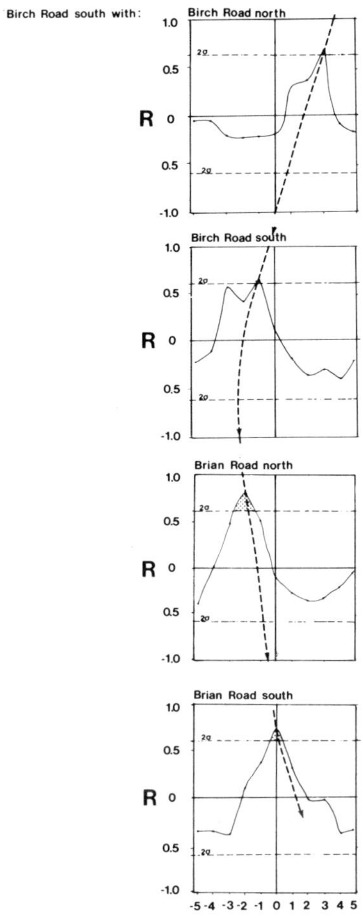

Figure 12.10. Curves of cross-correlation (for lags from 0 to ± 5 households of total coliform bacteria in wells on the south side of Birch Road, with those on other streets. Dotted lines delimit 95 percent confidence interval for theroretical zero correlation. Shaded portions of the curves indicate significant cross-correlation ($p < 0.05$). Arrow combines statistically significant peaks on cross correlogram and indicates probable direction of bacteria flow in the aquifer (21). Used by permission of Pergamon Press.

zero. This shows that bacterial propagation following ground contours occurred in a direction from northwest to southwest and then south (as shown by the dashed line on Fig. 12.10.)

Figure 12.11 shows a cross-correlation of nitrates between households on different streets. No systematic pattern was evident; all cross correlation estimates were well within 95 percent limits of that for theoretical zero correlation. The distribution of nitrates, therefore, appears to be random.

Figure 12.12 shows cross-correlation of nitrates and coliform bacteria. There was a significant correlation, $R = 0.7$, at lag zero between bacteria and nitrates in wells for the south side of Birch Road, but no significant correlation was observed for these on any of the other streets. This indicates that both nitrates and bacteria find access into drinking-water well on Birch Road, most likely around household 1, but from this point on downslope the correlation decreased due to differences in the way bacteria and nitrates move underground.

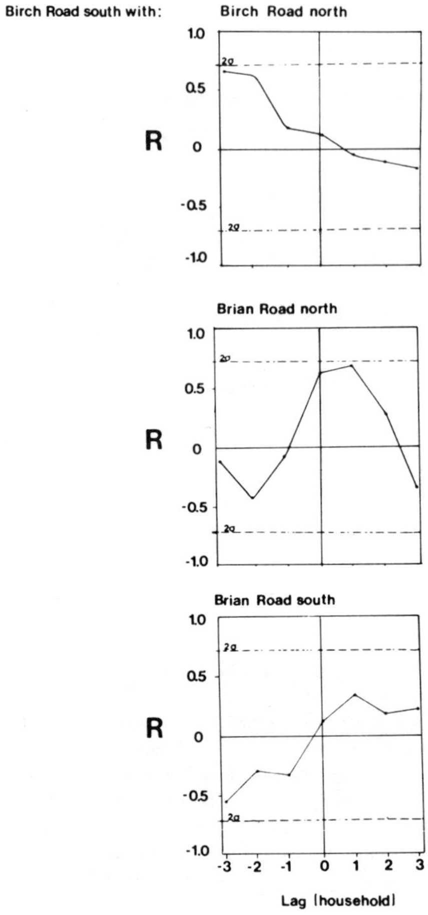

Figure 12.11. Curves of cross-correlation (for lags from 0 to ± 3 households) of nitrates wells on the south side of Birch Road with those of other streets. Dotted lines delimit 95 percent confidence interval for theoretical zero correlation.

The frequency diagram included in Fig. 12.14 shows that 28 percent of the total 66 wells sampled had coliform levels higher than 1 colony/100 ml required for potable water quality by State of Connecticut guidelines. Synagraphic representation of spatial distribution of bacteria and nitrate was done using SYMAP and SYMVU computer packages (13). Figure 12.13 gives a two-dimensional representation of the pattern of bacterial distribution in a given section of the city of South Windsor. Figure 12.14 provides a perspective graphic representation of bacterial contamination emphasizing the localized, channeled flow of bacteria in the general direction from northeast to southwest and then to the south. As to nitrates, synagraphic display did not reveal any special patterns and basically confirmed our previous conclusion that nitrates were essentially randomly distributed within this area.

The results suggest, therefore, that nitrate concentrations throughout most of the study area appear to be random with no persistent relation to any particular

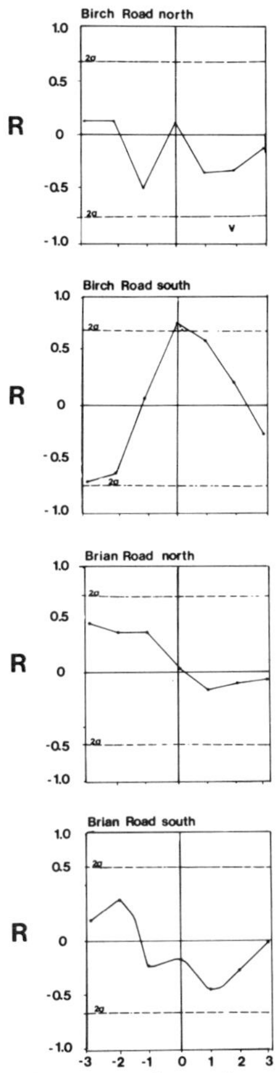

Figure 12.12 Curves of cross-correlation (for lags from 0 to ± 3 households) of total coliform bacteria with nitrates. Dotted lines delimit 95 percent confidence interval for theoretical zero correlation. Shaded portion of the graph on figure for Birch Road South corresponds to lag for which the correlation is significantly different from zero ($p < 0.05$) (21). Used by permission of Pergamon Press.

spatial point and no correlation with coliform bacteria except in the immediate vicinity of residence 1, where direct contamination of drinking water by sewage has been demonstrated. The concentration observed throughout the rest of the study area apparently reflects what is naturally occurring in this aquifer.

A definite pattern exists for coliform bacteria. Bacterial levels show strong spatial relations between certain wells. There appears to be a channeled flow of bacterial contamination through the area, with very little lateral dispersion away from a northeast to south and then to south line (Figs. 12.10, 12.13, and 12.14). It is unlikely that this pattern would be observed if the contamination entered the wells in a diffused fashion. The evidence is more in support of the conclusion that contamination entered the aquifer into which the wells were drilled in a single place and then propagated through some form of a permeable lens, or perhaps, a fracture system. Such a fracture could be either a natural property of the

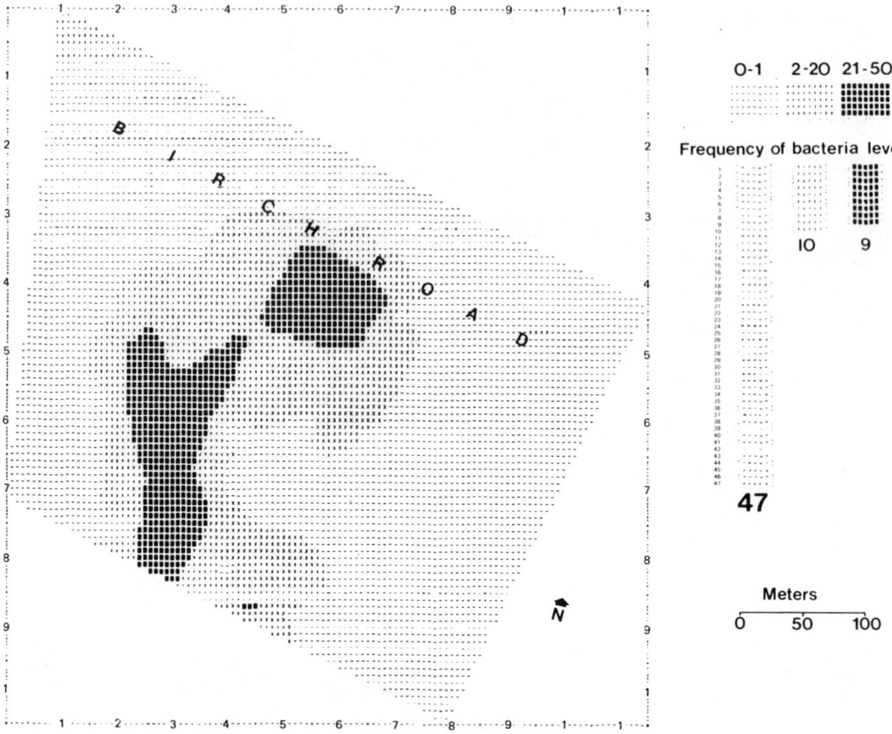

Figure 12.13. Two-dimensional computer display of total coliform levels in wells of the section of South Windsor, Connecticut (21). Used by permission of Pergamon Press.

underlying geological structure or a result of bedrock blasting which occurred during the extension of sewer lines.

All correlation coefficients of bacteria levels in wells were quite high throughout this entire path. Noteworthy is that the coefficient of correlation between Birch Road south and Kelly Road, which is the southernmost street in the study area, was just as high ($R = 0.75$) as the coefficient of correlation for the streets closer to Birch Road. This means that bacteria had traveled nearly 300 m underground with very little attenuation of bacterial concentration by the aquifer material. At this rate it is quite probable that contaminants could have traveled even farther beyond the study area.

In the studies of environmental problems, the serial correlation methodology has been applied primarily in the analysis of data representing time history. The variation brought by this study was the application of lagged correlations to a set of data collected at different locations, but at the same point in time. This was done in order to evaluate spatial patterns in the distribution of pollutants in search of a possible origin of the observed well contamination. Aside from this the mathematical treatment was the same as in time-series analysis.

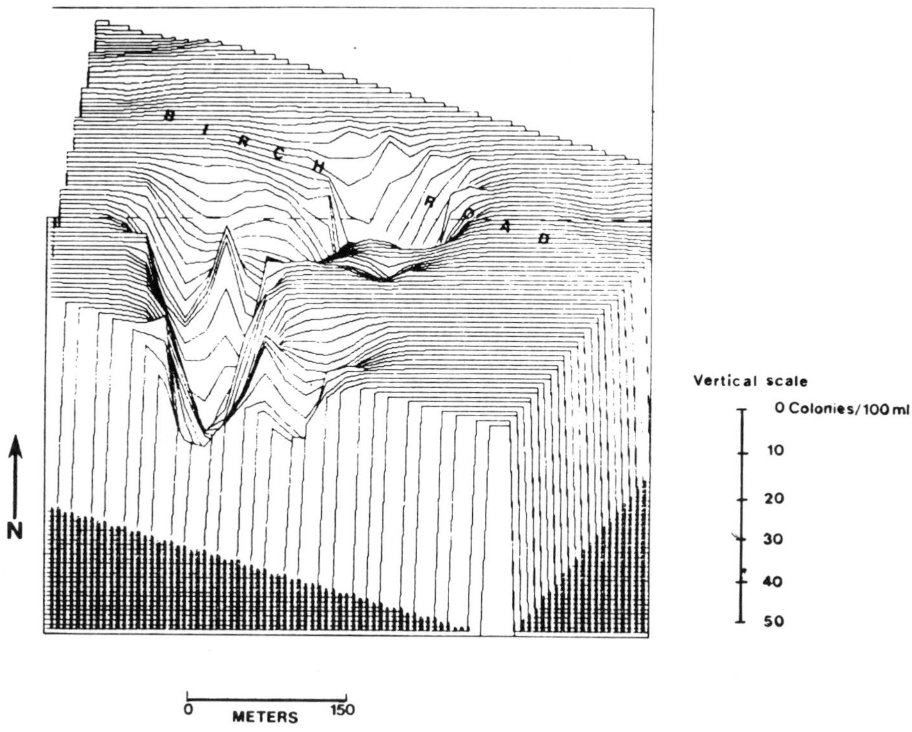

Figure 12.14. Perspective display of total coliform bacteria concentrations in wells of the section of South Windsor, Connecticut (21). Used by permission of Pergamon Press.

REFERENCES

1. R. D. Remington and M. A. Schork, *Statistics with Applications to the Biological and Health Sciences*, Prentice–Hall, Englewood Cliffs, N.J., 1970.
2. G. W. Snedecor and W. G. Cochran, *Statistical Methods*, The Iowa State University Press, Ames, Iowa, 1978.
3. C. P. Gerba, S. M. Goyal, I. Cech, and G. F. Bogdan, *Environ. Sci. Technol.* **15**, 940, (1981).
4. S. M. Goyal and C. P. Gerba, *Appl. Environ. Microbiol.* **38**, 241 (1979).
5. D. Brooks and I. Cech, *Water Res.* **13**, 33 (1979).
6. I. Cech, E. Davis, F. A. Gonzales, and D. Brooks, *Water Res. Bull.* **15**, 1691 (1979).
7. H. Comly, *J. Am. Med. Assoc.* **129**, 112 (1945).
8. P. Magee and J. Barnes, "Carcinogenic Nitroso Compounds," in A. Haddow and S. Weinhouse, Eds., *Advances in Cancer Research*, Vol. 10, Academic Press, New York, 1967, p. 163.
9. P. Bogovski, "The Importance of the Analysis of *N*-nitroso Compounds in International Research," in P. Bogovski, R. Preussman, and E. Walker, Eds. *N-Nitroso*

Compounds Analysis and Formation, International Agency for Research and Cancer, Scientific Publication, Lyon, France, 1972, p. 1.

10. W. George and W. Hastings, *Trans. Amer. Geophys. Union* **32,** 450 (1951).
11. Hach Chemical Company, *Hach Methods Manual,* Ames, Iowa, 1969.
12. Millipore Corporation, *Microbiological Analysis of Water*, Bedford, Mass., 1969.
13. Laboratory for Computer Graphics and Spatial Analysis, SYMAP, Harvard University, Cambridge, Mass., 1971.
14. B. J. Winer, *Statistical Principles in Experimental Design, 2nd ed., McGraw–Hill,* New York, 1971.
15. C. J. Hurst, C. P. Gerba, and I. Cech, *Appl. Environ. Microbiol.* **40,** 1067 (1980).
16. C. P. Gerba, S. M. Goyal, R. L. LaBelle, I. Cech, and G. F. Bogdan, *Am. J. Pub. Health* **69,** 1116 (1979).
17. C. P. Gerba, S. M. Goyal, I. Cech, and G. F. Bogdan, *J. Food Prot.* **43,** 99 (1980).
18. R. L. LaBelle, C. P. Gerba, S. M. Goyal, J. L. Melnick, I. Cech, and G. F. Bogdan, *Appl. Environ. Microbiol.* **39,** 588 (1980).
19. B. L. Portnoy, P. A. Mackowiak, C. I. Caraway, J. A. Walker, I. W. McKinley, and C. A. Klein, Jr., *J. Am. Med. Assoc.* **233,** 1065 (1975).
20. N. R. Draper and H. Smith, *Applied Regression Analysis*, John Wiley, New York, 1966.
21. H. H. Birden and I. Cech, *Water Res.* **15,** 291 (1981).
22. Connecticut State Department of Health. *Public Health Code of the State of Connecticut*, Hartford, Connecticut, 1971.
23. R. B. Blackman and J. W. Tukey, *The Measurement of Power Spectra*, Dover, New York, 1958.
24. G. M. Jenkins and D. G. Watts, *Spectral Analysis and Its Applications*, Holden–Day, San Francisco, 1969.
25. G. E. R. Box and G. M. Jenkins, *Time Series Analysis, Forecasting and Control*, Holden–Day, San Francisco, 1970.
26. C. Chatfield, *The Analysis of Time Series: Theory and Practice*, Chapman and Hall, London, 1975.

13

BIOCHEMICAL INDICATORS OF GROUNDWATER POLLUTION

Franklin R. Leach

Department of Biochemistry
Oklahoma State University, Stillwater, Oklahoma

13.1. INTRODUCTION

13.1.1. Scope

Several literature database files were searched for articles relevant to this chapter. Table 13.1 shows the results obtained using *Pollution Abstracts*, *Biosis*, and *Scisearch* for the indicated years. There were only 58 publications (duplicates not eliminated) found in the three files that had a groundwater, pollution, biochemical, and measurement term present in the title or the key word files. When those references were examined, very few proved to be relevent to the subject of this chapter. This means that very few articles have been published on the subject, and that my task is primarily to set the stage for future

Table 13.1. References to Biochemical Indicators of Groundwater Pollution

	Number of References		
	Pollution Abstracts 1970–1982	*Biosis* 1969–1982	*Scisearch* 1974–1982
1. Water	1408	2268	1255
Groundwater	1191	453	852
Ground water	323	1600	604
Water table	91	528	135
Subsurface water	48	77	36
2. Pollution	26,662	24,783	6135
Pollution	19,600	20,636	4556
Pollutant	11,199	5,159	1612
3. Type	6518	179,884	89,891
Biochemical	604	31,005	7878
Biological	3510	46,588	15,862
Bacterial	2230	93,876	30,629
Microbial	1116	20,144	35,247
4. Measurement	28,191	487,993	
Determination	8417	182,837	
Estimation	927	18,993	
Measurement	14,370	144,266	
Analysis	11,239	139,528	
Indicator	4461	114,211	
5. 1 × 2	598	409	60
6. 5 × 3	59	59	0
7. 6 × 4	36	22	

[a] The number do not add up, becuase of multiple use of terms in the title and the key word file for a single reference.

developments and to review accomplishments in potentially applicable approaches.

13.1.1.1. Groundwater and the Subsurface Environment

Because pollutants may reach groundwater either by percolation through the unsaturated soil areas above a water table or by direct discharge into the saturated zone without passage through the soil (1), biological activities in surface soil and the subsurface regions are pertinent. Any biological activity present may destroy, modify, or enhance the toxicity of a pollutant or convert a nontoxic compound to a toxic pollutant. The pollutants may modify the biological community present or change biological activity (either cellular or free). All these changes in biological activity, community composition, and individual organisms can be followed or determined through the use of biochemical procedures.

Much attention has been paid to the topmost layer of material, the soil, because of its agricultural importance. Much less attention has been given to the subsurface areas through which the water and pollutants pass by percolation to reach the groundwater. This passage requires time and thus gives the opportunity for any biological activity present to modify the pollutants. McNabb and Dunlap (1) reviewed the literature pertinent to the subsurface as a biological habitat. The environmental conditions range from those that we use in the laboratory to the extremes provided by the most hostile natural forces.

After reviewing the bacteriological quality of groundwater, Allen and Geldreich (2) concluded that improved bacterial detection methods were required. The apparent absence of coliform organisms because of the insensitivity of the assay does not preclude the presence of pathogens. The large bacterial populations present often suppress coliform detection. We are now aware of the fallacy of the following scenario: groundwater is a reliable and safe source of water since it is protected by a thick upper soil and subsurface mantle that removes potential pollutants by both filtration and biological activity as surface water travels to the groundwater (3).

Estuaries represent sites of transition from fresh to salt water which have been studied from the microbial and environmental viewpoints (4). Such studies may serve as models for the research on microorganisms in the groundwater environment. The range of environmental conditions occurring in estuaries are significant, but are not as extreme as the conditions under which microbial growth occurs. ZoBell (4) stated

> . . . my keynote message is that many environmental and biological factors are involved in estuarine microbial ecology. Cognizance must be taken of all of these in order to obtain the complete picture. Most urgently needed are quantitative data on the impact of microbial activities on the chemical components of estuaries, (and the) physiochemical properties of estuarine environments.

Substituting *groundwater* and *subsurface environment* for *estuarine* in the preceding quotation highlights the enormity of the problems of groundwater pollution microbiology.

13.1.2. Indicators and Criteria

13.1.2.1. Early Warning

The concept that it is easier and less expensive to prevent groundwater pollution than to clean it up afterward justifies the development of early warning systems. The ecosystem does not have an unlimited capacity to clean up water. Because of bioaccumulation or the nature of their toxicity, some pollutants produce adverse biological effects at concentrations that are not detected by the currently applied analytical techniques. Greater sensitivity in analysis must be developed.

Bioassays are required that make possible the determination of synergistic effects. Mixtures of materials obtained from wastes or dumps, for instance, are relevant samples. Because of the interaction of the various environmental conditions, biological populations, and mixtures of pollutants, knowledge of only the concentration of pollutants is insufficient to permit evaluation of the complete biological impact. Cairns and van der Schalie noted (5) that the search for a single method that will provide all the information necessary about an ecological system is useless. No one method ever will! A battery of tests will be required for detection and evaluation of potential hazards. There is no catholic attribute that is sufficient to define a biological system. Often an aquatic organism is used as a biological monitor; one fruitful area of research is in reducing the time required for bioassay. For a biological monitoring system to be applicable to wide use by municipal districts it should be automated. Cairns and van der Schalie (5) discussed such systems and their automation.

13.1.2.2. Biochemical Indicators

Escherichia coli and fecal coliforms are used as indicator organisms for pollution by pathogens (6). Since it is not possible to monitor all possible pathogens, indicator organisms are used to indicate the possible presence of pathogens. Dutka et al. (7) found that the current fecal coliform and *E. coli* enumeration procedures underestimate by up to 90 percent the true population of organisms present in a feces-inoculated water sample during a 56-day incubation. But the level of fecal coliforms found would have *indicated* fecal contamination of the water.

In a similar way, biochemicals, that is, chemical substances derived from or produced by living organisms, can be used as indicators of biological activity. The development of an arsenal of analytical techniques has led to the explosive development of biochemistry and molecular biology during the last four decades. Basic to the advancement of any science are techniques for quantitative measurements.

13.2. BACKGROUND INFORMATION

13.2.1. Review

Dermer et al. (8) comprehensively reviewed the literature through 1978 on *Biochemical Indicators of Subsurface Pollution.* There were 1185 references covered. They discussed the limits of analysis and the necessity of some form of amplication to make most techniques sensitive enough to be applicable to subsurface samples. Special emphasis was placed on the determination of enzymes because they act catalytically with an inherent amplification. Assays for the major macromolecules of nucleic acids, proteins, and the bacterial cell walls are covered. Other topics receiving special emphasis were ATP and GTP and the various coenzymes, vitamins, and electron carriers and acceptors.

Table 13.2. Sensitivity for Determination of ATP by Various Methods[a]

Technique	ng ATP
Spectrophotometric, 1 ml cell	100
Fluorimetric, direct enzymatic assay	50
Radioisotopic (Specific activity of commercial ATP and detect 100 cpm)	
$[^3H]$ label	0.01
$[^{14}C]$ label	0.5
$[^{32}P]$ label	0.012
Phosphate determination	2000
Coupled enzymatic assay, spectrophotometric	100
Enzymatic cycling reactions	
Spectrophotometric detection	0.5
Fluorimetric detection	0.05
Firefly luciferase	0.00005
HPLC, fluorimetric for 1-N^6-ethenoadenosine triphosphate	1.0

[a] Data from this laboratory except HPLC. Radioisotopic sensitivity was calculated from the specific activity of typical commercial preparations using the efficiency of our scintillation counter.

13.2.2. Sensitivities of Determination

Since the sensitivity inherent in the assay will determine the ultimate sensitivities obtainable, an idea of the range of sensitivities of various assays methods would be beneficial. Table 13.2 illustrates the sensitivities of various methods of ATP measurement. The choice of method influences whether or not a small amount of ATP present in the sample will be detected. Of course, the more sensitive assays generally require more sophisticated techniques and exotic equipment.

13.2.3. Magnitude of the Problem

Microorganisms occur in groundwater and the subsurface environment, but less is known about the microbiology of groundwater than any other part of the biosphere (3). The microorganisms present are extremely diverse and thus, the question ensues—which population should be enumerated (9)? The bacteria in the environment are not uniformly distributed and sensitive tests must be developed for sparse populations of interest that might occur in the samples.

Organisms occur in microhabitats. Those in the subsurface areas are subjected to the same variations in substrate concentration, mineral nutrients, the level of growth factors, moisture amount, atmosphere, pH and Eh, ionic strength, pressure, and temperature as surface samples. It was expected that the number of microorganisms present in groundwater and the subsurface might be lower by many orders of magnitude than in surface samples. That is not necessarily the situation.

13.2.4. Approach

Because of variations in microhabitats, ecological diversities, and interfering materials we must consider how to sample and extract the component to be analyzed. The interaction of the members of an ecological population with the pollutants might perturb some biochemical parameter. Because of the enormous surfaces and low-nutrient concentrations that exist in the groundwater and subsurface environments, these factors must be examined. The relevant analytical methods involve bioassays, measurement of growth, staining and microscopy, metabolic and physiological responses, measurement of enzyme activity, and determining cellular components. Any factor that influences the turnover of a component is important. Sensitive methods based on bioluminescence and chemiluminescence will be stressed. Other biochemical measures include fecal sterol levels and various miscellaneous methods.

13.3. SAMPLING

13.3.1. Methods

In Chapter 11 McNabb and Mallard cover sampling methods. Atlas and Bartha (10), in their text on *Microbial Ecology*, also discuss the sampling process. The importance of sampling cannot be overstressed—sensitive analytical techniques and expensive reagents must not be wasted on samples that were not taken with the greatest care using suitable techniques.

The Society of Applied Microbiology has devoted one of its technical publications to sampling and microbiological monitoring in the environment (11). Specific directions were given for the sampling and estimation of microbial populations in the aquatic environment (12). Another chapter (13) gave details on soil sampling. Because of the variety of microhabitats and unequal distribution of microorganisms within a soil sample, one must choose carefully the sample size with respect to the required data. Table 13.3 illustrates such a problem with subsurface core samples obtained from a depth of 25 ft. Part of the analysis for ATP was based on replicate 0.5 g samples. There is tremendous variation in the distribution of the microorganisms. When larger samples (100 g) were extracted by using a Waring blender and our extracting mixture, the amounts of ATP obtained from separate samples were in reasonable agreement. The size of the sample taken for analysis and whether or not the bulk sample has been homogenized influence the results.

Table 13.3. Effect of Sample Size on Analysis Results from a Single Core of Subsurface Material

Sample Number	Size (g)	ng ATP/g core
1	0.5	0.76
2	0.5	<0.1
3	0.5	100
4	0.5	11
5	0.5	<0.1
6	100	240
7	100	160

13.3.2. Storage

Once obtained, how long can samples be kept before analysis? As one might expect, the answer to this question depends upon the sample. With surface soils, the stability is good if the samples are refrigerated. Ross et al. (14) followed dehydrogenase activity in several soil samples for 77 days. Air drying of the samples and storage at 20° C were unsatisfactory. The dehydrogenase activity was retained in samples stored in either a refrigerator (4° C) or a deep-freezer (−20° C), up to 49 days with most samples.

Figure 13.1(A) shows the stability of ATP measurements during six weeks of storage of a surface soil sample obtained from Webster's garden. That this storage regime is not universally applicable is shown by Figure 13.1(B). There we

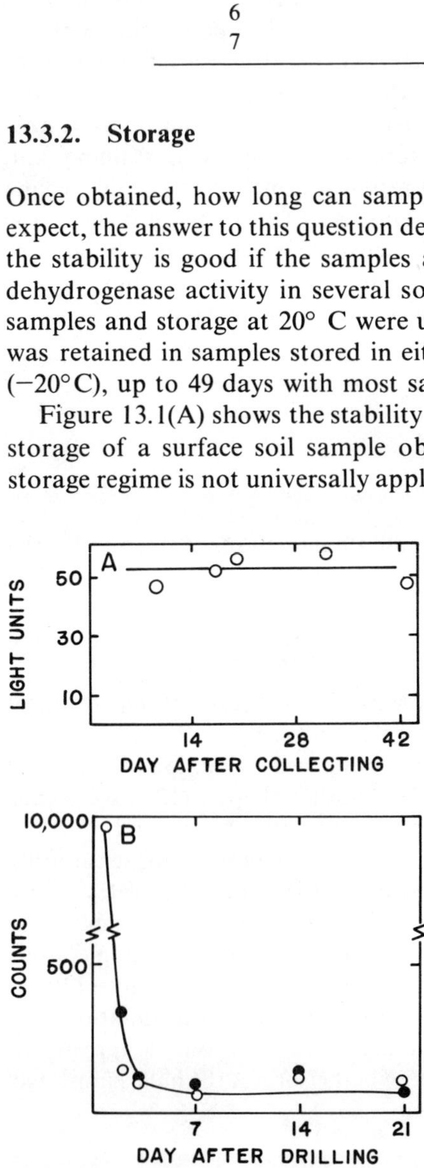

Figure 13.1 Effect of storage on ATP contents of soil and core samples. (A) Webster soil samples were stored at 4° C. At the indicated times duplicate samples were extracted and analyzed for ATP. (B) A core sample was obtained from Dr. John Wilson one day after drilling. The ATP content was determined that day, and the core was divided into several samples which were stored under different conditions. The • sample was stored at −20° C in a deep-freeze room while the ○ sample was stored at 4° C in the cold room. Duplicate samples were extracted at the indicated times and analyzed for ATP. Other experiments have shown that once extracted and neutralized ATP is stable.

see a rapid decrease in the ATP recoverable from core samples. Even storage in a super deep-freezer (−80°C) in the presence of glycerol was not satisfactory.

The variation of three biochemical mass measures (biomass carbon, ATP, and mineral N) was determined in four tussock topsoil samples taken at five sampling times over a 15-month period (14). There were seasonal variations, but the ratios of the biochemical parameters differed greatly at some of the sampling times. Yeager and Ward (15) have shown that the stability of bacteria is proportional to the moisture content of the sample. Removal of a sample from its environment represents a change in conditions that may be reflected in the measured biochemical parameter. All reactions modifying the measured parameter need to be quenched more rapidly than the time required for a measurable change to occur in the level of the biochemical parameter. Rapid freezing (freeze-clamping or immersion in liquid nitrogen) or chemicals such as an acid have typically been used in biochemistry. But are these methods applicable to the size of environmental samples required? Is there a change before the core sample is brought to the surface and quenched?

13.4. EXTRACTION

Significant numbers of microorganisms present in the subsurface materials are attached to surfaces. In aquatic environments, relatively high numbers of microorganisms are found in the first few centimeters of most sediments (16). The adsorption to surfaces may be even more important with respect to groundwater. All bacteria require water for growth, even those adsorbed to solid surfaces (17). To analyze those organisms by biochemical means they must be removed from the surfaces, that is, extracted. Sometimes the biochemical can be directly extracted without having to first remove the bacteria.

13.4.1. Release of Organisms or Material from Surfaces

Two recent volumes summarize information concerning adsorption of microorganisms to surfaces (18, 19). Corpe (20) reported a series of experiments on the release of marine bacteria from surfaces. The reagents that released the cells were 0.1 *N* NaOH, 0.1 percent EDTA, 0.1 percent 8-hydroxyquinoline, 0.1 percent sodium lauryl sulfate, 0.1 percent Triton X-100, 0.1 percent 2-mercaptoethanol, 0.01 *M* urea, and trypsin. To extract ATP from soil samples, we use a mixture containing EDTA, urea, DMSO, and Zwittergent 3–10, all of which should be active in removing the cells from particles and in lysing the cells.

An example of the rationale of extraction of ATP for analysis from sediments has been presented by Bulleid (21). The two major requirements for an extraction procedure are 1) release of all the ATP from the sample and 2) either no extraction of inhibitors or a subsequent removal of inhibitors before assay. There are interferences by anions in the luciferase assay. Luciferase has an anion-binding site and there are other ionic strength and solvent-related conformational changes in luciferase that influence its catalytic activity. There are sites in sediments that bind phosphate and that might bind to the phosphates in ATP. Heavy metals are extracted by acid and interfere with the luciferase assay by reacting with essential sulfhydryl groups and/or by changing the wavelength of light emitted. Bulleid (21) reported ATP recoveries of 95 percent from marine sediments and seawater using McIlvaine (phosphate-citrate), Tris, and sodium bicarbonate buffers. We find low recoveries (in the range of 10 percent) from soil using boiling buffers. To extract ATP from soil we use H_3PO_4 for the acid; this saturates the phosphate-binding sides of the soil material and insolubilizes some heavy metals. To saturate the possible aromatic binding sites which would yield the equivalent of base stacking we add adenosine. Stacking interactions between the flat aromatic surfaces of the bases contribute as much to stabilizing nucleic acids as the more familiar hydrogen bonding. If aromatic compounds are in the soil, they could bind the ATP. EDTA is added to prevent inhibition by heavy metal ions. Any possible hydrogen bond interactions of the released ATP are prevented by urea. Zwittergent 3-10 and DMSO are used to remove bacteria from the solid surfaces and to lyse them. Several of the components of the extracting mixture will inactivate any enzymes that would degrade the ATP.

This discussion of ATP extraction is given in detail to illustrate some of the precautions and types of factors that should be considered when designing an extraction procedure. I will not discuss extraction of enzymes or nucleic acids since that has been done (8).

The analysis of lipid components as performed by White and colleagues (see Section 13.11.1) has the advantage of the use of lipid solvents that extract, inactivate enzymes, and are selective for the lipids. This extraction and the ability to easily concentrate the organic solvents prior to analysis constitutes a facile procedure. Another example of improved procedures for the extraction of biochemicals from soils and sediments is the extraction and desalting of amino acids (22). Using a HF treatment prior to acid hydrolysis in 6 *N* HCl followed by anion-exchange chromatography (Dowex 1, fluoride form) increased the recovery from environmental samples from 32–75 percent. This example shows that sometimes an additional treatment of the material prior to extraction may greatly increase the recovery.

13.4.2. Purification of the Analyte

If interfering compounds are present or extracted by the procedure, purification of the compound to be determined is usually required. Again using ATP as an example, the acid extract is chromatographed on an activated charcoal column (23). The ATP is eluted with ethanol:water:ammonium hydroxide 150:95:1. Another means sometimes useful for circumventing this inhibition is a hundred-fold dilution before assay. I believe a selective extraction method would be preferable to using a second purification step.

13.4.3. Interfering Compounds

Pflug and Ziechmann (24) found that humic acid preparations inhibited the activity of malate dehydrogenase. If humic acids were extracted by the procedures used to extract enzymes and the enzymes were sensitive, inhibition would result and would make that particular enzyme unsuitable as a biomass measure. Or humic acid could inhibit the enzymes concerned with assay of the analyte.

Cunningham and Wetzel (25) found that fulvic acid interfered with ATP determination in sediments. The acid-soluble fulvic acid complexes with the extracted ATP. If, as Cunningham and Wetzel suggest, fulvic acid binds ATP by strong hydrogen bonds, a hydrogen-bonding-breaking-agent such as urea should circumvent this complication. A normal procedure for the extraction of nucleotides from biological materials uses perchloric acid followed by the removal of perchlorate by precipitating the potassium salt in the cold. When only small amounts of ATP are present ($<0.01 \text{m}M$), up to 50 percent of the ATP coprecipitated (26). The loss is negligible when the ATP concentration is 1–2 mM. This emphasizes the need that internal standards be used and that they include concentration of ATP in the range of that present in the sample. Internal standards can be added prior to extraction or analysis. Before extraction, either ATP or bacterial cells can be used as a source of ATP. I prefer the bacterial cells because they give an estimate of extraction efficiency. But you can always argue about the species and their metabolic condition. ATP added prior to assay would allow correction for interferences within the assay.

13.5. POPULATION AND ECOLOGICAL RELATIONSHIPS

13.5.1. Surface Interaction

As noted previously (Section 13.4), microorganisms are often associated with surfaces. Of the microorganisms present in seawater, between 14 and 60 percent are clumped (27). Jones and Jannasch (27) used surface-active agents such as Tween 80, MYRJ-52, Renex 31, and Dow Corning antifoam AF to disrupt the aggregates. Between 25 and 50 percent of the bacteria of salt marsh surface water were particulate-associated (28).

13.5.2 Growth Under Low-Nutrient Conditions

Most laboratory experiments do not reflect growth in natural environments (29). The microorganisms in the environment are not commonly in the exponential growth phase and do not have an abundance of every needed nutrient. The disparate nutritional conditions of laboratory cultures and of environmental samples might influence the biochemical parameters being measured. Walker, Webster, and Leach (unpublished) found a wide divergence among plate counts, light scattering measurement, and ATP levels during the germination and/or growth cycle of a spore-forming *Bacillus* strain. The ATP levels of an organism capable of becoming dormant seems to show greater variation with nutritional status than other types of organisms. Thus, the ATP levels will reflect the effect of environmental conditions on metabolism. This means that care must be taken in the selection and cultivation of the organism used as the standard. The microorganisms present in the subsurface environment and groundwater are probably adapted to growing in low-nutrient concentrations and possess the capacity for some kind of metabolic dormancy when times get really hard. Of course, when times are really hard, only the most adaptable survive, but that remnant is enough to replenish the population when better times occur.

Jannasch (30) reviewed the microbial ecology of aquatic low-nutrient habitats. The organisms faced with a low concentration of nutrients have developed high affinity transport systems for the minimum maintenance requirements. The other strategy to overcome poor environmental conditions is to produce progeny with the capacity of becoming dormant. The latter organisms consume the resources of the environment rapidly and then look for another habitat while the permanent residents use a habitat where there is a continual supply of nutrients even though at a low concentration.

Because there are so many parameters that influence growth potential in environmental samples, it is not always easy to specify a single factor that is limiting growth. In the subsurface environment bacteria are not uniformly distributed; they occur in microhabitats. Conditions within microhabitats may differ widely from one to another and these variations become evident with small sample sizes as we have seen from Table 13.3. The limiting factors for microbial growth and activity in soil have been reviewed by Dommergues et al. (31).

13.5.3. Stability of the System and Population Interactions

Measurement of changes in the level of a particular biochemical in a population of organisms can reveal an environmental perturbation. Jeffries (32) established that changes in fatty acid levels, despite inherent limitations, reflect changes in community structure. He found that as a community of plankton and benthos mature, the fatty acid composition becomes more complex. Transition from winter–spring to summer–fall species were reflected by a reorganization among the fatty acids. The changes formed an oval pattern over an annual cycle.

An excellent application of biochemical analysis techniques to detecting changes in community structure was described by White's group (33). They determined the effect of growth with and without light on the biomass and community structure of the estuarine microbiota. The biochemical factors analyzed included levels of glucosamine, muramic acid, adenosine nucleotides, lipid phosphate, chlorophyll a, neutral lipid, neutral carbohydrates, fatty acids, and incorporation of acetate into lipids, of sulfate, and of thymidine into DNA. The determination of several biochemicals can give insight into community structure.

13.6. GROWTH ASSAYS

13.6.1. Plate Counts

The plate count actually represents a biochemical amplification of the genetic information in the organism. The genetic information acting catalytically is responsible for the biosynthesis of the various systems required for growth.

The plate count in aquatic microbiology has been reviewed by Buck (34). There is a discussion of many complicating factors that influence plate count results. In spite of complications, the plate count technique is useful if the limitations are recognized. All plate count procedures are by nature selective according to the medium and conditions used.

13.6.2. Dividing Organisms

Hagstrom et al. (35) suggested the "frequency of dividing cells" as an indirect measure of the mean growth rate of a population. The cells were stained with acridine orange and viewed microscopically. Those cells that had a clear intervening zone between cells but were not yet separated were the dividing cells. These authors found a direct correlation between the frequency of occurrence of dividing cells and the specific growth rate. A statistically significant mathematical relationship between the frequency of dividing cells and the growth rate was confirmed by Newell and Christian (36).

13.6.3 Influence of Materials on Growth

Clay minerals influence the growth of organisms using various substrates that interact with clays. Marshman and Marshall (37) have studied the growth of bacteria on gelatin, bovine serum albumin, and lysozyme in the presence of montmorillonite and kaolinite. The proteins can be tightly bound to the clay minerals and their availability is determined by the clay mineral (type and amount), the protein (kind and amount), and the proteolytic enzymes concerned. The rate and extent of utilization of the protein for growth by the microorganisms depends on the protein/clay ratio. When the ratio is low, all the protein is adsorbed and not hydrolyzed. At intermediate ratios, the growth rate of the microorganisms is reduced but not the final yield. When the ratio is high, the presence of clay has no effect on either growth rate or yield.

Marshman and Marshall (38) went one step further in studying the effects of clays on growth of microorganisms by using a mixed culture of two organisms, *Bacillus* 3B-*Klebsiella* and *Bacillus* 3B-*Candida*. The starch in the medium was adsorbed by the montmorillonite. In the presence of the clay the rate of urea hydrolysis was reduced which limited growth; if ammonium sulfate is the nitrogen source, no such growth limitation was noted. These studies show the complicated interactions between soil materials, substrates, and organisms of a mixed culture that influence growth measured by plate counts and which will also occur in the subsurface environment.

13.7. STAINING AND DIRECT MICROSCOPIC ASSAYS

The most direct method of estimating biomass is by measuring the numbers and sizes of the organisms microscopically. As simple as the approach appears, there are complications due to the preparation of the samples, the amount of material that must be examined, and decisions concerning borderline structures. The distribution of dead, moribund, viable, and growing bacteria within the total population observed microscopically is an important question.

13.7.1. Epifluorescent Staining

An acridine orange-induced epifluorescence (indirect light fluorescence) technique for counting total (living and dead) bacteria in natural water has been described by Francisco et al. (39). Fluorescent staining of the bacteria makes it easier to differentiate bacteria from detritus.

The procedures for using epifluorescence were evaluated by Jones and Simon (40). The acridine-based fluorochromes were the best; changes in methodology could cause significant differences in the results. The largest errors were produced by changes in the volume of sample filtered and in the pore size of membrane used.

Daley and Hobbie (41) introduced a new light filter combination and used Sartorius cellulose membrane filters and a distilled water rinse. Field fixation with formalin allowed storage of samples for two weeks without change in bacterial numbers. In a later paper, Hobbie et al. (42) report that Nucleopore filters were superior to cellulose filters. Before use the polycarbonate Nucleopore filter was dyed with Irgalan black. Bacterial counts from lake and ocean waters were doubled by the switch to Nucleopore filters.

A technique for rendering microorganisms directly observable in sediment samples has been described by Polonenko et al. (43). They percolated 8-anilino-1-naphthalenesulfonic acid (a protein stain) through the sediment samples to be examined on a specially designed incubation chamber four hours before observation. A cover slip over the compartment allows microscopic examination of the microorganisms at weekly intervals. The fluorescent bacteria were counted. Protein staining is probably less clear cut than nucleic acid staining.

A method developed by Kogure et al. (44) yields a direct count of viable bacteria. The sample is supplemented with 0.025 percent yeast extract and 0.002 percent nalidixic acid and incubated for 6 hr. The bacterial cells did not divide but formed easily recognizable elongated cells. This technique allows detection of cells that grow on 0.025 percent yeast extract. However, with a natural population of microorganisms, it was not possible to completely suppress cell division of some of the organisms. A further application of the nalidixic acid method has been made by Peele and Colwell (45) using different substrates. A sixfold increase in the number of swollen cells occurred with the increase of yeast extract concentration from 50 to 500 μg/ml. There was a threefold decrease in cell number when the concentration of yeast extract was further increased from 500 to 1000 μg/ml. The number of substrate-responsive bacteria varied up to ninefold with various substrate concentrations. This again illustrates the perverseness of natural organisms.

Daley (46) presented arguments suggesting that direct epifluorescence is the best available technique for estimating the total number of native aquatic bacteria. With epifluorescence, 10 to 10^5 times as many bacteria are detected in samples as by conventional growth techniques (plate counts) or transmitted light microscopy. Because of the variation in size of microorganisms, the direct count of acridine-orange epifluorescent cells cannot be converted to total biomass.

Fluorimetric methods are extremely sensitive and have been used to detect single enzyme molecules or single bacterial cells (for a discussion, see Ref. 8). Could epifluorescence methods be automated and interfaced with video analysis units and a computer for automated quantitation? An example of automation of fluorimetric procedures is the development of a scanning microfluorometer for micro-test plates (47). A microcomputer-assisted biomass determination has been developed (48) to calculate the cell volume shown by scanning electron micrographs. Krambeck et al. (48) reduced the time required for a determination from 1 day to 2.5 hr.

13.7.2. Metabolic Staining

By combining either a radioisotopic incorporation procedure or a stain for an active enzyme with microscopy, the number of actively metabolizing cells can be determined. Meyer–Reil (49) combined radioautography and epifluorescence to determine actively metabolizing cells. About 30 percent of the cells in natural water (Baltic Sea) were metabolically active.

When the electron transport system of respiring organisms reduces 2-(*p*-iodophenyl)-3-(*p*-nitrophenyl)-5-phenyltetrazolium chloride (INT), INT-formazan (INTF) is formed and accumulated intracellularly as dark red spots. These spots may be intense enough to be viewed visually. Zimmerman et al. (50) used INTF and epifluorescence to differentiate between respiring and nonrespiring organisms. About 8 percent of the bacteria in Baltic Sea water were respiring while about 20 percent of the bacteria from lakes and ponds were respiring.

Tabor and Neihof (51) found that the immersion oil caused a loss of 70 percent in the number of visible INTF deposits. When the stained cells were transferred from the filter to a gelatin film, a two- to tenfold greater number of respiring organisms were found as compared to the previous method.

13.7.3. Immunochemical Staining

It would be useful to quantitate specific organisms within a population. The antigen–antibody reaction can be used to detect specific organisms. Henry et al. (52) detected *Leptospira* in both soil and water using antiserum that had been labeled with fluorescein isothiocyanate. Bohlool and Brock (53) used an immunofluorescence approach to study *Thermoplasma acidophilum*, a thermophilic, acidophilic organism lacking a cell wall. The immunofluorescence approach is useful for ecological studies.

A membrane-filter fluorescent-antibody test has been used to identify fecal group D streptococci (54). Using a fluorescent antibody techique, Apel et al. (55) established that *Thiobacillus ferrooxidans* was responsible for the production of the acid in coal refuse piles.

Fluorescence immunoassay is a promising alternative to the familiar radioimmunoassay (56). A systems approach has resulted in the development of a microprocessor-based automated instrument with a precision and accuracy of measurement of 1–3 percent. Competitive binding assays are possible.

The use of immunofluorescence to enumerate specific populations of microorganisms has been reviewed by Stanley et al. (57). There are, however, several drawbacks to immunofluorescent assay: 1) there is a limit to the volume of sample that can be passed through a filter without complications; 2) there is a problem of specificity because the assay may become too specific if the laboratory strain of organism used for antibody production is slightly different from the organisms present in the ecological samples (same strain but different serotype), and 3) the procedure does not differentiate between viable and nonviable cells.

Baker and Mills (58) combined fluorescent antibody and INT staining to determine the number of respiring *Thiobacillus ferroxidans* in water samples. They called the combined procedures FAINT. Between 3 and 72 percent (depending on the specific sample) of the cells that were counted by immunofluorescence were also counted by the FAINT procedure. They concluded that significant numbers of nonviable (nonrespiring) cells can be present in environmental samples.

13.8. TURNOVER

One of the difficulties noted in the preceding discussion is the presence of dead microorganisms or components thereof. If a component is to be a measure of a living cell it should turn over rapidly (disappear) upon the cell's death.

Holms et al. (59) found that ATP was a constant function of growth of *E. coli* when five different substrates were used. They found that the ATP turned over about 300 times/min. The turnover rate for ATP has been calculated by Atkinson (60) to be 1 s or less. The metabolic activity of ATP suggests that it will not remain for more than a few minutes unless all of the utilizing enzymes are quenched.

Since lipids are considered precursors to petroleum, their persistence in soil has been investigated (61). An average of about 40 percent of the lipids were degraded within a 4-week period. The amount of degradation increased linearly with time, and, of course, there was variation in the rates depending upon the soil sample used. The soils themselves contained between 200 and 3500 mg of unbound lipid per kg. Since lipids do not disappear from the environment as rapidly as ATP, they are not as suitable a living biomass indicator as ATP.

13.9. METABOLIC AND PHYSIOLOGICAL ASSAYS

13.9.1. Fumigation

A sample of soil is treated with a volatile fumigant such as $CHCl_3$ which kills most of the organisms. The C derived from the killed organisms comes in compounds that are readily utilized by the surviving organisms and evolved as CO_2. The amount of CO_2 produced is proportional to the biomass. The fumigation technique has been reviewed (62) in detail by its originator, D. S. Jenkinson.

Lynch and Panting (63) found that sieving of soil decreased the apparent biomass in winter but increased it in summer. They ascribe this observation to the presence of different substrates and different microbial populations. However, they offer no experimental evidence for the differences. This illustrates the changes that can occur during processing of samples.

The microbial biomass in nine topsoil samples was determined by three biochemical methods (64). The soils ranged in pH from 4.4 to 6.2 and had organic C contents of 2.5–20 percent. All three estimates, chloroform fumigation for biomass C (range 530–2780 $\mu g/g$), mineral -N (range 59–167 $\mu g/g$), and ATP (range 2.2–10.7 $\mu g/g$) were sigificantly correlated with each other, organic C, total N, and CO_2 production. The correlations were: ATP and biomass C, $r = 0.90$; ATP and mineral -N, $r = 0.67$; biomass C and mineral -N, $r = 0.57$. ATP was correlated with soil moisture, $r = 0.86$; organic C, $r = 0.85$; total N, $r = 0.93$; CO_2 produced by unfumigated soil, $r = 0.67$. There is variation among the biochemical estimates that must await elucidation. In the absence of direct counts, there is no estimation of which procedure most accurately reflects the actual biomass content.

The effect of storage of soil samples on the biomass measured by the three biochemical methods was determined (65). Ross et al. found that storage at 4°C was most suitable, while storage at 25°C was unsatisfactory. The biomass C and ATP contents were better retained during storage at −20°C, but variable results were obtained with mineral -N contents.

Because the activity of much soil biomass is nutrient limited, Sparling et al. (66) supplemented four soil samples with glucose. The samples were incubated up to 14 days at 22°C and the biomass was determined at various times and by various means: chloroform fumigation, substrate-induced respiration, and levels of ATP, amylase, and dehydrogenase. The increases in biomass, ATP, and enzyme activity were related to the amounts of glucose added. They found no generally significant correlations between ATP, biomass, or enzyme activities.

13.9.2. Respiration

A method has been described by Anderson and Domsch (67) for estimating the amount of carbon in the living, nonresting microbial biomass. When the microbial population was supplied with an excess of carbon and energy, the initial respiration response was proportional to the microbial biomass. A dozen soil samples were used with glucose as the substrate; the concentration of glucose used depended upon the particular soil sample. A correlation coefficient of $r =$ 0.96 was obtained with Jenkinson's fumigation technique (see Section 13.9.1) on 50 soil samples. The respiratory method provided reproducible estimates of soil biomass within 3 hr after glucose addition. Combination of the method with selective inhibitors can determine the relative contributions by bacteria and fungi.

The relationship between oxygen consumption and electron transport system activity has been established in five species of marine bacteria by Christensen et al. (68). The respiration/electron transport system ratios were fairly constant with a value of 5.02 for growing populations and 0.43 for senescent populations. With the termination of growth, the respiration rate decreased.

Ross (69) measured oxygen uptake rather than carbon dioxide evolution when he applied the respiration method of Anderson and Domsch (67). The oxygen uptake as stimulated by glucose was correlated with biomass carbon (0.84), mineral N (0.78), and ATP (0.75). The suitability of the method was thus confirmed for estimating soil microbial biomass in tussock grassland soils.

13.9.3. Radioactive Isotopes

Compounds labeled with radioisotopes can be used to follow metabolism. The limitations are the compound has to be transported into the cell and metabolized. Individual organisms can be examined using microautoradiography or populations by measuring gas release on radioisotope incorporation. Table 13.4 summarizes the use of radioisotopic techniques with environmental samples.

13.9.4. Bacterial Bioluminescence

An extremely sensitive physiological assay using light emission by bioluminescent bacteria has been developed for toxic components in water (90). Freeze-

Table 13.4. The Use of Radioisotopes to Ascertain Microbial Metabolic Activity in Environmental Samples

Isotope	Compound	Reference	Sample	Remarks
[^{3}H]				
	Thymidine	70	Water	Active metabolism; 2-hr labeling period;
	Glucose	70	Water	direct count: plant count: glucose labeled = 100:45:1.
	Thymidine	71	Soil	Incorporation into acid insoluble material; bulk population measured.
	Amino acids	72	Water	60 percent of the bacteria in the Baltic Sea were active in protein synthesis.
	Thymidine	73	Sediment	Four-fold variation in nine Lake Erie samples.
	Thymidine	74	Water	30 percent of the bacteria tooks up thymidine; generation time on the order of days.
	Adenine	75–78	Water	DNA/RNA synthesis rate is proportional to growth rate; determine specific activity of precursor pools.
[^{14}C]				
	Lactose	79	Water	100 *E. coli* cells detected in 1 hr.
	Formate	80	Water	Detect 20 cells.
	Glucose	81	Water	A measurable response from 10 bacteria in 10 hr.
	CO_2	82	Soil	Pyrolysis gas chromatography and acidification–wet oxidation used for detection.
	Lactose	83	Water	A response to 1–10 bacteria in 6 hr.
	Glucose	84	Soil	Determines potential microbial activity.
	Various	85	Water	Benzoic, salicylic, 2,4-D, Dicamba, glucose were nutrients; applied in a most-probable-number method.
[^{35}S]				
	SO_4	86	Water	Measures microbial biomass.
	SO_4	87	Water	C:S ratio 100–200.
	SO_4	88	Soil	Specific extraction of microbial population.
	SO_4	89	Water	75 percent of the S is in protein.

dried luminescent bacteria are incubated with various solutions and the light output is measured using a photometer. Reduction in light output is proportional to the concentration of the toxic compound that depends on the nature of the compound. For example, a linear response was obtained for toluene from 7.5 to 75 mg/liter. The bacteria are aquatic and therefore constitute a natural assay system. Comparison with the 96-hr fish bioassay (90) showed that the sensitivity of the bioluminescent bacteria was sufficient and constituted a quicker assay.

Bacterial luciferase uses a reduced flavin, a long-chain aldehyde, and oxygen to produce light. The reactions involve a branch of the electron transport chain. Since the flow of electrons in the respiratory chain is an indication of the metabolic state of the cell, many different compounds will inhibit light production. We (91) found a correlation coefficient of 0.90 with rat toxicity and 1.0 with fish toxicity for seven different inhibitors.

The EC_{50} can be calculated within 30 min of sample collection (92). The bioluminescent data obtained in a 5-min test are comparable with the data obtained in a 96-hr fish assay. Application of the Microtox test to fossil-fuel process waters has been described (93). The authors concluded that inhibition of bacterial bioluminescence is of value for predicting the toxicity of complex effluents.

The measurement of bioluminescence is a sensitive assay reflecting the physiological state of the bacteria and is extremely sensitive to many pollutants. This assay is therefore useful in measuring toxicity present in water samples.

13.10. ENZYMES

Because they are easy to detect and act catalytically, enzymes have been objects of assays to determine environmental parameters. Soil scientists have made many measurements in attempting to develop a relationship between soil enzyme levels and soil fertility. Our prior discussion (8) on determination of enzymes and the significance of their concentrations should be consulted. That review summarizes the extensive data that were available through 1978.

When the enzyme to be determined does not produce an easily measurable product, another enzyme or enzymes may be added to yield a coupled system that produces an easily measurable product. Whole systems of analysis can be based on the determination of single products such as the pyridine nucleotide coenzymes. Cleland (94) has developed equations that make possible minimization of cost and calculation of the maximum practical rates for coupled assays involving two or more coupling enzymes.

The presence of certain enzymes in environmental samples indicates the presence of specific organisms. The amount of enzyme present may be changed by an environmental perturbation or pollutant. The patterns of enzymes found can be just as indicative of a particular organism as fingerprints. The demand for assay of multiple samples for enzyme activity has led to the development of automatic systems. Many types are available and automatic enzyme determination has been reviewed by Moss (95). Table 13.5 summarizes studies using mea-

Table 13.5. Use of Enzyme Determinations for Measuring Microorganisms in Environmental Samples

Enzyme Assayed	Substrate	Reference	Sample	Remarks
Glutamic acid decarboxylase	Glutamic acid	96	Water	Detect 1 bacterium/100 ml of water in 13 hr.
Dehydrogenases	Glucose or yeast extract	97	Soil	37° C used because this condition gave greater growth, but it is not the environmental temperature. Used 2,3,5-triphenyl-tetrazolium chloride as the electron acceptor.
Lipolytic activity	Cottonseed oil	98	Water	Found in many water samples, psychrophiles.
Phosphatase	*p*-Nitrophenyl phosphate	99	Soil	Phosphate-deficient soil gave $r = 0.84$ for ATP and biomass.
Urease	Urea	99	Soil	Urease and biomass $r = 0.94$
β-Galactosidase	*o*-Nitrophenyl-D-galactoside	100	Water	820 cells in 9.5 hr or 5 cells in 17 hr.
Protoelytic	Hide powder azure	101	Water	Must be assayed above 15°.
Esterase	Fluorescein diacetate	102	Soil	Spectrophotometric method
Lipase	4-Methylumbelliferone	103	Soil	Extract enzymes from soil with 0.1 *M* sodium pyrophosphate.

surement of enzyme activity for evaluation of biomass or microbial activity in environmental samples.

13.11. CELLULAR COMPONENTS AND METABOLIC PRODUCTS

Central to the science of biochemistry is an inventory and quantitative accounting of the cellular components and metabolic products. The development of analytical procedures continues as the importance of accurately determining specific compounds at low concentrations becomes increasingly relevant to agriculture, medicine, and ecology. Just as one can recognize a pathological condition by a change in one or more of the biochemical components in a diseased organism, the changes brought about by a pollutant in the environment of an organism should be similarly recognizable. Organisms themselves or their products can be biologically detrimental to others and they should be biochemically determinable. In general, organisms are fairly conservative with their resources because of the close regulation of the biochemical processes that tend to limit intermediates and balance synthesis. Thus, the quantity of a cellular component should reflect the number and mass of organisms present.

13.11.1. Lipids

Kowalenko and McKercher (104) found that hexane-acetone gave a better extraction than $CHCl_3$-acetone or hexane-benzene of soil phospholipids. Sediments from Boston Harbor and vicinity were analyzed for lipids by gas chromatography. Differences were found in the lipid contents that Shaw attributed to the complexities of the estuarine sedimentation process (105).

White et al. (106) used lipid synthesis (from both [^{14}C]- and [^{32}P]- labeled precursors), alkaline phosphatase activity, ATP levels, respiration, and α-mannosidase activity to characterize microbial activity in estuarine detritus. Lipid synthesis increased with time of incubation and was proportional to the mass of detritus. The presence of penicillin, streptomycin, nystatin, and cycloheximide reduced lipid synthesis by 40 percent. There was a hundredfold reduction in lipid biosynthesis by 2 percent chloroform. Because of the nonrandom distribution of microbes on detrital surfaces, the use of randomly sampled disks from a pooled collection of leaves was necessary to reduce the variance. The incorporation of phosphate into phospholipid, ATP levels, and the phosphates paralleled each other during the six-week period. The following correlation coefficients were found: mass of detritus–lipid synthesis, $r = 0.98$; alkaline phosphatase activity–lipid synthesis, $r = 0.64$; ATP–lipid synthesis, $r = 0.41$; α-mannosidase activity–lipid synthesis, $r = 0.66$; oxygen utilization–lipid synthesis, $r = 0.82$.

White et al. (107) proposed lipid phosphate as a microbial biomass indicator for marine and estuarine sediments. The assay is simple to perform on chloroform-methanol extracts. Lipid phosphate correlated with ATP ($r = 0.84$) and the rate of DNA synthesis ($r = 0.99$). Phospholipids are the major lipids of the bacterial membrane and are not storage products. There is active metabolism of the phospholipids ($t_{1/2}$ of 2 and 24 days) and quantitative recovery of phospholipids from the sediments can be achieved. The pattern of lipid compositions define the microbial community composition. Thus phospholipid content, like ATP content, represents a relatively nonspecific measure of biomass.

Fatty acid analysis has been used to characterize the benthic microbial community structure (108). Gas–lipid chromatographic analysis of the methyl esters of the fatty acids was reproducible and highly sensitive. The recovery of fatty acids was quantitative. By addition of mass spectral fragmentation analysis an ultrasensitive diagnostic analysis was possible. Fatty acid analysis is superior since staining and epifluorescence microscopy requires release of the organisms (not quantitative) or assumptions about the concentrations on the hidden faces.

Nickels et al. (109) have applied their analytical techniques to an investigation of microfouling of aluminum and titanium exposed to seawater. Cleaning of the surfaces after fouling left a residual population of organisms that was different from the original. They fouled the metal at a much greater rate and secreted a twofold larger amount of extracellular polymer than did the free-fouling community. After removal of the metal samples from the seawater, total organic carbon was determined and the surface was examined by scanning electron microscopy. The lipid portion of the chloroform/methanol extract was examined for lipid phosphate and fatty acids. The aqueous phase of the extract was analyzed for adenine nucleotides by high-pressure liquid chromatogrpahy (detected fluorimetrically). Table 13.6 presents the various biochemical markers and describes what these mean in terms of both biomass and community structure. This represents a key for the interpretation of lipid analysis.

Silica grains with different microtopography but the same size and water pore space bear different biomasses and community structures after eight-week exposure to running seawater (110). Biochemical measures were used to show the differences in biomass and community structure. In soil and sediments, most of the microorganisms are located on the surfaces and not in the water (111). Different properties are sometimes shown by enzymes and organisms that are attached to surfaces than when they are free in solution (112). Microbial attachment involves chemotaxis, reversible then irreversible binding, and extracellular polymer synthesis. Often there is specificity in the attachment interactions. Bacteria are the primary colonizers of new surfaces.

An ultramicro determination (1 μl–1 nl) of triglycerides by a bioluminescent assay has been developed (113) and illustrates how specific enzymes can be used for biochemical analysis when coupled with firefly luciferase. Such techniques might be applicable to environmental analysis with suitable modifications. This procedure shows the extreme sensitivity that can be obtained using bioluminescent assays rather than conventional spectrophotometric determination. A distinguishable signal is obtained from 60 μg of triglyceride.

Smith et al. (114) have determined the effect of oil- and gas-well drilling fluids on sand-colonizing microbiota. The well-drilling fluids increase a portion of the bacterial population and microfauna when either overlying on or mixed with sand. Shading the microbiota from light depressed the microflora without

Table 13.6. Summary of Biochemical Measures of Microfouling Biomass and Community Structure[a]

Biochemical Marker	Measures
Biomass	
Lipid phosphate	Phospholipid; total "membranes" biomass.
Total palmitic acid	Total lipid; membrane + triglyceride biomass.
Total organic carbon	Total mass; exopolymer + cellular carbon.
Total *a* + *i* 15:0	Bacteria; a measure of some bacteria.
Total 15:0	Bacteria; one component of the bacteria.
Total Δ 17:0	Bacteria; another component of the bacteria forming cyclopropane fatty acids.
Total 18:1ω7	Bacteria; a third component of the bacteria with the anaerobic desaturation pathway.
Total 18:2ω6	Linoleic acid; filamentous bacteria, blue-green algae (cyanobacteria), fungi.
Total polyenoics >20	Microeucaryotes; protozoa, fungi, algae.
Total 24:0	Long straight-chain; microeucaryotes.
Community structure	
a + *i* 15:0/15:0	Proportion of total bacteria of one type.
a + *i* 15:0/16:0	Proportion of that type of bacteria in the total microfouling organisms.
Δ17:0 + Δ19:0/16:0	Proportion of the second type of bacteria in the total microfouling community.
18:1ω7/16:0	Proportion of third type of bacteria in the total microfouling community.
18:1ω7/18:19	Proportion of the third type of bacteria in the portion of the community containing oleic acid (bacteria and microeucaryotes).
18:2ω6/16:0	Proportion of filamentous, gliding, or blue-green algae in the total film.
Total polyenolics >20/16:0	Proportion of the total microeucaryotes in the total film.
Polyenoic ω6/ω3	Proportion of protozoa (animals) to algae, diatoms, and fungi (plants) in the microeucaryotes.

[a] *a* is anteiso; *i* is isobranching; Δ is cyclopropane ring. This table is reproduced, slightly modified from J.S. Nickels, et al., *Appl. Environ. Microbiol.*, **41**, 1442 (1981) with permission.

changing the biomass, and there were some such shading effects of the well-drilling fluids. This paper showed how the lipid bioanalysis procedures developed by White's laboratory can be used to determine the effect of environmentally significant chemicals on biomass and community structure. The analysis was based on quantification of 14 different lipid components and the calculation of 12 ratios. Part of the components and ratios determined are shown in Table 13.6.

13.11.2. Lipopolysaccharides

The horseshoe crab, *Limulus polyphemus*, contains in its amebocytes, the only circulating blood cells, all the components required for clotting. One of the agents responsible for formation of a clot is the endotoxin, that is, the lipopolysaccharides (LPS), from gram-negative bacteria. When the *Limulus* blood is shed, the amebocytes are transformed so that they aggregate and the clotting materials are released during degranulation. The clot that forms helps confer resistance to bacterial infection.

A total of 48 water samples from an advanced waste treatment plant in Dallas, Texas were tested by the *Limulus* endotoxin assay, standard plate counts, and total coliform counts (115). The *Limulus* assays were simpler and quicker. The correlation coefficients for each of the counts and endotoxin were in the range of $r = 0.7$. Jorgensen et al. (115) think that poor correlation may be due to alteration of the normal ratio of viable gram-negative cells to endotoxin produced by the treatments. Because the effluent had been chlorinated, exposed to ultraviolet irradiation, and treated with ozone, a majority of the bacterial population were killed but the endotoxin was not necessarily destroyed. When samples were taken before treatment, the correlation coefficient between plate count and total endotoxin was $r = 0.945$.

Watson et al. (116) compared epifluorescence, transmission electron microscopy, and the *Limulus* amebocyte lysate test to measure bacterial numbers and biomass in the marine environment. A correlation coefficient of $r = 0.948$ was obtained for 188 seawater samples comparing epifluorescence and LPS measurements.

When the *Limulus* lysate test was applied to stream water, it reflected the degree of contamination as measured by coliform, enteric, gram-negative, and heterotrophic bacteria (117). The spectrophotometric method of assay was more sensitive than the firm-clot method. Bound endotoxin was a better measure of the endotoxin associated with bacterial cells than was total endotoxin.

Watson and Hobbie (118) reviewed measurement of bacterial biomass as LPS. They found a correlation coefficient of $r = 0.82$ between ATP measurements and LPS measurements for 12 samples of seawater. The correlation between LPS and epifluorescence count was $r = 0.96$ and that between the ATP method and epifluorescence count was $r = 0.81$.

Between 90 and 95 percent of endotoxin was removed when sewage traveled through 100 to 250 cm of loamy sand soil (119). Addition of distilled water to simulate rainfall mobilized the endotoxin, which traveled as a concentrated band. In some groundwater samples from land treatment sites, as much as 480 ng of LPS/ml were found. Because of the chemical nature of LPS, its behavior should be similar to that of some viruses that have lipoprotein-containing coats. Goyal et al. (119) concluded that there is potential for contamination of groundwater with endotoxin.

Alkaline hydrolysis destroys LPS and lysozyme hydrolyzes peptidoglycans; these reactions can be used to distinguish between specific and nonspecific *Limulus* lysate reactions (120). Sykora et al. (120) recommended the use of at least two techniques in conjunction with *Limulus* lysate test for drinking-water quality measurements.

13.11.3. Nucleic Acids

The genetic material of most organisms is composed of a linear sequence of purine and pyrimidine nucleotides. DNA determinations have been based on the reaction of diphenylamine with deoxyribose, on the formation of complexes with antibiotics, and complexes of fluorescent or dye substances with DNA. Many of these methods have been discussed previously (8), so the emphasis here will be placed on the more sensitive fluorimetric techniques and those techniques that have been applied to environmental samples. Anderson (121) suggested microbial DNA must survive for some time after decomposition of the cell structure. At least when DNA is in aqueous solution, its transforming activity is rapidly lost in the environment (122). Components of the soil may stabilze the DNA or it may be of sufficient length to be detected chemically, but its genetic information is not retained because it is degraded to short pieces.

Nucleic acid bases, that is, the purine and pyrimidine building blocks, adsorb to clays and the adsorption decreases with an increase in pH (123). A fulvic acid–montmorillonite complex adsorbed less than montmorillonite alone, but the adsorption was still substantial. This adsorption would complicate measurement of biomass based on DNA measurement with subsurface samples. Table 13.7 summarizes DNA determinations.

13.11.4. β-Hydroxybutyrate

Poly-β-hydroxybutyrate (PHB) is an endogenous reserve polymer widely found in both gram-negative and gram-positive bacteria. The sensitivity of the quantity and metabolism of poly-β-hydroxy-butyrate to environmental perturbation led Herron et al. (127) to determine the levels of this compound in estuarine detrital microflora. Since PHB content of detrital microbial communities does

Table 13.7. DNA Determination in Environmental Samples

Reagent	Reference	Sample	Remarks on Sensitivity
3,5-Diaminobenzoic acid	124	Soil	8×10^8 cells required for measurement.
Mithramycin	124	Soil	4×10^9 cells required for measurement.
4,6-Diamidion-3-phenylindole	125	Culture	10^6 cells detected.
Hoechst dyes	126	Water	50 ng of DNA, correlated with acridine orange $r = 0.90$.

not correlate with microbial biomass (lipid phosphate), Herron et al. concluded that PHB more appropriately reflects metabolic status rather than biomass.

13.11.5. Coenzymes

Procedures for the extraction and estimation of the intracellular nicotinamide nucleotides of *E. coli* have been developed (128). Such assays of coenzymes might offer another measure of biomass or means of examining the metabolic status of bacteria.

13.11.6. Nucleotides

13.11.6.1. ATP

ATP occurs ubiquitously in living organisms, where it functions as an allosteric factor, as a group-carrying coenzyme, and as a substrate. To run the biochemical machinery of life, cells require 1) a source of reducing power, which is NADPH, 2) a source of energy, which is ATP, and 3) several biosynthetic building blocks (only 10 different compounds). ATP plays the central role in metabolism and given the closely regulated biochemical mechanisms ATP should be found at a fairly constant concentration. Atkinson (129) has also defined the concept of energy charge.

$$\text{Energy charge} = \frac{[\text{ATP}] + \frac{1}{2}[\text{ADP}]}{[\text{ATP}] + [\text{ADP}] + [\text{AMP}]}$$

The energy charge is a linear measure of the metabolic energy stored in the adenylate nucleotide system. The energy charge is also closely regulated and its determination most closely measures the metabolic health of the organism.

The love life of fireflies has yielded a byproduct that enables analysts to determine ATP with great sensitivity and through coupling to other enzymes it is possible to determine each of the adenylates. Two symposia (130, 131) have been devoted to ATP methodology. I have recently reviewed in detail ATP determination using firefly luciferase (132).

In 1966, Holm–Hansen and Booth (133) suggested ATP analysis as a means for estimating biomass. The properties of a reliable biomass indicator are:

1. That it be a constituent of all living cells.
2. That a fairly constant amount of the indicator is present per cell weight.
3. That it rapidly disappears upon the death of the organism.
4. That the concentration within the cell is independent of environmental conditions.
5. That the concentration in the cell is independent of growth rate.
6. That a sensitive assay for measurement of the indicator exists.

ATP does not meet Numbers 2 and 4 of these specifications, but also consider how the other proposed indicators qualify.

Many different procedures and reaction conditions have been used for the determination of ATP with firefly luciferase. The situation is confusing since there have been few direct considerations of the influences of the different conditions on the sensitivity and specificity of the determination. Therefore, a discussion of the technology of ATP determination with firefly luciferase follows.

An improved method for the determination of ATP in soil has been developed by Eiland (134). Extraction of ATP from soil or sediments presents special problems that are not encountered with aqueous samples. 1) There is a lower concentration of ATP. 2) There are substances that interfere with extraction. 3) There are substances that interfere with the enzymatic assay for ATP. Eiland found that none of the extraction methods described in the literature were satisfactory for measurement of small amounts of ATP in soil. The method he developed entails shaking the soil with 1 to 1.5 *N* H_2SO_4, filtering, and chromatography on an Amberlite IR-120 column. The ATP was eluted with water and adjusted to pH 9.8 with 5 *N* ethanolamine. The recovery of internal ATP standards was 81 percent.

Factors important in the determination of ATP in aqueous samples have been reviewed by Deming et al. (135). They believe that instantaneous peak height measurement of the light production is a more valid measure than the integration mode. Using a purified luciferase, they can detect 100 fg of ATP. The stability of firefly luciferase is increased by the addition of 1 *mM* dithiothreitol. Several different commercially manufactured instruments are available and suitable for ATP determination. Better extraction of ATP was achieved with inorganic acids (they use 0.1 *N* HNO_3). They concluded that ATP assays provide a unique measurement of naturally occurring aquatic microorganisms, but that a single analysis is difficult to interpret in terms of biomass or cell number.

Davis and White (136) have developed a high-pressure liquid chromatography technique with fluorescent detection for the measurement of adenosine, adenine, cAMP, AMP, NAD^+, ADP, and ATP. They found that rapid freezing in a dry-ice–acetone bath within 15 s preserved the adenine nucleotides with the same energy charge and the same ATP concentration. They extracted sediment samples using an aqueous buffer, methanol, and chloroform combination. The organic phase was used for lipid phosphate determination while the various adenine-containing compounds were in the aqueous phase. The 1-N^6-etheno derivatives of the adenine-containing compounds were formed. They demonstrated quantitative extraction and derivatization. The high pressure liquid chromatography method was shown equivalent to the firefly luciferase assay of ATP.

A method to correct for the influence of cations from soil samples on ATP measurement with luciferase has been developed by Eiland and Nielsen (137). Ion exchange (Dowex 50) was used to reduce the concentrations of Fe, Al, and other cations in the extract. They found different patterns of light emission depending on the kind of soil involved. Measurement of the light production at one given time interval such as 1 or 60 s will not reflect the true ATP content. The true content is represented by the integrated curves.

There are many variations in procedures in the analysis for ATP using firefly luciferase. I have summarized over 100 of the procedures (Table VII, Ref. 132). The temperature used for the assay range from 0°–37°C, but the optimum is 25°C. The pH values varied from 7 to 8, but the optimum is pH 7.8. Even though arsenate inhibits firefly luciferase, 64 percent of the reaction mixtures contain it.

Because of the variations in buffers used, we (138) tested several and found Tricine superior. We compared a dozen different commercial firefly luciferase preparations (139, 140). Lantern extracts (the dried firefly tails are extracted with buffer and not further purified) have high inherent (without ATP addition) light emission and contain contaminating enzymes that permit light production from GTP, ADP, and AMP. The purified preparations have lower inherent light emissions and are more specific because the contaminating enzymes have been mostly removed. Crystalline preparations (Sigma, Boehringer–Mannheim, and this laboratory) all show better specificity and greater response. We found that all except a LKB preparation were markedly stimulated by additional luciferin (of course, the crystalline preparations must have luciferin added). By changing the luciferase and/or luciferin concentration in the reaction mixture, the sensitivity of the assay can be modified (see Fig. 2, Ref. 139). Figure 13.2 shows the effect of additional luciferin on the activity of the Firelight luciferase preparation. Since these data were published, several companies have modified their formulations or have developed new preparations. Table 13.8

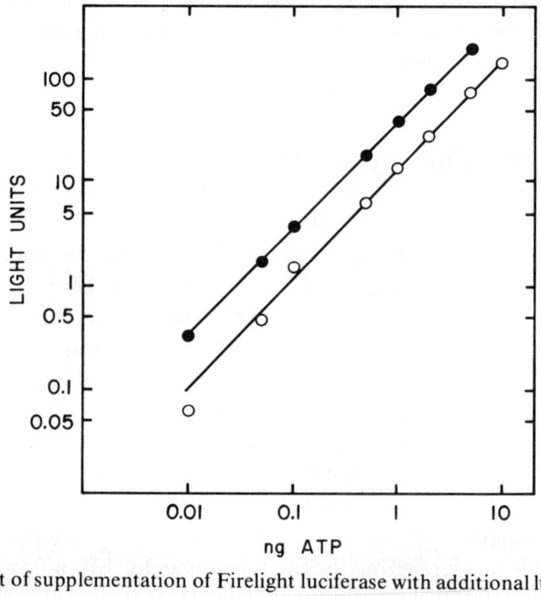

Figure 13.2. Effect of supplementation of Firelight luciferase with additional luciferin on detection sensitivity. An addition of 50-μg luciferin (●) per assay was added to Firelight and the bioluminescence produced by various concentrations of ATP was compared to that of unsupplemented Firelight (○).

Table 13.8. Properties of Some New Firefly Luciferase Preparations[a]

	Light Units					
	Inherent		1 ng ATP		50 ng AMP	50 ng GTP
(Luciferin 50 μg)	−	+	−	+	+	+
			Sample			
Analytical Lumin.	0	0.4	17.4	38.6	0	2.2
Boehringer-Mann.	0	0	0	28.1	0	3.8
Diagnostics	0	0.2	38.9	16.8	0	3.2
LKB	2.8	0.5	35.3	22.1	0	4.7
Lumit PM	0.1	0	66.4	188.7	0.1	9.0
Sigma L-1759	0	0.1	0	152.7	0.2	22.9
Sigma L-9124	0	0	30.0	241.4	0.1	5.0
Sigma L-0633	0	0	11.2	62.4	0	1.1
Sigma L-4130	0.1	0	55.0	228.1	0.2	14.6
Sigma L-9009	0	0	0	57.6	0	8.0

[a] Some of the luciferase preparations were purchased. We thank Diagnostic Sicence, Inc., 3M Medical Products Division for the Lumit, and Sigma for their donated samples.

summarizes the characterization of those preparations while Fig. 13.3 shows standard curves. Again let me stress that the relative positions of the lines can be changed by using different amounts of luciferase and luciferin.

We (141) optimized firefly luciferase assay of ATP using commercially available equipment and reagents. The pH optimum was 7.8 in 0.025 *M* Tricine buffer. With concentration of buffer >0.1 *M* considerable inhibition was observed. Mg^{2+} reacts with ATP to form the actual substrate for the reaction, the optimum Mg^{2+} concentration being 5 m*M*. Because many of the commercial preparations contain both luciferase and luciferin, their dilution does not produce a linear reduction in light production. An additional 50 μg per reaction of luciferin was used to assure maximum sensitivity. Oxygen is also a required substrate, but to demonstrate an oxygen requirement we had to enzymatically deplete the O_2 prior to the addition of ATP. Because the possibility of contaminating heavy metals is present, we routinely make each reaction mixture 0.5 m*M* EDTA. Dithiothreitol, 0.5 m*M*, and 100 μg of bovine serum albumin final concentrations are added to each reaction mixture.

There are marked effects of assay volume size and container on the light production measured. Figure 13.4 shows that using Beckman Biovials instead of glass scintillation vials caused more light output to be measured. A new holder for the vial must be made. Reductions in the assay volume from 1 to 0.5 and then to 0.2 ml increase the measured light production. The geometry within the light measuring instrument is an important factor in determining sensitivity.

Some applications of ATP determination to environmental samples will now be described. This discussion will be divided into four areas: seawater, sediments, soil, and populations.

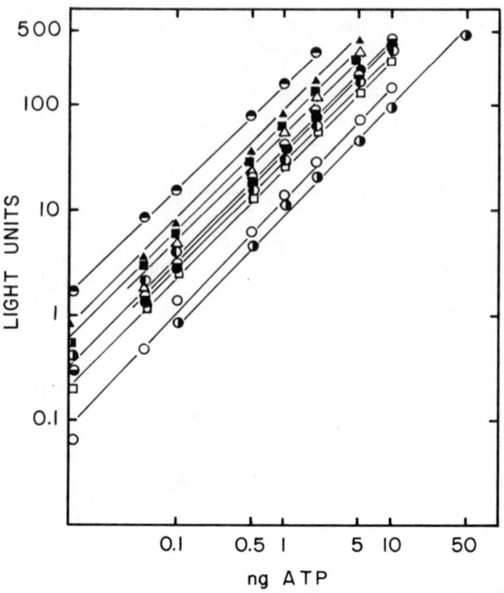

Figure 13.3. Comparison of the bioluminescence produced by various ATP concentrations with selected commercially available luciferase preparations. The various firefly luciferase preparations were constituted to the manufacturer's recommended volume in Tricene. The luciferase preparations are: ◒, Sigma L-1759·▲ , Lumit PM; ■ , Sigma L-9009; △ , Sigma L-4130; ◒ , DSI; ◐ , Sigma L-9134; ● , LKB; □ , Boehringer–Mannheim; ○, Analytical Luminescence; ◑, Sigma L-0633. By changing the amount of luciferase and luciferin, the positions of the curves can be changed. Therefore, the order in which the various preparations occur does not necessarily reflect their true limiting sensitivity.

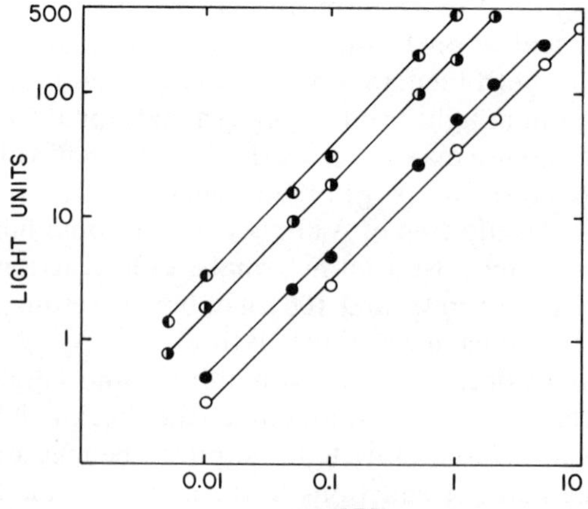

Figure 13.4. Influence of reaction container and volume on the measured light output. A SAI Model 3000 ATP photometer was used to measure light production. The light production by various amounts of ATP was determined in glass 20-ml liquid scintillation vials containing 1 ml, ○ ; in a Beckman Biovial containing 1 ml, ● ; Biovial containing 0.5 ml, ◑ ; Biovial containing 0.2 ml, ◐ .

An effect of the sample size filtered on the values obtained for the ATP content of seawater was noted by Sutcliffe et al. (142). Filtration of several different sized samples will allow detection of this complication.

An interesting example is the measurement of ATP and organic carbon in the Cariaco Trench (143). ATP was found to a depth of 1200 m. The ATP and particulate organic carbon values were between 4 and 12 times those previoulsy reported from oxygenated oceanic environments. The higher ATP content found in an anoxic ocean zone suggests that the anoxic conditions permit the development of a large population of bacteria (144). This finding may be important in the microbiology of groundwater and the subsurface area because of their possible anoxic conditions.

Christian et al. (145) measured the vertical distribution of microbial ATP in salt marsh sediments. Three-inch core samples were taken in triplicate for each region and then divided into six segments by depth. Boiling 0.1 M $NaHCO_3$ (pH 8.5) was the extracting reagent. *Enterobacter aerogenes* cells were used to determine recovery which was "always about 100 percent." The concentration of ATP was greatest in the first centimeter of sediment and decreased with depth. ATP concentrations were highest in July and lowest in February. In four sediment samples from widely varying sources the ATP concentration varied between 1 and 10 $\mu g/g$.

Just as work with seawater had been slowed by a lack of a good biomass indicator the same situation exists with soil biomass. Ausmus (146) compared seven methods for the extraction of ATP from soil. The best extractants were butanol and cold 0.6 N H_2SO_4. He concluded that ATP is a particularly useful tool in studying terrestrial decomposition of organic mater.

ATP measurements have been established as a reliable parameter for the measurement of microbial activity in coal strip-mine spoils (147). As in marsh sediments, the ATP concentration within core samples varied with depth. The greatest level of ATP was at the top. The spoils contained less ATP than native range soils. The distribution of ATP within core samples of native range soils were different. Thus, by comparing the distribution of ATP, the two soil systems (native range soil and coal strip-mine spoils) may be differentiated.

Jenkinson and Ladd (62) have reviewed the application of ATP analysis as a measure of the biomass of the soil. The difficulties with the application of ATP analysis techniques to soil are varying extraction efficiencies and the variation in the amount of ATP per unit of biomass. Many of the complications that Jenkinson and Ladd noted can be overcome with improved extraction methods, addition of internal standards, and the use of purified firefly luciferase.

The ATP contents of seven soil samples were determined after air-drying, freeze-drying, storage, and incubation with glucose or water (148). Air drying reduced the ATP content. An increase in ATP content was produced by wetting but this may not represent new ATP synthesis. Ahmed et al. (148) concluded that determination of ATP as a biomass measure has great potential since it can show changes in a few minutes and is suitable for fractions of soils that cannot be incubated.

Bobbie et al. (149) note that the use of multiple parameters for microbial biomass and activity measurement makes possible better comparisons between substrates and populations of organisms than does measurement of a single parameter. Among the parameters they measured are weight change of the detritus, ATP, oxygen uptake, scanning electron microscopy, and levels of alkaline phosphatase, phosphodiesterase, muramic acid, and glycoesterase. Using biodegradable pine needles, they found two-to ten-fold higher values of extractable ATP, rate of oxygen utilization, activities of alkaline phosphatase and phosphodiesterase, and muramic acid in the microbiota than on similar sized needles of polyvinyl chloride.

Cavari (150) concluded that even though ATP concentrations may vary within an organism depending upon its nutritional state, it is more reliable as a biomass indicator than classical counting techniques. He suggested that ATP determinations may be used to indicate nutrient limitations in natural aquatic environments.

Because nucleotides are found in all cells, the ATP assay is nondifferentiative (151). There are many assay procedures and experimental variations that lead to differences in results. It was concluded that the advantages of using ATP as an index of all microbial populations include: 1) sparse communities are detectable, 2) all physiological types of microbes are included, 3) only living cells give the response, 4) the analysis is quick, and 5) the measurements are precise.

Davis and White (136) concluded that ATP is a precarious biomass measure under some conditions. The variation in conditions that they used such as anoxia and addition of a substrate followed by development of anoxia produced a significant metabolic shock to organisms endogenously respiring. The shock was absorbed by an adjustment in the ATP concentration and thus, the ATP level reflected the metabolic status of cells more than it does biomass. If metabolic conditions were held constant, then ATP could be related to biomass.

13.11.6.2. Energy Charge

The adenylate energy charge yields information concerning the metabolic status of the cells. Firefly luciferase is used to determine the three adenylate nucleotides in three separate reaction mixtures. First, ATP is determined using firefly luciferase. Addition of phosphoenolpyruvate and pyruvate kinase converts the ADP to ATP and the measured light production in the sum of ATP + ADP. When phosphoenolpyruvate, pyruvate kinase, and adenylate kinase are incubated, the light produced is the sum of ATP + ADP + AMP. From these values the energy charge is calculated.

Wiebe and Bancroft (152) proposed use of the energy charge ratio as a means of determining the growth state of natural communities and the effect of environmental changes on them. Actively growing and dividing cells have energy charges in the range of 0.8–0.95; stationary cells maintain an energy charge of 0.6; senescent cells have an energy charge below 0.5. Other measures such as oxygen consumption and substrate uptake should also be studied.

Karl (153) has reviewed cellular nucleotide measurements and their applications in microbial ecology. He states, "No single approach to the study of microbial ecology is universally accepted." The approaches vary from reductionist to synthesist. Karl presented the first comprehensive review on cellular nucleotides and energy charge as indicators in microbial ecology. He pointed out the need for an extensive study of methods of sampling, transporting, storing, and other experimental manipulating of environmental materials. There is a summary table presenting 40 different extracting conditions. An average of about 6 nmol of ATP/mg dry weight with an energy charge of 0.8 is found for representative samples of procaryotic microorganisms.

A modified method for measuring energy charge in natural samples was presented by Romano and Daumas (154). Dead cells that still contain AMP but not ATP can interfere with energy charge determinations. There are three "ranges" of energy charge data with water samples. Values as low as 0.4 are associated with nonliving or severely stressed organisms. When the value is 0.4 to 0.6, the system has either a low-structure or a pollution-perturbed ecosystem. Values >0.6 represent oligotrophic water and a mature ecosystem. They concluded that energy charge is a useful ecophysiological indicator for the study of cellular metabolic activity as it relates to energy flow.

Spielmann et al. (155) found that the determination of the three adenine ribonucleotides was no longer limited by the purity of firefly luciferase with the advent of the partially purified commercial preparations. The limiting factors now are the levels of ribonucleotides in pyruvate kinase and adenylate kinase. Exclusion of the ribonucleotides and removal of ammonium sulfate is required.

13.11.6.3. GTP

Karl (156) found an interference by nonadenine nucleotides with Sigma FLE-50 luciferase. This Sigma FLE-50 is a crude extract of firefly tails and contains adenylate kinase and nucleoside diphosphate kinase. The responsible nucleotides were GTP and UTP. He found that GTP/ATP ratio in *Serratia marinorubra* increased in direct proportion to the rate of cell growth. He concluded that determining the GTP/ATP ratio in environmental systems would be useful for measuring microbial growth.

13.11.7. Proteins and Amino Acids

In determining amino acid composition of proteins in subsurface soils, Cheng et al. (157) found that a HF treatment to destroy hydrated clay materials was required prior to hydrolysis with 6 *N* HCl. An ion-exchange resin in the fluoride form was used to remove the salt which interferes with the analysis.

Nanogram quantities of protein can be determined after hydrolysis by reaction with *o*-phthalaldehyde or enzymatic determination of glutamate (158). We (manuscript in preparation) applied the method and found a detection limit of 6 ng of protein instead of the 3 ng reported by Butcher and Lowry (158). The major problem we encountered was the background of protein present even on new test tubes. The test tubes were prepared by three cycles of acid hydrolysis using 6 *N* HCl and 110°C for 5 hr followed by extensive washing with water. Without using acid-treated test tubes and pipette tips, the detection limit varied between 30 and 100 ng.

13.11.8. Muramic Acid

The hexosamine muramic acid (3-*O*-2-carboxyethyl-*D*-glucosamine) is a constituent of the cell walls of both bacteria and blue-green algae. Millar and Casida (159) found muramic acid in concentrations of 0–150 μg/g of soil. This concentration indicated that there were 10^2–10^3 more bacteria present than were platable in 33 different soil samples. A method for estimating the biomass of bacteria in aqueous sediments by enzymatic measurement of the lactic acid produced by acid hydrolysis of muramic acid has been published by Moriarty (160). Lactic acid is released from the muramic acid by alkaline hydrolysis and is measured using lactic acid dehydrogenase. The recovery of muramic acid from sediments was quantitative.

The sensitivity of the measurement of muramic acid was increased by two orders of magnitude by assaying the NADH (produced by oxidation of lactic acid by lactic acid dehydrogenase) using bacterial luciferase (161). This more specific assay procedure revealed that the previously reported amounts of muramic acid were high. [^{14}C]-labeled muramic acid of *Bacillus subtilis* cell walls disappeared from a marine sediment in 48 hr.

Samples of estuarine and marine sediments were subjected to acid hydrolysis to cleave the peptide linkage involving muramic acid, the muramic acid was purified by thin layer chromatography, lactate was released by alkaline hydrolysis, and then lactate was determined colorimetrically (162). Enzymatic assays gave a tenfold greater level. King and White (162) felt that the nonspecificity of lactic dehydrogenase was responsible. Lactic acid dehydrogenase oxidizes many other compounds. The metabolic turnover of muramic acid was found to be bipartite with half lives of 3.2 and 78.3 hr. Thus, muramic acid furnishes a measure of detrital or sedimentary procaryotic biomass.

Fazio et al. (163) improved muramic acid measurements by a gas-liquid chromatographic method. The time required for analysis has been reduced to 17 hr with only 3 hr of continuous attention.

13.11.9. Iron Porphyrins

Protein-bound iron atoms act as electron carriers, as in cytochromes, or as catalytic sites in enzymes, such as catalase and horseradish peroxidase. Iron porphyrins can be quantitated by the chemiluminescent oxidation of luminol (5-

amino-2,3-dihydro-1,4-phthazinedione). Metals such as Co^{2+}, Fe^{2+}, and Cu^{2+} can also act as catalysts.

Oleniacz et al. (164) have devised two automated systems using luminol for the detection of microorganisms in water. Between 10^3 and 10^5 viable bacterial cells/ml were detectable. Either sodium perborate or sodium pyrophosphate peroxide was used as reactant. They used a bacterial filter to allow establishment of the background level of metals present in the water which can catalyze the light production.

The luminol luminescence method for detection of bacteria has been adapted to an Auto Analyzer and used to detect bacteria (165). The analysis takes only 2 min. There were <1 percent false negatives and 7 percent false positives as compared with actual counts. A linear response was obtained between 10^4 and 10^6 bacteria/ml.

Ewetz and Thore (166) modified the luminol-perborate assay by treatment of the sample with 0.1 *M* NaOH 5 min before addition to the alkaline luminol reagent. Perborate was added 10 s later to initiate light production.

Miller and Vogelhut (167) claimed that the limit of detection of bacteria using chemiluminescence is no lower than 10^5 to 10^6 bacteria/ml. Ewetz and Thore (168) concluded that the luminol assay can be used as a quantitative method for determination of protohemin in microorganisms. We (manuscript in preparation) found, using a Pico-Lite photometer and peak height measurements, that the minimum number of detectable *E. coli* cells is 1×10^3.

13.12. BIOLUMINESCENCE, CHEMILUMINESCENCE, AND FLUORESCENCE

Light produced during a reaction can be measured with great sensitivity and accuracy. Bioluminescent, chemiluminescent, and fluorescent labels can be measured at extremely low concentrations and may replace ratioisotopes.

DeLuca (169) has edited Volume 57 of *Methods in Enzymology* which is devoted to bioluminescent and chemiluminescent methods. Two international symposia have been held on the analytical applications of bioluminescence and chemiluminescence (170, 171). Luminescence methods were covered in Glick's (172) review of microchemical analytical techniques of potential clinical interest. Both Gorus and Schram (173) and Whitehead et al. (174) have stressed the role of analytical bio- and chemiluminescence in the clinical laboratory. A monograph on *Bioluminescence in Action* (175) was edited by P. Herring. The fundamentals and biomedical applications of bioluminescent and chemiluminescent analysis have been reviewed (176). All of these reviews have been published since 1977 and illustrate the growing importance of bioluminescence and chemiluminescence. The analytical procedures of the future will be based upon these techniques.

Femtogram levels of peroxidase can be measured through the reaction of luminol and H_2O_2 (178). A pholad luciferase in the absence of H_2O_2 gives a method of extreme sensitivity. Immunological reactions can be used, for example, peroxidase-coupled antibodies, glucose oxidase-coupled antibodies, and bacterial luciferase-coupled antibodies, using light emission techniques.

Jablonski and DeLuca (178) have immobilized bacterial luciferase on arylamine glass beads. The immobilized enzyme could detect 0.2 pmol of NADH per sample. When glucose 6-phosphate dehydrogenase was co-immobilized, 1 pmol of glucose 6-phosphate could be quantitated. The immobilized enzymes can be reused and are more stable.

Methods for the use of bacterial luciferase to assay 1-nmol to 1-pmol amounts of NADH have been described by Golden and Katz (179). Among the substrates that can be determined are: glucose 6-phosphate, fructose 6-phosphate, dihydroxyacetone phosphate, fructose 1,6-diphosphate, glycerol 3-phosphate, malate, UDP-glucose, pyruvate, aspartate, citrate, and glutamate.

A highly sensitive modified ELISA (enzyme-linked immunosorbent assay) technique using a fluorescent substrate has been developed for *Rhizobium* strain identification (180). With a 2-hr incubation, they can detect 10^4 cells/ml.

An example of bacterial luciferase immunoassay has recently been provided by Wannlund and DeLuca (181). While the antigens used were dinitrophenol and trinitrotoluene, the procedures are suitable for measuring any antigen. The level of sensitivity is 10 amol (10^{-17} mol).

A new, highly sensitive chemiluminescent enzyme immunoassay for cortisol has been established (182). Horseradish peroxidase was the enzyme label and its activity was measured using luminol and hydrogen peroxide. The chemiluminescence was measured in a photon counter. Comparison of results with the radioimmunoassay gave a correlation coefficient of $r = 0.913$. The detection limit was 10 pg of cortisol.

Kather et al. (183) have described a sensitive bioluminescence determination of glycerol using NADH-linked, bacterial luciferase. The assay was 100 times as sensitive as the conventional spectrophotometric assay.

13.13. FECAL STEROLS

Coprostanol (5β-cholestan-3β-ol) is one of the five major sterols in feces of humans and higher animals and therefore, it is an indicator of fecal pollution (184). Cholesterol is not as specific because it can also come from eggs, milk, lard, and wool grease. A method was proposed for solvent extracting, hydrolyzing, purifying by thin-layer chromatography, and determining this sterol by gas-liquid chromatography. The finding of two coliform bacteria in 100 ml of water is equivalent to detecting coprostanol in the extract of 2 liters.

Dutka et al. (185) improved the analytical techniques, but failed to find a consistent, significant relationship between bacterial densities and fecal sterols. They concluded that fecal sterols should be considered on their own merits as indicators. Later Dutka and El-Shaarawi (186) found a significant relationship between coprostanol and cholesterol, a significant relationship between fecal sterols and fecal coliforms at two sampling sites, a positive relationship between fecal sterols and the fecal coliform/fecal streptococci ratio, and significant A.M.

vs. P.M. sample differences for coliform, fecal coliform, and fecal streptococci, but not so for coprostanol and cholesterol.

An extensive review of coprostanol as an indicator of fecal pollution has just appeared (187). The water solubility of coprostanol has not been reported, but that for cholesterol is 19 ppb at 30° C. While about 30 percent of the coprostanol in human feces was esterified, only 8 to 15 percent was esterified in sewage. It appears that coprostanol is a metabolite of only higher animals. Coprostanol is 40–60 percent of the total neutral sterols that humans excrete. It appears to be a "near absolute" indicator of fecal pollution. Coprostanol is biodegradable and removed by an adequate sewage treatment facility.

13.14. MISCELLANEOUS METHODS

13.14.1. Impedimetric

Gencer and Mutharasan (188) have developed a cell concentration measurement technique based on capacitance measurement. The results are useful when the conductance is low, but if the conductivity of the suspending medium is high, as when ionizable salts are present, the method fails.

Fecal coliforms growing in a selective medium at 44.5° C generate a change in the electrical impedance when the population reaches 10^6–10^7/ml (189). The detection time for one fecal coliform ranged between 8.7–11.4 hr. The detection time was dependent upon the inoculum size, as expected. The measurement process can be automated. Munoz and Silverman (190) also developed a rapid, single-step, most-probable-number method based on electrical impedance measurements. The test can be completed in 18 hr compared to the standard 72 hr.

The impedance-measuring techniques have been applied to quantitating coliforms in meat (191). Such a determination can be accomplished in 24 hours with a single sample dilution and four replicates. Seventy-nine percent of the impedimetric estimates fell within the 95 percent confidence limits of the standard most-probable-number test. Some 88 percent of the coliforms gave a positive test while only 15 percent of the noncoliforms did.

13.14.2. Uronic Acids

Extracellular polysaccharide polymers bind microbes to surfaces, confer virulence on pathogens, impede desiccation, regulate transport, and physically modify the microenvironment. Since uronic acids are components of the extracellular films that are most concentrated in an extramembrane location, White's group (192) developed a quantitative assay for them. The uronic acids were methylated and reduced with sodium borodeuteride so that they could be recovered quantitatively. The dideuterated sugars were hydrolyzed, the resulting neutral sugar derivatized, and separated and assayed by glass capillary gas–

liquid chromatography. The uronic acid-containing polymers are important in stabilizing marine sediments.

13.14.3. Calorimetric

Sparling (193) supplemented soils with 5 mg of glucose/g of soil and found an immediate increase in microbial activity. The rates of heat output increased 18 percent and respiration increased 23 percent in 30 min at 22° C. The heat output rate remained stable for 6 hr and correlated with CO_2 production. Sparling (194) found that the microcalorimetry correlated with basal respiration (0.99), ATP content (0.96), biomass respiration (0.81), and amylase activity (0.97).

13.14.4. Computer-Assisted Biochemical Tests

Willemse–Collinet et al. (195) developed a computer-assisted identification technique to identify *Bacillus* species. Twenty-four biochemical tests were used with 18 species. Some 30 test cultures were used and in 29 of 30 cases the identification was at a probability level of one.

13.15. COMPARATIVE STUDIES

There are four compilations or comparative reviews of the applications of biochemical methods which will be considered now. As Strange (196) put it, the problem is rapid detection and assessment of sparse microbial populations. There is no trouble in analysis if the concentration of the organisms is high or if immediate or rapid results are not needed. Microbial cells can be determined by use of a Coulter counter, but they must be concentrated and extraneous particulate material interferes. Two broad-spectrum methods based on bioluminescence and chemiluminescence are determination of ATP using firefly luciferase and heme iron using luminol. Strange concluded that detection of a *few* microbes in a *few* minutes is a difficult problem for which a general solution has not been found.

Jones (197) gave methods for direct counts on membrane filters, the use of counting chambers, autoradiographic fluorescent antibody assay, colony counts, most-probable-number, ATP, DNA, photosynthetic pigments, microbial cell wall materials, lipopolysaccharide assay, luminol, and chemiluminescence. These are detailed laboratory procedure descriptions, but no actual results or standard curves are given.

With a goal of reducing the time required for coliform analysis from 24 hr to <1 hr, increasing sample output, and increasing selectivity, Cundell (198) reviewed the rapid coliform-counting procedures. Table 13.9 summarizes many analytical methods with their sensitivity expressed in terms of numbers of *E. coli* cells detectable. Cundell's goals included specificity for fecal coliforms, 1-hr test

Table 13.9. Sensitivity in Terms of Numbers of *E. coli* Cells and the Time Required for Various Biochemical Determinations

Method	Number of *E. coli* Cells	Time[a]
$[^{14}C]0_2$ production from lactate	1–10	6 hr
Limulus amebocyte lysate for LPS	1–100	1 hr
H_2 evolution	1	8 hr
Impedance	$1–10^5$	10 or 2 hr
Electron transport	10^8	15 min
Glutamic dehydrogenase	1	10 hr
β-galactosidase, colorimetric	10^3	9 hr
β-galactosidase, fluorimetric	10^5	1.5 hr
Gas chromatography measuring ethanol	5	9 hr
ATP	10^3	10 min
Iron porphyrins	10^3	2 hr
Immunoenzymatic	10^2	1 hr
Muramic acid	3×10^8	17 hr
Protein, Lowry	10^8	2 hr
Protein, bromosulfalein	10^7	3 hr
Protein, *o*-phthalaldehyde	3×10^4	10 hr
DNA, diphenylamine	10^8	2 hr
DNA, DAPI	3×10^4	30 min
Lactate dehydrogenase	1×10^6	2 hr
Alkaline phosphatase	9×10^4	24 hr
Catalase	10^7	2 hr
Adenylate kinase	4×10^5	2 hr
Lipid phosphate	10^8	30 hr
Poly-β-hydroxybutyrate	10^5	5 hr

[a] When time was not mentioned in the procedure, it was estimated.

time, results as good as or better than those of present methods, good reproducibility and accuracy, the ability to detect a single coliform, susceptibility to automation, use of modestly priced equipment and reagents, technical simplicity, and utility on line or in the field. The stage is set for such developments but additional work is needed. An excellent discussion of the application of some (not all) methods to the measure of biomass in soil has been presented by Jenkinson and Ladd (62).

13.16. DISCUSSION AND SUMMARY

A myriad of methods have been presented and each has its proponents. Table 13.10 summarizes many of the techniques discussed with some complications listed.

Many are applying various biochemical methods without sufficient understanding of the limitations and pitfalls of the methods. Critical evaluation of the

Table 13.10. Summary of Bases and Some Complications of Determinations

Method	Basis	Complications
Growth		
Plate count	Growth to form a visible colony	Media and conditions of incubation are selective.
Microscopic		
Epifluorescence	Nucleic acid or protein staining.	Subjective visual measurement. Unable to differentiate dead, moribund, viable, and growing cells.
Epifluorescence for dividing cells	Cells with visible septum	Subjective visual measurement and recognition of a small structure.
Epifluorescence with nalidixic acid	Nalidixic acid inhibits DNA synthesis but cell increases in size	Visual observation. Not all organisms are susceptible to nalidixic acid.
Immunochemical staining	Antibody to specific organisms react	Visual observation. Strain differences in reactivity.
Metabolic staining	A specific reaction such as oxidation deposits a substance that can be visually detected	Visual observation. Permeability of the cell to the acceptor.
Metabolic		
Fumigation	Kill most organisms and the cellular material that remains is used first as substrates by the remaining cells	Different metabolic capabilities of the organisms.
Respiration	Measurement of gas	Substrate specificity and transport specificity.
Radioisotope incorporation	Measure biosynthesis	Variation in rates, pool sizes, and transport ability.
Bioluminescence	Measure the integrity of the light producing system in bioluminescent bacteria	Measures a physiological response that has a range of sensitivities depending upon the substance that is inhibiting.
Enzymatic activity	Enzymatic assay	The enzyme content varies with the conditions of the organism and the available substrates.

Table 13.10. (*Cont.*)

Method	Basis	Complications
		To be differentiative special substrates are used and this produces all the usual problems of specificity.
Impedimetric	Electrical measurement during growth	Ionizable salts interfere. Growth is required. Selective media.
Calorimetric	Heat output during metabolism	Growth required. Selective medium and substrate utilization.
Components		
Uronic acids	Chemical	Extracellular product and not proportional to growth.
Muramic acid	Enzymatic or chemical	Present in dead bacteria. Enzyme not specific. Amount varies with organism.
LPS	*Limulus* amebocyte lysate coagulation	Indicative of gram-negative organisms only.
Lipids	Chemical analysis after extraction	Do not rapidly turnover in the environment. Some found lipid phosphate to be nonspecific biomass indicator like ATP.
β-Hydroxybutyrate	Chemical	A storage compound and the amount varies depending upon conditions. Reflects metabolic state.
ATP	Light production by firefly luciferase or chemical	Amount varies between organisms and metabolic states.
Energy charge	Enzymatic determination of ATP, ADP, AMP, or chemical	Requires a rapid quench. Measures metabolic state.
Nucleic acids	Fluorimetric	Variation in content.
Iron porphyrins	Luminol chemiluminescence	Metal ion interference.
Fecal sterols	Chemical	Not necessarily correlated with bacterial numbers.

procedure with standards and controls is required. White's work is an example of the attention required in developing and validating a procedure. It is an example of the care required for quantitation in microbial ecology.

To repeat the take-home message—no single method is going to be sufficient. A suite of methods with high analytical precision applied in a sophisticated and careful manner will give insights into biomass and biochemicals in groundwater pollution microbiology. Many of the methods are in their developmental stages and extensive optimization and validation with environmental samples is required.

ACKNOWLEDGMENTS

The original work that I report was supported in part by EPA Grant R804613, the National Center for Ground Water Research, OWRT project A-097-Okla, and Oklahoma Agricultrual Experiment Station projects 1640 and 1806. Although the research described in this article has been funded wholly or in part by the United States Environmental Protection Agency through contract or grant R804613 and the National Center for Ground Water Research to Oklahoma State University, it has not been subjected to the agency's required peer and policy review and therefore does not necessarily reflect the views of the agency and no official endorsement should be inferred. Ginger Hampton obtained the firefly luciferase results presented herein. This is manuscript # J-4205 of the Oklahoma Agricultural Experiment Station. Appreciation is expressed to Jeralyn Jackson for the drawings and to Sue Heil, Pam Barnes, and Sandy Shamas for typing. Dr. Otis Dermer and Dr. John Wilson critically reviewed the manuscript and made many suggestions for improvement. The manuscript had to be prepared in less than a month and I apologize to those researchers whose work was not included. If the journal or a reprint was not immediately available, there was no way it could be obtained in time for coverage. Ms. Carolyn Warmann of the Edmon Low Library did the computer literature research and the Xerox-copying staff of the Physical Science section made up for the early (August) closing of the library at 6 P.M.

REFERENCES

1. J. F. McNabb and W. J. Dunlap, *Ground Water* **13,** 33 (1975).
2. M. J. Allen and E. E. Geldreich, *Ground Water* **13,** 45 (1975).
3. C. P. Gerba and J. F. McNabb, *ASM News* **47,** 326 (1981).
4. C. E. ZoBell, In L. H. Stevenson and R. R. Colwell, Eds., *Estuarine Microbial Ecology*, Univ. South Carolina Press, Columbia, South Carolina, 1973, p. 9.
5. J. Cairns and W. H. van der Schalie, *Water Res.* **14,** 1179 (1980).
6. P. A. Krenkel and V. Novotny, *Water Quality Management*, Academic Press New York, 1980, p. 71.

7. B. J. Dutka, S. Kuchma, and K. K. Kwan, *Water Air Soil Pollut.* **11,** 349 (1979).
8. O. C. Dermer, V. S. Curtis, and F. R. Leach, *Biochemical Indicators of Subsurface Pollution*, Ann Arbor Science, Ann Arbor, MI, 1980, 203 pp.
9. J. W. Costerton and G. G. Geesey, *ASTM/STP* **695,** 1979, p. 7.
10. R. M. Atlas and R. Bartha, *Microbial Ecology*, Addison-Wesley, Reading, MA, 1981, Chapter 4.
11. R. G. Board and D. W. Lovelock, Eds., *Sampling-Microbiological Monitoring of Environments*, Acedemic Press, New York, 1973, 267 pp.
12. V. G. Collins, J. G. Jones, M. S. Hendrie, J. M. Shewan, D. D. Wynn-Williams, and M. E. Rhodes, in R. G. Board and D. W. Lovelock, Eds., *Sampling-Microbiological Monitoring of Environments*, Academic Press, New York, 1973, p.77.
13. S. T. Williams and T. R. G. Gray, in R. G. Board and D. W. Lovelock, Eds., *Sampling-Microbiological Monitoring of Environments*, Academic Press, New York, 1973, p.111.
14. D. J. Ross, K. R. Tate, A. Cairns, and K. F. Meyrick, *Soil Biol. Biochem.*, **13,** 109 (1981).
15. J. G. Yeager and R. L. Ward, *Appl. Environ. Microbiol.* **41,** 1117 (1981).
16. R. M. Atlas and R. Bartha, *Microbial Ecology*, Addison-Wesley, Reading, MA, 1981, Chapter 6.
17. K. C. Marshall, In R. C. W. Berkeley, J. M. Lynch, J. Melling, P. R. Rutter, and B. Vincent, Eds., *Microbial Adhesion to Surfaces*, Ellis Horwood Ltd, Chichester, 1980, Chapter 9.
18. G. Bitton and K. C. Marshall, Eds., *Adsorption of Microorganisms to Surfaces*, John Wiley, New York, 1980.
19. R. C. W. Berkeley, J. M. Lynch, J. Melling, P. R. Rutter, and B. Vincent, Eds., *Microbial Adhesion to Surfaces*, Ellis Horwood Ltd., Chichester, 1980.
20. W. A. Corpe, *Develop. Indust. Microbiol* **15,** 281 (1975).
21. N. C. Bulleid, *Limnol. Oceanogr.* **23,** 174 (1978).
22. C.-N. Cheng, *Soil Biol. Biochem.* **7,** 319 (1975).
23. R. E. Hodson, O. Holm–Hanson, and R. Azam, *Mar. Biol.* **34,** 143 (1976).
24. W. Pflug and W. Ziechmann, *Soil Biol. Biochem.* **13,** 293 (1981).
25. H. W. Cunningham and R. G. Wetzel, *Limnol. Oceanogr.* **23,** 166 (1978).
26. S. Wiener, R. W. Wiener, M. Urivetzky, and E. Meilman, *Anal. Biochem.* **59,** 489 (1974).
27. G. E. Jones and H. W. Jannasch, *Limnol. Oceanogr.* **4,** 269 (1959).
28. R. W. Harvey and L. Y. Young, *Appl. Environ. Microbiol.* **39,** 894 (1980).
29. T. D. Brock, *Bacteriol. Rev.* **35,** 39 (1971).
30. H. W. Jannasch, in M. Shilo, Ed., *Strategies of Microbial Life in Extreme Environments*, Verlag Chemie, Berlin, 1979, p. 243.
31. Y. R. Dommergues, L. W. Belser, and E. L. Schmidt, *Adv. Microbial Ecol.* **2,** 49 (1978).
32. H. P. Jeffries, *Am. Nat.* **113,** 643 (1979).
33. R. J. Bobbie, J. S. Nickels, G. A. Smith, S. D. Fazio, R. H. Findlay, W. M. Davis, and D. C. White, *Appl. Environ. Microbiol.* **42,** 150 (1981).
34. J. D. Buck, *ASTM/STP* **695,** 19 (1979).

35. A. Hagstrom, U. Larsson, P. Horstedt, and S. Normark, *Appl. Environ. Microbiol.* **37,** 805 (1979).
36. S. Y. Newell and R. R. Christian, *Appl. Environ. Microbiol,* **42,** 23 (1981).
37. N. A. Marshman and K. C. Marshall, *Soil Biol. Biochem.* **13,** 127 (1981).
38. N. A. Marshman and K. C. Marshall, *Soil Biol. Biochem.* **13,** 135 (1981).
39. D. E. Francisco, R. A. Mah, and A. C. Rabin, *Trans. Am. Micros. Soc.* **92,** 416 (1973).
40. J. G. Jones and B. M. Simon, *J. Appl. Bacteriol.* **39,** 317 (1975).
41. R. J. Daley and J. E. Hobbie, *Limno. Oceanogr.* **20,** 875 (1975).
42. J. E. Hobbie, R. J. Daley, and S. Jasper, *Appl. Environ, Microbiol.* **33,** 1225 (1977).
43. D. R. Polonenko, C. I. Mayfield, and W. E. Inniss, *Water Air Soil Pollut.* **11,** 237 (1979).
44. K. Kogure, U. Simidu, and N. Taga, *Can. J. Microbiol.* **25,** 415 (1979).
45. E. R. Peele and R. R. Colwell, *Can. J. Microbiol.* **27,** 1071 (1981).
46. R. J. Daley, *ASTM/STP* **695,** 29 (1979).
47. J. E. deJosselin de Jong, J. F. Jongkind, and H. R. Ywema, *Anal. Biochem.* **102,**120 (1980).
48. C. Krambeck, H. J. Krambeck, and J. Overbeck, *Appl. Environ. Microbiol.* **42,** 142 (1981).
49. L.-A. Meyer-Reil, *Appl Environ. Microbiol.* **36,** 506 (1978).
50. R. Zimmermann, R. Iturriaga, and J. Becker-Birck, *Appl. Environ. Microbiol.* **36,** 926 (1978).
51. P. S. Tabor and R. A. Neihof, *Appl. Environ. Microbiol.* **43,** 1249 (1982).
52. R. A. Henry, R. C. Johnson, B. B. Bohlool, and E. L. Schmidt, *Appl. Microbiol.* **21,** 953 (1971).
53. B. B. Bohlool and T. D. Brock, *Appl. Microbiol.* **28,** 11 (1974).
54. A. P. Pugsley and L. M. Evison, *J. Appl. Bacteriol.* **38,** 63 (1975).
55. W. A. Apel, P. R. Dugan, J. A. Filppi, and M. S. Rheins, *Appl. Environ. Microbiol.* **32,** 159 (1976).
56. R. E. Curry, H. Heitzman, D. H. Riege, R. V. Sweet, and M. G. Simonsen, *Clin. Chem.* **25,** 1591 (1979).
57. P. M. Stanley, M. A. Gage, and E. L. Schmidt, *ASTM/STP* **695,** 46 (1979).
58. K. H. Baker and A. L. Mills, *Appl. Environ. Microbiol.* **43,** 338 (1982).
59. W. H. Holms, I. D. Hamilton, and A. G. Robertson, *Arch. Mikrobiol.* **83,** 95 (1972).
60. D. E. Atkinson, *Cellular Energy Metabolism and Its Regulation*, Academic Press, New York, 1977, p. 86.
61. J. Moucawi, E. Fustec, P. Jambu, and R. Jacquesy, *Soil Biol. Biochem.* **13,** 461 (1981).
62. D. S. Jenkinson and J. N. Ladd, in E. A. Paul and J. N. Ladd, Eds., *Soil Biochemistry*, **5,** Dekker, New York, 1981, Chapter 10.
63. J. M. Lynch and L. M. Panting, *Microb. Ecol.* **7,** 229 (1981).
64. D. J. Ross, K. R. Tate, A. Cairns, and E. A. Pansier, *Soil Biol. Biochem.* **12,** 375 (1980).

65. D. J. Ross, K. R. Tate, A. Cairns, and K. F. Meyrick, *Soil Biol. Biochem.* **12,** 369 (1980).

66. G. P. Sparling. B. G. Ord, and D. Vaughan, *Soil Biol. Biochem.* **13,** 99 (1981).

67. J. P. E. Anderson and K. H. Domsch, *Soil Biol. Biochem.* **10,** 215 (1978).

68. J. P. Christensen, T. G. Owens, A. H. Devol, and T. T. Packard, *Mar. Biol.* **55,** 267 (1980).

69. D. J. Ross, *New Zealand J. Sci.* **23,** 229 (1980).

70. A. J. Ramsey, *J. Gen. Microbiol.* **80,** 363, (1974).

71. D. R. Thomas, J. A. Richardson, and R. J. Dicker, *Soil Biol. Biochem.* **6,** 293 (1974).

72. H.-G. Hoppe, *Mar. Biol.* **36,** 291 (1976).

73. R. S. Tobin and D. H. J. Anthony, *Limnol. Oceanogr.* **23,** 161 (1978).

74. J. A. Fuhrman and F. Azam, *Mar. Biol.* **66,** 109 (1982).

75. D. M. Karl, *Appl. Environ. Microbiol.* **42,** 802 (1981).

76. D. M. Karl, *Appl. Environ. Microbiol.* **38,** 850 (1979).

77. D. M. Karl, C. D. Winn, and D. C. L. Wong, *Mar. Biol.* **64,** 1 (1981).

78. D. M. Karl, C. D. Winn, and D. C. L. Wong, *Mar. Biol.* **64,** 13 (1981).

79. G. V. Levin, V. R. Harrison, W. C. Hess, and H. C. Gurney, *Am. J. Pub. Health* **46,** 1405 (1956).

80. G. V. Levin, V. L. Strauss, and W. C. Hess, *J. Water Pollut. Cont. Fed.* **33,** 1021 (1961).

81. J. R. Waters, *Appl. Microbiol.* **23,** 198 (1972).

82. R. M. Atlas and J. S. Hubbard, *J. Microb. Ecol.* **1,** 145 (1974).

83. U. Bachrach and Z. Bachrach, *Appl. Microbiol.* **28,** 169 (1974).

84. R. R. Christian and W. J. Wiebe, *Limnol. Oceanogr.* **23,** 328 (1978).

85. L. G. Lehmicke, R. T. Williams, and R. L. Crawford, *Appl. Environ. Microbiol.* **38,** 644 (1979).

86. R. H. Monheimer, *Can. J. Microbiol.* **20,** 825 (1974).

87. M. J. Jordan and B. J. Peterson, *Limnol. Oceanogr.* **23,** 146 (1978).

88. W. Saggar, J. R. Bettany, and J. W. B. Stewart, *Soil Biol. Biochem.* **13,** 493 (1981).

89. R. L. Cuhel, C. D. Taylor, and H. W. Jannasch, *Arch. Microbiol.* **130,** 8 (1981).

90. A. A. Bulich, *ASTM/STP* **667,** 98 (1979).

91. J. C. Chang, P. B. Taylor, and F. R. Leach, *Bull. Environ. Contam. Toxicol.* **26,** 150 (1981).

92. A. A. Bulich, M. W. Greene, and D. L. Isenberg, *ASTM/STP* **737,** 338 (1981).

93. M. E. Lebsak, A. D. Anderson, G. M. De Graeve, and H. L. Bergman, *ASTM/STP* **737,** 348 (1981).

94. W. W. Cleland, *Anal. Biochem.* **99,** 142 (1979).

95. D. W. Moss, *Adv. Clin. Chem.* **19,** 1 (1977).

96. P. A. Trinel, N. Hanoune, and H. Leclerc, *Appl. Environ. Microbiol.* **39,** 976 (1980).

97. L. E. Casida, *Appl. Environ. Microbiol.* **34,** 630 (1977).

98. C. R. Blaise and J. B. Armstrong, *Appl. Microbiol.* **26,** 733 (1973).

99. P. Nannipieri, R. L. Johnson, and E. A. Paul, *Soil Biol. Biochem.* **10,** 223 (1978).

100. L. S. Warren, R. E. Benoit, and J. A. Jessee, *Appl. Environ. Microbiol.* **35,** 136 (1978).
101. J. E. Little, R. E. Sjogren, and G. R. Carson, *Appl. Environ. Microbiol.* **37,** 900 (1979).
102. R. Swisher and G. C. Carroll, *Microb. Ecol.* **6,** 217 (1980).
103. A. B. Cooper and H. W. Morgan, *Soil Biol. Biochem.* **13,** 307 (1981).
104. C. G. Kowalenko and R. B. McKercher, *Soil Biol. Biochem.* **2,** 269 (1970).
105. D. G. Shaw, *Environ. Sci. Technol.* **7,** 740 (1973).
106. D. C. White, R. J. Bobbie, S. J. Morrison, D. K. Oosterhof, C. W. Taylor, and D. A. Meeter, *Limnol. Oceanogr.* **22,** 1089 (1977).
107. D. C. White, W. M. Davis, J. S. Nickels, J. D. King, and R. J. Bobbie, *Oecologia* **40,** 51 (1979).
108. R. J. Bobbie and D. C. White, *Appl. Environ. Microbiol.* **39,** 1212 (1980).
109. J. S. Nickels, R. J. Bobbie, D. F. Lott, R. F. Martz, P. H. Benson, and D. C. White, *Appl. Environ. Microbiol.* **41,** 1442 (1981).
110. J. S. Nickels, R. J. Bobbie, R. F. Martz, G. A. Smith, D. C. White, and N. L. Richards, *Appl. Environ. Microbiol.* **41,** 1262 (1981).
111. K. C. Marshall, *Interfaces in Microbial Ecology*, Harvard University Press, Campridge, MA, 1976.
112. K. C. Marshall, *Annu. Rev. Phytopathol.* **13,** 357 (1975).
113. M. Werner, D. G. Gabrielson, and J. Eastman, *Clin. Chem.* **27,** 268 (1981).
114. G. A. Smith, J. S. Nickels, R. J. Bobbie, N. L. Richards, and D. C. White, *Arch. Environ. Contam. Toxicol.* **11,** 17 (1982).
115. J. H. Jorgensen, J. C. Lee, G. A. Alexander, and H. W. Wolf, *Appl. Environ. Microbiol.* **37,** 928 (1979).
116. S. W. Watson, T. J. Novitsky, H. L. Quinby, and F. W. Valois, *Appl. Environ. Microbiol.* **33,** 940 (1977).
117. T. M. Evans, J. E. Schillinger, and D. G. Stuart, *Appl. Environ. Microbiol.* **35,** 376 (1978).
118. S. W. Watson and J. E. Hobbie, *ASTM/STP* **695,** 82 (1979).
119. S. M. Goyal, C. P. Gerba, and J. C. Lance, *Appl. Environ. Microbiol.* **39,** 544 (1980).
120. J. L. Sykora, G. Keleti, R. Roche, D. R. Volk, G. P. Kay, R. A. Burgess, M. A. Shapiro, and E. C. Lippy, *Water Res.* **14,** 829 (1980).
121. G. Anderson, *Soil Biol. Biochem.* **11,** 213 (1979).
122. P. O. Clark, S. R. Hull, and F. R. Leach, *Curr. Microbiol.* **2,** 119 (1979).
123. J. Cortez and M. Schnitzer, *Soil Biol. Biochem.* **13,** 173 (1981).
124. V. L. Torsvik and J. Goksoyr, *Soil Biol. Biochem.* **10,** 7 (1978).
125. J. L. Howard, Ph.D. Dissertation, Oklahoma State University, Stillwater, OK 1980.
126. J. H. Paul and B. Myers, *Appl. Environ. Microbiol.* **43,** 1393 (1982).
127. J. S. Herron, J. D. King, and D. C. White, *Appl. Environ. Microbiol.* **35,** 251 (1978).
128. E.-M. Lilius, V.-M. Multanen, and V. Toivonen, *Anal. Biochem.* **99,** 22 (1979).

129. D. E. Atkinson, *Cellular Energy Metabolism and Its Regulation*, Academic Press, New York, 1977, 293.

130. G. A. Borun, Ed., *ATP Methodology Seminar*, SAI Technology Co., San Diego, CA, 1975, pp. 616.

131. G. A. Borun, Ed., *2nd Bi-Annual ATP Methodology Symposium*, SAI Technology, San Diego, CA, 1977, pp. 657.

132. F. R. Leach, *J. Appl. Biochem.* **3,** 473 (1981).

133. O. Holm–Hansen and C. R. Booth, *Limno. Oceanogr.* **11,** 510 (1966).

134. F. Eiland, *Soil Biol. Biochem.* **11**, 31 (1979).

135. J. W. Deming, G. L. Picciolo, and E. W. Chappelle, *ASTM/STP* **695,** 89 (1979).

136. W. M. Davis and D. C. White, *Appl. Environ. Microbiol.* **40,** 539 (1980).

137. F. Eiland and B. S. Nielsen, *Microb. Ecol.* **5,** 129 (1979).

138. J. J. Webster, J. C. Chang, E. R. Manley, H. O. Spivey, and F. R. Leach, *Anal. Biochem.* **106,** 7 (1980).

139. J.J. Webster, J. C. Chang, J. L. Howard, and F. R. Leach, *J. Appl. Biochem.* **1,** 471 (1979).

140. J. J. Webster, J. C. Chang, J. L. Howard, and F. R. Leach, in M. A. DeLuca and W. D. McElroy, Eds., *Bioluminescence and Chemiluminescence*, Academic Press, New York, 1981, p. 755.

141. J. J. Webster and F. R. Leach, *J. Appl. Biochem.* **2,** 469 (1980).

142. W. H. Sutcliffe, E. A. Orr, and O. Holm–Hansen, *Limnol. Oceanogr.* **21,** 145 (1977).

143. D. M. Karl, P. A. LaRock, and D. J. Shultz, *Deep-Sea Res.* **24,** 105 (1977).

144. P. A. LaRock, R. D. Lauer, J. R. Schwartz, K. K. Watanabe, and D. A. Wiesenburg, *Appl. Environ. Microbiol.* **37,** 466 (1979).

145. R. R. Christian, K. Bancroft, and W. J. Wiebe, *Soil Sci.* **119,** 89 (1975).

146. B. S. Ausmus, *Bull. Ecol. Res. Commun.* **17,** 223 (1973).

147. L. E. Hersman and K. L. Temple, *Soil Sci.* **126,** 350 (1978).

148. M. Ahmed, J. M. Oades, and J. N. Ladd, *Soil Biol. Biochem.* **14,** 273 (1982).

149. R. J. Bobbie, S. J. Morrison, and D. C. White, *Appl. Environ. Microbiol.* **35,** 179 (1978).

150. B. Cavari, *Limnol. Oceanogr.* **21,** 231 (1976).

151. L. H. Stevenson, T. H. Chrzanowski, and C. W. Erkenbrecher, *ASTM/STP* **695,** 99 (1979).

152. W. J. Wiebe and K. Bancroft, *Proc. Nat. Acad. Sci., USA* **72,** 2112 (1975).

153. D. M. Karl, *Microbiol. Rev.* **44,** 739 (1980).

154. J. C. Romano and R. Daumas, *Mar. Biol.* **62,** 281 (1981).

155. H. Spielmann, U. Jacob–Muller, and P. Schulz, *Anal. Biochem.* **113,** 172 (1981).

156. D. M. Karl, *Appl. Environ. Microbiol.* **36,** 349 (1978).

157. C.-N. Cheng, R. C. Shufeldt, and F. J. Stevenson, *Soil Biol. Bochem.* **7,** 143 (1975).

158. E. C. Butcher and O. H. Lowry, *Anal. Biochem.* **76,** 502 (1976).

159. W. N. Millar and L. E. Casida, *Can. J. Microbiol.* **16,** 299 (1970).

160. D. J. W. Moriarty, *Oecologia* **20,** 219 (1975).

161. D. J. W. Moriarty, *Oecologia* **26,** 317 (1977).

162. J. D. King and D. C. White, *Appl. Environ. Microbiol.* **33,** 777 (1977).

163. S. D. Fazio, W. R. Mayberry, and D. C. White, *Appl. Environ. Microbiol.* **38,** 349 (1979).

164. W. S. Oleniacz, M. A. Pisano, M. H. Rosenfeld, and R. L. Elgart, *Environ. Sci. Technol.* **2,** 1030 (1968).

165. L. Ewetz and K. Strangert, *Acta Path. Microbiol. Scand., Sec. B.* **82,** 375 (1974).

166. L. Ewetz and A. Thore, *Anal. Biochem.* **71,** 564 (1976).

167. C. A. Miller and P. O. Vogelhut, *Appl. Environ. Microbiol.* **35,** 813 (1978).

168. L. Ewetz and A. Thore, *Appl. Environ. Microbiol.* **36,** 790 (1978).

169. M. A. DeLuca, Ed., *Methods in Enzymology*, Vol. 57, Academic Press, New York, 1978.

170. E. Schram and P. Stanley, Eds., *International Symposium on Analytical Applications of Bioluminescence and Chemiluminescence*, State Printing and Publishing, Westlake Village, CA, 1979.

171. M. A. DeLuca and W. D. McElroy, Eds., *Bioluminescence and Chemiluminescence*, Academic Press, New York, 1981.

172. D. Glick, *Clin. Chem.* **23,** 1465 (1977).

173. F. Gorus and E. Schram, *Clin. Chem.* **25,** 512 (1979).

174. T. P. Whitehead, L. J. Kricka, T. J. N. Carter, and G. H. G. Thorr, *Clin. Chem.* **25,** 1531 (1979).

175. P. Herring, Ed., *Bioluminescence in Action*, Academic Press, New York, 1978.

176. W. R. Seitz, *CRC Crit. Rev. Anal. Chem.* **13,** 1 (1981).

177. K. Puget, A M. Michelson, and S. Avrameas, *Anal. Biochem.*, **79,** 447 (1977).

178. E. Jablonski and M. DeLuca, *Clin. Chem.* **25,** 1622 (1979).

179. S. Golden and J. Katz, *Biochem. J.* **188,** 799 (1980).

180. S. J. Morley and D. G. Jones, *J. Appl. Bacteriol.* **49,** 103 (1980).

181. J. Wannlund and M. DeLuca, *Anal. Biochem.* **122,** 385 (1982).

182. H. Arakawa, M. Maeda, and A. Tsuji, *Anal. Biochem.* **97,** 248 (1979).

183. H. Kather, F. Schroder, and B. Simon, *Clin. Chim. Acta* **120,** 295 (1982).

184. J. J. Murtaugh and R. L. Bunch, *J. Water Pollut. Contr. Fed.* **39,** 404 (1967).

185. B. J. Dutka, A. S. Y. Chau, and J. Coburn, *Water Res.* **8,** 1047 (1974).

186. B. J. Dutka and A. El-Shaarawi, *Can. J. Microbiol.* **21,** 1386 (1975).

187. R. W. Walker, C. K. Wun, and W. Litsky, *CRC Crit. Rev. Environ. Cont.* **12,** 91 (1982).

188. M. A. Gencer and R. Mutharasan, *Biotech. Bioengr.* **21,** 1097 (1979).

189. M. P. Silverman and E. F. Munoz, *Appl. Environ. Microbiol.* **37,** 521 (1979).

190. E. F. Munoz and M. P. Silverman, *Appl. Environ. Microbiol.* **37,** 527 (1979).

191. S. B. Martins and M. J. Selby, *Appl. Environ. Microbiol.* **39,** 518 (1980).

192. S. A. Fazio, D. J. Uhlinger, J. H. Parker, and D. C. White, *Appl. Environ. Microbiol.* **43,** 1151 (1982).

193. G. P. Sparling, *Soil Biol. Biochem.* **13,** 373 (1981).

194. G. P. Sparling, *Soil Biol. Biochem.* **13,** 93 (1981).

195. M. F. Willemse–Collinet, Th. F. J. Tromp, and T. Huizinga, *J. Appl. Bacteriol.* **49,** 385 (1980).

196. R. E. Strange, *Adv. Microb. Physiol.* **8,** 105 (1972).

197. J. G. Jones, *A Guide to Methods for Estimating Microbial Number and Biomass in Fresh Water*, Freshwater Biological Association, Scientific Publication No. 39, 1979.

198. A. M. Cundell, *Adv. Appl. Microbiol.* **27,** 169 (1981).

14

U.S. FEDERAL LEGISLATION PERTAINING TO GROUNDWATER PROTECTION

Jay H. Lehr, David M. Nielsen, and John J. Montgomery

National Water Well Association
Worthington, Ohio

14.1. INTRODUCTION

Groundwater has been described as one of our nation's most valuable resources. Americans have long depended on groundwater for many uses, but primarily for drinking water. More than 100 million Americans—nearly 50 percent of the population—depend on groundwater for their drinking water. Over 75 percent of the nation's public water supplies, more than one-third of the nation's largest cities, and nearly 95 percent of all domestic water needs in rural areas are served by groundwater resources. In addition, groundwater is used extensively in the western states for irrigation, in the northern states for residential and commercial heating, in the southern states for air conditioning, and across the nation for various industrial purposes.

National reliance on groundwater has increased dramatically over the years and will continue to increase as consumption and use of water increases in the future. This reliance will be underscored if surface water shortages caused by prolonged droughts continue to occur.

Despite the overwhelming evidence of the vital importance of groundwater, until very recently it has been provided with little legal protection. Specifically, it has been unprotected from a vast and complex array of organic and inorganic contaminants generated by a number of different sources including solid and hazardous waste landfills, surface and subsurface mining operations, subsurface waste injection wells, wastewater stabilization and sludge lagoons, oil and gas-well drilling operations, septic-tank leachfields, beef cattle feedlots, unregulated open dumps, and industrial waste storage areas. The massive national pollution cleanup efforts associated with the landmark environmental legislation of the early 1970s largely ignored groundwater. In fact, ironically, by strictly regulating discharges into surface water and air, some of the earlier environmental laws encouraged land disposal and underground injection of waste (wastes tend to gravitate for disposal purposes to the point of least regulation), thus hastening the contamination of groundwater supplies. Only since 1976 or so has the nation become aware of the dangers of groundwater contamination and of the many ways in which groundwater can become contaminated.

Probably a major reason groundwater has received so little attention in legislation is that it is "out of sight" and it is not readily understood by most people. Many people still believe that groundwater occurs in *veins* or *underground rivers* that transport water over great distances. A number of people believe that groundwater can be found by *water witching* with a forked stick or divining rod. These beliefs were prevalent when the first court decisions addressing groundwater were handed down, as the following passage from an Ohio Supreme Court case, *Frazier vs. Brown* (1861) attests:

> The existence, origin, movement and course of such waters, and the causes which govern and direct their movements, are so secret, occult, and concealed, that an attempt to administer any set of legal rules in respect to them would be, therefore, practically impossible.

Unfortunately, these beliefs persist even today in some states, when hydrogeology exists as a sophisticated science.

That the nature and occurrence of groundwater is misunderstood is especially regrettable because, in many ways, preventing groundwater contamination is far more crucial than policing surface waters. Groundwater contamination is particularly serious because it is difficult and expensive to cleanup after the fact. Unlike surface water, groundwater cannot be "flushed" clean. If an aquifer is contaminated with certain nonbiodegradable, long-lived chemicals, it may remain so for years. However, recent technological advances have made it possible to reclaim contaminated groundwater in many situations.

The most expedient approach to protecting groundwater from contamination is to regulate those activities that may have an adverse impact on the quality of groundwater. The regulatory techniques utilized over the years to protect groundwater and water quality in general are many and highly varied. As the following history of water-quality legislation points out, concern for groundwater protection has been quite deficient until just recently.

14.2. THE WATER POLLUTION CONTROL ACT OF 1948

Conscious effort and legislation toward comprehensive water pollution control began in 1948, when the first Water Pollution Control Act was passed into law. This initial legislation, like much of that to follow, focused primarily on protecting surface water. But, within this law were the roots of groundwater protection. The Water Pollution Control Act directed the Surgeon General to:

> ... prepare or adopt comprehensive programs for eliminating or reducing the pollution of interstate waters and tributaries thereof and improving the sanitary conditions of surface and underground waters.

No specific regulations intended to protect groundwater were promulgated as a result of this legislation, however.

14.3. THE NATIONAL ENVIRONMENTAL POLICY ACT

Following the 1948 Water Pollution Control Act, the concern of Congress for water pollution control lay dormant until 1969, when the National Environmental Policy Act (NEPA) was passed. NEPA mandates the preparation of environmental impact statements for all Federally funded projects that may have a "significant adverse impact" on water quality. However, while the potential is clearly there for an environmental impact statement to be an effective tool in groundwater protection, historically this has not been the case. The degree of analysis of groundwater considerations varies in environmental impact statements on a case-by-case basis, but again the focus has primarily been on surface water, even though NEPA does list groundwater protection as a significant parameter in determining the need for an environmental impact statement. A directive from the Administrative branch could assist in expanding the degree of consideration of groundwater in review procedures.

14.4. THE FEDERAL WATER POLLUTION CONTROL ACT

The Federal Water Pollution Control Act Amendments (also known as the Clean Water Act) of 1972 (Public Law 92-500) and 1977 (Public Law 95-217) established a major course change in water pollution control law. The objective of the FWPCA was to ". . . restore and maintain the chemical, physical, and biological integrity of the Nation's waters." However, the language of the Act stresses the need to protect "navigable waters," thus, the major emphasis and the thrust of enforcement has been toward surface waters. Nevertheless, several provisions of the FWPCA have served to enhance groundwater protection.

The most effective potential means for controlling groundwater contamination is found in Section 208, which provides for statewide and areawide planning for pollution control. The water-quality management program required by Section 208 has served as a catalyst for the development of several State groundwater management programs. The most powerful means for controlling groundwater contamination under Section 208 requires a plan to include "a process to control the disposal of pollutants on land or in subsurface excavations to protect ground- and surface-water quality."

Section 208, while it is part of a Federal regulatory scheme, provides the agency assigned to promulgate the regulations required under the Act, the U.S. Environmental Protection Agency (EPA), with no effective power to force action to protect groundwater. The primary responsibility for preparing plans and implementing programs is in the hands of state, regional, and local agencies. It is within the EPA's power to withhold approval of a plan or program that does not adequately provide for groundwater protection, but not within its power to act if the groundwater provisions of the plan are not implemented.

Another Section of PL 92-500 that holds potential for protecting groundwater is Section 304, which states that ". . . the Administrator shall develop and publish criteria for water quality and on the kind and extent of all identifiable effects on health and welfare which may be expected from the presence of pollutants in any body of water, including groundwater." Furthermore, this section requires the EPA to develop and issue guidelines for identifying and evaluating the nature and extent of nonpoint sources of pollutants and processes, procedures and methods to control pollution resulting from, among other things, ". . . the disposal of pollutants in wells or in subsurface excavations, saltwater intrusion resulting from reduction of freshwater flow from any cause, including extraction of groundwater, and changes in the movement, flow, and circulation of any navigable waters or groundwaters." Again, this section is not enforceable as law, as the guidelines can only serve as information to be used at the discretion of the Federal, State, regional, or local agencies.

The National Pollutant Discharge Elimination System (NPDES), described under Section 402, empowers the EPA to issue permits for the discharge of any point source pollutant or combination of pollutants to navigable waters. Individual states may issue NPDES permits if they develop programs and are authorized by the EPA to do so. One requirement for approval of state programs is that the states must provide for the control of the disposal of pollutants into wells. However, wells under this section of the Act have been interpreted by the EPA to mean only deep-waste injection wells, a term that is poorly defined. Additionally, Section 502 goes on to state that

> . . . pollutant does not mean water, gas, or other material which is injected into a well to facilitate production of oil and gas, or water derived in association with oil or gas production and disposed in a well if the well is approved by authority of

the State and if the State determines that such injection or disposal will not result in the degradation of ground- or surface-water resources.

This makes for a very narrow coverage of this portion of the Act. Other sections of the FWPCA that apply in some way to groundwater include Sections 104 and 106, which provide for the establishment and funding of national and state programs to equip and maintain both surface water and groundwater surveillance systems. While the authority exists for the use of funds provided under Section 106 for groundwater pollution control, most money has been channeled to surface water programs at the state level.

Until the passage of the Federal Water Pollution Control Act, the only viable legal remedy for groundwater pollution was suing on the basis of common-law nuisance. A nuisance is anything that disturbs an individual in the enjoyment of his property or that renders its use or physical occupation uncomfortable. The difficulty with a nuisance lawsuit is that it must be tried according to the common law, a body of unwritten laws and legal practices adopted in England at the time of the American Revolution. Since common law is derived from legal practices that are several hundred years old, applying it to modern scientific legal issues can often be an exercise in futility. In cases of groundwater pollution, a causal connection must be shown between the damage and the act of pollution. Since groundwater is "out of sight," this has always been extremely difficult to do.

After the passage of the Federal Water Pollution Control Act, it was believed that a new tool existed to deal with water pollution. Although that may have been the case with surface water, a controversy arose as to the Act's applicability to groundwater. At issue was what Congress meant when it used the term "water."

The objective of the Federal Water Pollution Control Act, as stated earlier, is to "restore and maintain the chemical, physical, and biological integrity of the Nation's waters," thereby eliminating pollution of navigable waters. Navigable waters is defined as being "waters of the United States." While the term "navigable waters" is generally considered to mean surface waters, "waters of the United States" is a term that can encompass all types of waters, including groundwater. Although four out of five sections of the Act refer specifically to groundwater, the enforcement section does not. This gap caused some defendants in pollution cases to argue that the Federal Water Pollution Control Act had no jurisdiction over groundwater pollution. A series of court cases throughout the United States showed that even the judiciary could not agree on what the Act covered.

In *United States vs. GAF Corporation*, the court decided that the Federal Water Pollution Control Act did not cover groundwater. The GAF Corporation was in the process of constructing two deep-injection wells to dispose of organic chemical wastes when the U.S. EPA demanded that GAF receive EPA approval for the wells. The court reviewed the legislative history of the Federal Water Pollution Control Act and concluded that the ". . . irrebuttable language and unambiguous action found in the legislative history . . ." indicated Congress did not intend to include groundwater in the Act.

In another case, *Exxon Corporation vs. Train*, another court analyzed the legislative history and found a clear pattern of congressional intent with respect to groundwater. The court concluded that ". . . the pattern is one of federal information gathering and encouragement of state efforts to control groundwater pollution—but not of direct federal control over groundwater pollution."

In another deep-well case, *United States Steel Corporation vs. Train,* a federal circuit court reviewed the same legislative history as the other courts and decided that Congress had intended to include groundwater within the meaning of the Act. In fact, the court specifically referred to the legislative history used in the GAF case to demonstrate congressional intent not to include groundwater. However, the U.S. Steel court interpreted the same material to imply inclusion of groundwater in the Federal Water Pollution Control Act.

The issue of whether groundwater is included in the Federal Water Pollution Control Act is now mostly of historical interest. Although Congress was ambiguous about the Federal Water Pollution Control Act's authority over groundwater, the laws passed since that time are not. The Safe Drinking Water Act, the Resource Conservation and Recovery Act, and the Surface Mining Control and Reclamation Act all specifically include groundwater within their jurisdiction. These laws have eliminated the potential gaps in the Federal Water Pollution Control Act and have demonstrated Congress' intent to protect groundwater.

14.5. THE SAFE DRINKING WATER ACT

With the passage of the Safe Drinking Water Act (SDWA) in December 1974 (Public Law 93-523), the void in groundwater protection began to be filled. The SDWA was the first major piece of legislation involving water management to recognize groundwater as a fundamental component of the nation's water resources.

The goal of the SDWA, as its name implies, is to assure the provision of safe drinking water to all Americans served by public water supply systems. Several mechanisms have been developed to ensure that this goal is met.

Promulgation of drinking water-quality standards to protect public health is required by Section 1412 of the SDWA. Two separate sets of standards have been developed by the U.S. EPA to meet this requirement. The first set of standards, known as the Primary Drinking Water Standards, became effective in June 1977 as "interim" standards. These are health-related standards that set maximum contaminant levels (MCL's) for bacteria, turbidity, inorganic constituents, pesticides, and radionuclides in public water supplies (Table 14.1). The second set of standards, the Secondary Drinking Water Standards, set guidelines for aesthetic characteristics of public drinking water supplies (Table 14.2). These became effective in July 1979.

Although the U.S. EPA was required under the SDWA to develop both primary and secondary standards, enforcement is limited to primary standards, as

Table 14.1. U.S. EPA Primary Drinking Water Standards[a]

Constituent	Maximum Contaminant Level (mg/liter unless otherwise specified)	Effects
Arsenic	0.05	Linked with skin cancer; recognized carcinogen
Barium	1.0	Toxic to heart, blood vessels, nervous system
Cadmium	0.010	Bioaccumulative; recognized carcinogen
Chromium	0.05	Ulcers after long-term exposure
Lead	0.05	Bioaccumulative; causes paralysis
Mercury	0.002	Highly toxic to humans
Nitrate (as N)	10.0	Methemoglobinemia, diarrhea
Selenium	0.01	Similar to arsenic
Silver	0.05	Fatal at high concentrations
Fluoride	1.4–2.4[b]	Fluorosis
Turbidity	1–5 turbidity units	Aesthetically displeasing; interferes with disinfection
Coliform	1 per 100 ml (mean)	Most are nonpathogenic
Endrin	0.002	Neurotoxin; suspected carcinogen, bioaccumulative
Lindane	0.004	Suspected carcinogen; bioaccumulative
Methoxychlor	0.10	Fatal at high concentrations
Toxaphene	0.005	Neurotoxin
2, 4 D	0.10	Suspected carcinogen and mutagen
2, 4, 5 TP (Silvex)	0.01	Bioaccumulative; suspected carcinogen
Radium 226 and 228	5 pCi/liter	Somatic and genetic damage
Gross alpha activity	15 pCi/liter	Somatic and genetic damage
Gross beta activity	4 millirem/year	Somatic and genetic damage

[a] *Source*: U.S. EPA Primary Drinking Water Regulations as outlined in the *Federal Register*, Vol. 45, No. 168, August 27, 1980.
[b] Based on average annual temperature.

the secondary standards are essentially only advisory in nature. In addition, the courts have specifically directed EPA to adopt regulations to limit the levels of organic chemical contaminants, including trihalomethanes and synthetic organic chemicals, in drinking water supplies. In recent years, studies of water quality in some public systems have indicated that drinking water may be contaminated with organic chemicals. Increased awareness of the presence of synthetic organic contaminants in groundwater supplies has heightened interest in this area.

Both the Primary and Secondary Drinking Water Standards apply only to public water supply systems, or those which supply water to 25 or more people, or have more than 15 service connections, whether the source of supply is surface or groundwater. It is estimated that there are 50,000 community water supplies in the United States that serve 180 million people. The standards do not apply to single-family private water systems, the majority of which are wells tapping groundwater. This means that nearly 45 million people who rely on groundwater from domestic wells are not protected by the standards.

Table 14.2. U.S. EPA Secondary Drinking Water Standards[a]

Constituent	Level (mg/liter unless otherwise specified)	Effects
Color	15 color units (ACU)	Aesthetically displeasing
Odor	3 threshold odor number	Aesthetically displeasing
pH	6.5–8.5 pH	High pH causes halogen reactions; increased corrosivity; unpleasant taste
Corrosivity	Noncorrosive	Damages plumbing; leaches toxic materials from plumbing
Foaming agents	0.5	Aesthetically displeasing; unpleasant taste
Total dissolved solids	500	Unpleasant taste; deteriorates plumbing
Chloride	250	Objectionable taste; corrosion in pipes
Copper	1.0	Liver damage; objectionable taste
Iron	0.3	Objectionable taste; discolors and stains
Manganese	0.05	Objectionable taste; discolors and stains
Sulfate	250	Scale formation; unpleasant taste; physiological effects
Zinc	5.0	Physiological effects; unpleasant taste

[a] *Source*: U.S. EPA publication no. EPA-570/9-76-000, July 1979.

While the U.S. EPA was required to develop the drinking water standards, primary enforcement responsibility is delegated to individual states, if the state adopts adequate procedures to enforce the federal standards and provides adequate records and reports. Most states require that utilities obtain operating permits if they provide water for human consumption and define the responsibilities of utility owners, including water-quality monitoring, recordkeeping, and public notice of failure to comply with a maximum contaminant level.

In addition to requiring that all public water supply systems meet drinking water standards, the SDWA also imposes controls on the underground injection of contaminated water, through the Underground Injection Control (UIC) program. Part C (Section 1424) of PL 93-523 was intended to protect current and future sources of drinking water (defined as all groundwater with a Total Dissolved Solids content of less than 10,000 mg/liter), from endangerment by underground injection practices.

The basic concept of the UIC program is to prevent contamination of freshwater aquifers by ensuring that injected fluids are confined within the wells and in the intended injection zone. The need for such a program becomes compelling when one considers the types and volumes of fluids currently being injected, the number of injection facilities in operation, and the complexity and diversity of geology in areas where underground injection is practiced. The EPA estimates that there are more than 600,000 wells injecting more than 850 billion gallons of fluid per year below the surface.

To ensure effective regulation of injection wells, standards have been set for each of the five types or classes of injection wells (Table 14.3) based on sound engineering practices accepted and used by the affected industries. The UIC regulations establish minimum standards for injection well design, construction, operation, monitoring, and abandonment procedures, and State program requirements. Wells of Classes I–III come under rigid permit requirements; Class IV wells are forbidden; Class V wells are to be permitted or forbidden on a case-by-case basis.

Under the UIC program, aquifers that do not currently serve as a source of drinking water or could not in the future because they are either (1) mineral, hydrocarbon, or geothermal-energy producing; (2) situated at a depth or location that makes recovery of water for drinking economically or technologically impractical; or (3) so contaminated that it would be infeasible to render the water fit for human consumption, are exempted from protection. Moreover, in keeping with SDWA policy, only groundwater that supplies or could supply in the future any public water system is protected. Consequently, the UIC Program does not apply to either groundwater used for purposes other than drinking or groundwater supplying nonpublic water systems.

Another provision of the SDWA related to protection of groundwater is the Sole-Source Aquifer program, also known as the "Gonzales Amendment." This program provides local, regional, or state agencies with a legal mechanism to protect the recharge zones of specially designated aquifers. It establishes a procedure whereby the EPA, either on its own initiative or upon petition, may designate an aquifer as the sole or principal source of drinking water for an area.

Table 14.3. Injection-Well Classifications[a]

Class of Well	Description
Class I	Wells used by generators of hazardous wastes or owners/operators of hazardous waste management facilities to inject hazardous waste. Also, industrial and municipal disposal wells used to inject fluids beneath the lowermost formation containing, within ¼ mile of the well, an underground source of drinking water.
Class II	Wells that inject fluids that are brought to the surface in connection with conventional oil and natural gas production, those used for enhanced recovery of oil or natural gas, and those used for storage of liquid hydrocarbons.
Class III	Wells that inject for the purpose of extracting minerals or energy, including solution mining wells, wells used for *in-situ* combustion of fossil fuel, wells used for recovery of geothermal energy, and wells used in the mining of sulfur by the Frasch process.
Class IV	Wells used by generators of hazardous wastes or radioactive wastes, by owners/operators of hazardous waste management facilities, or by owners/operators of radioactive waste disposal sites to dispose of hazardous wastes of radioactive wastes into or above a formation that, within ¼ mile of the well, contains an underground source of drinking water.
Class V	Any injection well not included in Classes I–IV, including: 1. Air-conditioning return-flow wells used to return to the supply aquifer the water used for heating or cooling in a heat pump. 2. Cesspools or other devices that receive wastes, which have an open bottom and sometimes have perforated sides. The UIC requirements do not apply to single family residential cesspools. 3. Cooling-water return-flow wells used to inject water previously used for cooling. 4. Drainage wells used to drain surface fluid, primarily storm runoff into a subsurface formation. 5. Dry wells used for the injection of wastes into a subsurface formation. 6. Recharge wells used to replenish the water in an aquifer. 7. Saltwater intrusion barrier wells used to inject water into a freshwater aquifer to prevent the intrusion of saltwater into the freshwater. 8. Sand backfill wells used to inject a mixture of water and sand, mill tailings, or other solids into mined out portions of subsurface mines. 9. Septic system wells used: a. To inject the waste or effluent from a multiple dwelling, business establishment, community or regional business establishment septic tank, or b. For a multiple dwelling, community, or regional cesspool. The UIC requirements do not apply to single family residential waste disposal systems.

Table 14.3. (Cont.)

Class of Well	Description
	10. Subsidence control wells (not used for the purpose of oil or natural gas production) used to inject fluids into a nonoil or gas producing zone to reduce or eliminate subsidence associated with the overdraft of freshwater
	11. Wells used for the storage of hydrocarbons which are gases at standard temperature and pressure.
	12. Geothermal wells used in heating and aquaculture.
	13. Nuclear disposal wells.

[a] *Source*: U.S. EPA Underground Injection Control Program regulations as outlined in the *Federal Register*, Vol. 45, No. 123, June 24, 1980.

After such designation, no Federal financial assistance may be granted to a project that the EPA determines could contaminate the aquifer through its recharge zone so as to create a significant hazard to public health.

For the purposes of this program, the term "significant hazard to public health" is defined as:

> . . . any level of contaminant which causes or may cause the aquifer to exceed any maximum contaminant level set forth in any national drinking water standard at any point where the water may be used for drinking water purposes, or which may otherwise adversely affect the health of persons, or which may require a public water system to install additional treatment to prevent such adverse effects.

Upon petition, EPA had made 12 sole-source aquifer designations as of June 1982:

1. The Edwards Aquifer, Texas.
2. The Spokane Valley Aquifer, Washington–Idaho.
3. The Long Island Aquifer (Magothy) underlying Nassau and Suffolk Counties, New York.
4. The Helena Valley Aquifer, Montana.
5. The Maryland Piedmont Aquifer.
6. The Upper Rockaway Basin Aquifer, New Jersey.
7. The Delaware Basin Aquifer, Texas and New Mexico.
8. The Verdegris Valley Aquifer, Oklahoma.
9. The Biscayne Aquifer in southeastern Florida.
10. The Aquifer beneath Fresno, California.
11. The Buried Valley Aquifer System, western Essex and southeastern Morris Counties, New Jersey.
12. The groundwater System of Guam.

A limiting factor of the Sole-Source Aquifer provision is that it protects aquifer recharge zones only from Federally funded projects that might contaminate an aquifer. Nonfederally financed projects are not regulated. Although it is not extensively used, this program is the only EPA program that uses recharge zone protection as a groundwater management tool.

The only case involving the designation of a sole-source aquifer was *Montgomery County vs. United States Environmental Protection Agency*, involving the Maryland Piedmont Aquifer. Two citizens' groups petitioned the EPA to designate a 40-square-mile area as a sole-source aquifer. The groups were opposed to a landfill that had been proposed for the area. The EPA modified the area of the designated aquifer area to include seven drainable basins that drained into the original 40 square miles. This resulted in a 180-square-mile sole-source aquifer region.

Montgomery County appealed this enlargement, arguing the inclusion of the seven drainable basins was unreasonable, arbitrary, and capricious because each basin acted independently as a separate and distinct hydrogeologic unit. In denying the County's claims, the court pointed out that each basin contributed groundwater to the affected area. The inclusion of all the basins was therefore not arbitrary and capricious.

14.6. THE RESOURCE CONSERVATION AND RECOVERY ACT

In October 1976, Congress enacted the Resource Conservation and Recovery Act (RCRA) (Public Law 94-580), thereby setting into motion the largest, most complex regulatory program ever conceived by the Federal government. RCRA, which is the U.S. EPA's main tool for managing solid and hazardous waste, has three primary objectives:

1. The improvement of solid-waste disposal practices to protect public health and the environment.
2. The regulation of hazardous wastes, from generation through disposal (cradle-to-grave).
3. The establishment of resource conservation as the preferred approach to solid and hazardous waste management.

Section 1003 of RCRA, which outlines these objectives, clearly indicates the applicability of the Act to groundwater protection, as does Section 1004, which defines the terms used in the Act. Congress was acutely aware of the link between waste disposal and groundwater when they enacted RCRA, as its definition of "disposal" evidences a conscious recognition of the groundwater pollution problem. RCRA defines disposal as:

> . . . the discharge, deposit, injection, dumping, spilling, leaking, or placing of any solid waste or hazardous waste or any constituent thereof that may enter the environment or be emitted to the air or discharged into any water, including groundwaters.

For solid-waste management, RCRA authorizes grants and technical assistance to State and local government agencies developing solid-waste management plans that meet Federal guidelines. In addition, the Act provides Federal standards for disposal of solid municipal waste (Section 4004), but these are not Federally enforceable. The EPA's Criteria for Classification of Solid-Waste Disposal Facilities, issued in September 1979, contains a provision that a solid-waste facility "shall not contaminate an underground drinking water source beyond the solid waste boundary." Contamination is defined as exceeding certain maximum contaminant levels specified in the regulation or as causing any increase in the concentration of those contaminants where the background level of the contaminant is already at or above the maximum contaminant limit. The maximum contaminant levels are based on the Primary Drinking Water Standards established under the Safe Drinking Water Act. Further, an underground drinking water source is defined as an aquifer currently supplying drinking water for human consumption or an aquifer containing less than 10,000 ppm of total dissolved solids.

For hazardous waste management, a much more strict program has been authorized by Congress through RCRA, since hazardous wastes pose a more significant and immediate threat to public health and the environment than do nonhazardous wastes. The objective of the hazardous waste management program developed under RCRA is to control hazardous wastes from "cradle-to-grave" through a complex system of standards applicable to generators and transporters of hazardous waste and to owners and operators of hazardous waste treatment, storage, and disposal facilities. In addition, RCRA establishes a permitting system applied to sites utilized for the treatment, storage, or disposal of hazardous waste.

The regulatory program to control the generation, transportation, treatment, storage, and disposal of hazardous waste was partially outlined in May 1980. Known as Subtitle C, the hazardous waste regulations developed by the EPA are broken down into 11 separate sections. Section 3004 has by far the most impact on groundwater protection, as it sets forth standards for owners/operators of hazardous waste treatment, storage, or disposal facilities (TSDF's).

With specific reference to groundwater, the EPA has prepared performance criteria that apply to most forms of land disposal, including dumps, landfills, pits, ponds and lagoons, and landspreading of wastewater and sludge. The siting, design, and operating specifications developed for landfills and other hazardous waste facilities require that the facility owner/operator employ natural geologic or engineering design features and waste management practices that minimize adverse effects on groundwater and surface water. The basic purposes of these requirements are to minimize the production of leachate and to avoid situations that could compromise the integrity of the facility's liner and final cover (landfills) or its natural ability to ameliorate waste migration (land treatment facilities).

In addition to prescribing standards for siting, design, and operation of hazardous waste TSDF's, Section 3004 also sets forth requirements for the installation and operation of groundwater monitoring systems as a means of evaluating the performance of such facilities. These groundwater monitoring requirements outline procedures for (1) installing groundwater monitoring systems; (2) preparing a groundwater quality assessment plant; (3) developing a groundwater sampling and analysis program. Exempt from these requirements are those TSDF's that can demonstrate that there is a low potential for migration of hazardous waste from the facility via the uppermost aquifer to water-supply wells or surface water. Such a facility may apply for a waiver from groundwater monitoring requirements.

The RCRA regulations for groundwater monitoring at hazardous waste facilities are the most specific regulations with regard to groundwater protection ever promulgated by an agency of the Federal government. As stated in the regulations, each hazardous waste TSDF must have installed groundwater monitoring wells both hydraulically upgradient and hydraulically downgradient of the limit of the waste management area. The number, location, depth and construction details of the upgradient wells must be sufficient to yield groundwater samples that are representative of background water quality in the uppermost aquifer beneath the facility. The number, location, depth and construction details of downgradient wells must ensure that they detect any wastes that migrate from the waste management area to the uppermost aquifer. Both upgradient and downgradient wells must be cased in a manner that maintains the integrity of the monitoring well borehole, screened and packed with gravel or sand to enable the collection of groundwater samples, and the annular space above the sampling depth sealed to prevent contamination of samples and the groundwater. Regulations require a minimum of one upgradient well and three downgradient wells.

In addition to requiring the installation of a groundwater monitoring system, RCRA regulations under Section 3004 require owners/operators of hazardous waste TSDF's to outline procedures that are to be followed to assess the extent to which wastes may have entered the uppermost aquifer underlying the facility if the facility is suspected to be contaminating groundwater. Implementing the groundwater quality assessment plan would be required in the event that water quality analysis results indicate a statistically significant difference between downgradient and upgradient water quality. This will have been determined by sampling and analysis conducted in accordance with the sampling and analysis plan.

The groundwater sampling and analysis plan developed for compliance with RCRA regulations must include procedures and techniques for sample collection, sample preservation, analytical procedures, and chain-of-custody control. For a period of one year after the owner/operator has installed the groundwater monitoring system, he must monitor groundwater in all wells to establish background water quality. Samples must be taken quarterly and analyzed for three separate sets of indicator parameters:

1. Parameters characterizing the suitability of the groundwater as a drinking water supply, including all parameters mandated for analysis under the Safe Drinking Water Act (see Table 14.1).

2. Parameters establishing groundwater quality, including chloride, iron, manganese, phenols, sodium, and sulfate.
3. Parameters used as indicators of groundwater contamination, including pH, specific conductance, total organic carbon, (TOC) and total organic halogens (TOX).

After the first year, all monitoring wells must be sampled and the samples analyzed such that parameters used to establish groundwater quality are sampled and analyzed annually and parameters used as indicators of groundwater contamination are sampled and analyzed semiannually.

In order to determine whether the hazardous waste TSDF is performing up to standards, the owner/operator must compare the results of the indicator parameter analysis with the computed background by use of the Cochran's Approximation to the Behrens-Fisher Student's t-test. If a statistically significant difference exists (at the 0.05 confidence interval) between measured and established background water-quality parameters, the owner/operator must then follow a complex sequence of events to assess the extent of the problem. This sequence is depicted in Fig. 14.1.

If it is determined during the course of monitoring that the facility is contaminating ground water, the facility owner/operator must, in addition to assessing the extent of the problem, take corrective action to remedy the problem within a "reasonable time period" after the contamination is detected. The corrective action must restore the contaminated groundwater to the point at which concentration limits of hazardous constituents are not exceeded in samples from the downgradient monitoring wells. This portion of RCRA is unique in federal environmental legislation.

Despite the extensive consideration of groundwater in the regulations promulgated under RCRA, this legislation still falls short of providing complete control of at least one class of potential groundwater pollution sources. The major drawback is that RCRA applies only to facilities that were active as of November 1980. Those facilities that were inactive or abandoned prior to this date, which may number in excess of 50,000, are not subject to these regulations. In addition, the regulations do not specifically address the need for remedial actions that may be necessary at these facilities. This task has been left for later legislation.

14.7. THE TOXIC SUBSTANCES CONTROL ACT

The Toxic Substances Control Act (Public Law 94-469), signed into law in October 1976, provides the U.S. EPA with the authority to require premarket testing of a wide range of chemicals in order to evaluate the health effects that they may cause and to regulate chemical substances. While this Act has no direct impact on groundwater protection, it has the potential to be used as a tool in protecting groundwater. Manufacturers and processors of chemical substances may be required to keep records identifying chemicals and their uses. To be

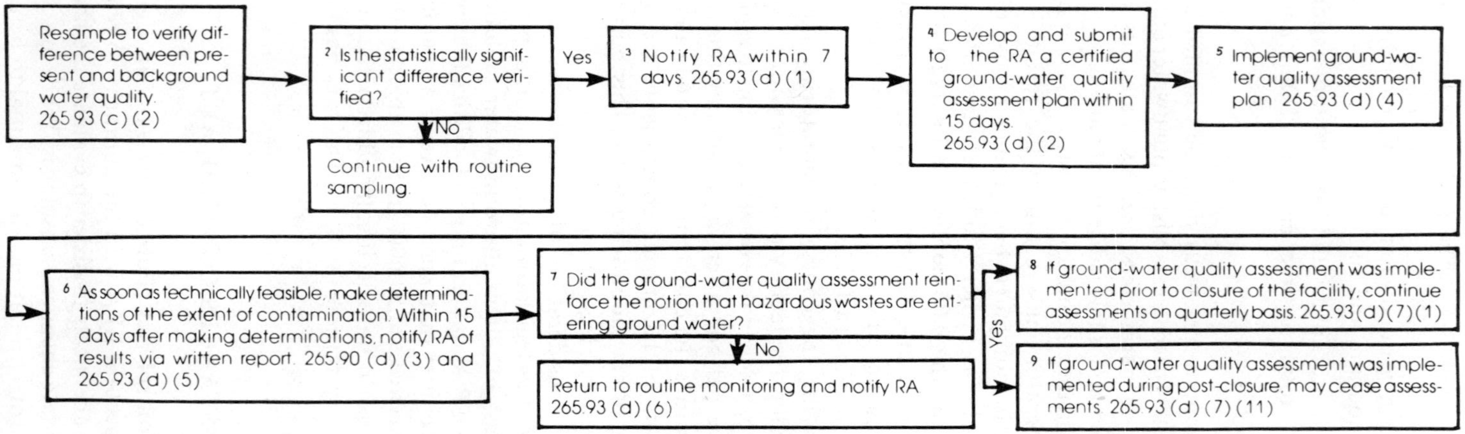

Figure 14.1. Required procedures for owners/operators of hazardous waste TSDF's whose sampling and analysis programs indicate that hazardous wastes are entering groundwater. Source: Graves, L. S., Ground Water Monitoring Requirements of RCRA; *Ground Water Monitoring Review*, Vol. 1, No. 1, 1981, pp. 34–36.

included on these records are data such as the amounts of chemicals produced, how and where the chemicals are stored and transported, any known or projected occupational exposures, and the methods used to dispose of the chemicals. The EPA does have the power to prohibit or regulate the manufacture, processing, distribution, use, or disposal of certain chemicals found to "pose an unreasonable risk to public health or the environment." Regulation of chemical substances may range from simple labeling requirements to a strict ban.

However, the Act is somewhat narrow in its application, as it addresses only new chemicals produced for the first time and existing chemicals used in a new way. Thus, TSCA is not applicable to most of the toxic substances presently in distribution and offers protection from only a limited number of toxic materials.

14.8. THE SURFACE MINING CONTROL AND RECLAMATION ACT

The Surface Mining Control and Reclamation Act (SMCRA) of 1977, under the administration of the Department of the Interior, provides authority for various levels of government to control environmental impacts resulting from all mining activities although the title of the Act refers only to surface mining. Among the purposes of the Act are:

To establish a nationwide program to protect society and the environment from the adverse effects of surface mining operations.

To assure that surface mining operations are so conducted as to protect the environment.

To assure that adequate procedures are taken to reclaim surface areas as contemporaneously as possible with the surface mining operations.

To promote the reclamation of mined areas left without adequate reclamation prior to the enactment of this Act and which continue, in their unreclaimed condition, to substantially degrade the quality of the environment, prevent or damage the beneficial use of land or water resources, or endanger the health or safety or the public.

Several sections of SMCRA deal specifically with groundwater.

For new mines, permit applications described under Section 507 must include a determination of the probable hydrologic consequences of the mining and reclamations proposed. Of particular concern is the determination of the impact of mining and reclamation on the quantity and quality of water in both surface water and groundwater systems. All permit applications must be accompanied by geologic cross sections and maps of the land to be affected, showing among other things the location of aquifers and estimated water levels.

Reclamation plan requirements outlined in Section 508 compel mine operators to provide a detailed description of the measures to be taken during the mining and reclamation process to assure the protection of the quality of surface water and groundwater from the adverse effects of the mining and reclamation process. In addition, the operator must recognize the rights of the present users to this water and must assure the protection of the quantity of surface and groundwater from the mining and reclamation operation, or provide alternative sources of water to affected users where such protection cannot be assured. This requirement could potentially have far-reaching effects, as it could set a precendent for other environmental statutes to follow. To date, SMCRA is the only piece of Federal legislation to provide users with protection of groundwater from a specific threat.

Section 515 of SMCRA outlines general environmental protection performance standards applicable to mining and reclamation operations that require the operation to " . . . minimize the disturbances to the prevailing hydrologic balance at the mine site and to the quality and quantity of water in surface and groundwater systems both during and after mining operations and during reclamation." Water quality is to be preserved by avoiding acid or other toxic mine drainage through such means as preventing or removing water from contact with toxic-producing deposits, treating drainage to reduce its toxic content, or burying or otherwise disposing of acid-forming or toxic materials in a manner designed to prevent contamination of both groundwater and surface water. Mine operators are to maintain the hydrologic balance of the area by restoring the recharge capacity of the mined area to approximate premining conditions and by preserving, throughout the mining and reclamation process, the essential hydrologic functions of alluvial valley floors (this applies mainly to arid and semiarid areas).

Under Section 517, the Department of the Interior (specifically the Office of Surface Mining) may require operators to install, use, and maintain groundwater monitoring systems. For those mining and reclamation operations that disturb aquifers, the regulatory authority has the power to specify monitoring sites to record the level, amount, and quality of groundwater in aquifers affected or potentially affected by the mining and those directly below the lowermost coal seam to be mined. The Department of the Interior has set forth standards and procedures for the collection and analysis of data generated by the groundwater monitoring programs.

Section 522, which outlines the means for designating areas unsuitable for mining, states that a surface area may be declared unsuitable if the mining operation will "affect renewable resource lands in which such operations could result in a substantial loss or reduction of long-range productivity of water supply." The "renewable resource lands" referred to in this passage are construed to include aquifers and aquifer recharge areas.

In summary, the Surface Mining Control and Reclamation Act has significant potential to protect both groundwater and the users of groundwater in mining areas from the environmental effects of mining and reclamation operations. The language within the Act is significantly different from that contained within any other similar legislation and could serve as a precedent for requiring broad protection of both groundwater quality and availability.

14.9. THE COMPREHENSIVE EMERGENCY RESPONSE, COMPENSATION, AND LIABILITY ACT

The second phase of the national attack on hazardous waste is the implementation of the Comprehensive Emergency Response, Compensation, and Liability Act (CERCLA), also referred to as the "Superfund." This legislation, which was passed in December 1980, supports the cleanup of abandoned hazardous waste sites that may be contaminating groundwater Superfund, which is funded mostly by an industry hazardous waste surcharge, is meant to serve both as a deterrent to environmentally unsound waste-disposal practices and as a means of recovering cleanup costs incurred by Federal, State, and local governments. CERCLA provides the Federal government with both the authority and the resources to clean up abandoned sites and permits speedy action prior to long judicial proceedings to prevent and abate situations that may present an " . . . imminent and substantial danger to public health." In addition, the Act imposes joint, several, and strict liability on those who caused or are causing a groundwater contamination problem.

Superfund offers an opportunity to close the gap in the enforcement of the Resource Conservation and Recovery Act, if the U.S. EPA chooses to utilize the Act in that manner. However, to date, few cleanup operations have been attempted under the provisions of CERCLA and it is unclear as to how effective this legislation will be in policing hazardous waste problem areas.

14.10 CONCLUSIONS

There are eight major pieces of Federal environmental legislation that either directly or indirectly contribute to the cause of groundwater protection in the United States. At present, no national program exists to tie together all of the various mechanisms utilized in each of the Acts into a comprehensive groundwater protection effort. However, the more recent legislation, including FWPCA (CWA), SDWA, RCRA, SMCRA, TSCA, and CERCLA, provide a framework for such a comprehensive national program to be constructed. Each of these Acts offers its own unique contribution to groundwater protection and, in combination, appear to be sufficiently comprehensive to provide broad protection. Their somewhat diverse objectives and means of attaining them, however, require that the Acts be coordinated more closely to maximize their usefulness in a groundwater protecton effort. A comprehensive National Ground Water Protection Strategy, utilizing existing legislation, is needed to accomplish this objective.

INDUSTRIAL POLLUTION CONTROL—Volume I: Agro-Industries
E. Joe Middlebrooks

BREEDING PLANTS RESISTANT TO INSECTS
Fowden G. Maxwell and Peter Jennings, Editors

NEW TECHNOLOGY OF PEST CONTROL
Carl B. Huffaker, Editor

THE SCIENCE OF 2,4,5-T AND ASSOCIATED PHENOXY HERBICIDES
Rodney W. Bovey and Alvin L. Young

INDUSTRIAL LOCATION AND AIR QUALITY CONTROL: A Planning Approach
Jean-Michel Guldmann and Daniel Shefer

PLANT DISEASE CONTROL: Resistance and Susceptibility
Richard C. Staples and Gary H. Toenniessen, Editors

AQUATIC POLLUTION
Edward A. Laws

MODELING WASTEWATER RENOVATION: Land Treatment
I. K. Iskandar

AIR AND WATER POLLUTION CONTROL: A Benefit Cost Assessment
A. Myrick Freeman, III

SYSTEMS ECOLOGY: An Introduction
Howard T. Odum

INDOOR AIR POLLUTION: Characterization, Prediction, and Control
Richard A. Wadden and Peter A. Scheff

INTRODUCTION TO INSECT PEST MANAGEMENT, Second Edition
Robert L. Metcalf and William H. Luckman, Editors

WASTES IN THE OCEAN—Volume 1: Industrial and Sewage Wastes in the Ocean
Iver W. Duedall, Bostwick H. Ketchum, P. Kilho Park, and Dana R. Kester, Editors

WASTES IN THE OCEAN—Volume 2: Dredged Material Disposal In the Ocean
Dana R. Kester, Bostwick H. Ketchum, Iver W. Duedall and P. Kilho Park, Editors

WASTES IN THE OCEAN—Volume 3: Radioactive Wastes and the Ocean
P. Kilho Park, Dana R. Kester, Iver W. Duedall, and Bostwick H. Ketchum, Editors

LEAD AND LEAD POISONING IN ANTIQUITY
Jerome O. Nriagu